HERIOT-WATT UNIVERSITY

38007 00005255 6

D1388114

Inorganic Compounds with Unusual Properties

R. Bruce King, EDITOR

The University of Georgia

A symposium sponsored by
the Division of Inorganic
Chemistry of the American
Chemical Society, held at
the University of Georgia,
Athens, Ga., Jan. 6-8, 1975.

ADVANCES IN CHEMISTRY SERIES **150**

AMERICAN CHEMICAL SOCIETY

WASHINGTON, D. C.   1976

K32114c
546

Library of Congress CIP Data

Inorganic Compounds with unusual properties.
(Advances in chemistry series; 150 ISSN 0065-2393)

Includes bibliographical references and index.
  1. Chemistry, Inorganic—Congresses.
  I. King, R. Bruce. II. American Chemical Society.
Division of Inorganic Chemistry. III. Series: Advances in
chemistry series; 150.

QD1.A355 no. 150 [QD146] 540'.8 [546] 76-13505
ISBN 0-8412-0281-8   ADCSAJ 150 1-437 (1976)

Copyright © 1976

American Chemical Society

All Rights Reserved. No part of this book may
be reproduced or transmitted in any form or by
any means—graphic, electronic, including photo-
copying, recording, taping, or information storage
and retrieval systems—without written permission
from the American Chemical Society.

PRINTED IN THE UNITED STATES OF AMERICA

Advances in Chemistry Series

Robert F. Gould, *Editor*

Advisory Board

Kenneth B. Bischoff

Jeremiah P. Freeman

E. Desmond Goddard

Jesse C. H. Hwa

Phillip C. Kearney

Nina I. McClelland

John B. Pfeiffer

Joseph V. Rodricks

Aaron Wold

FOREWORD

ADVANCES IN CHEMISTRY SERIES was founded in 1949 by the American Chemical Society as an outlet for symposia and collections of data in special areas of topical interest that could not be accommodated in the Society's journals. It provides a medium for symposia that would otherwise be fragmented, their papers distributed among several journals or not published at all. Papers are refereed critically according to ACS editorial standards and receive the careful attention and processing characteristic of ACS publications. Papers published in ADVANCES IN CHEMISTRY SERIES are original contributions not published elsewhere in whole or major part and include reports of research as well as reviews since symposia may embrace both types of presentation.

CONTENTS

v

PREFACE

In January 1975, a symposium on "Inorganic Compounds with Unusual Properties" was held at the Georgia Center for Continuing Education of the University of Georgia as one of the biennial symposia of the Division of Inorganic Chemistry of the American Chemical Society. The purpose of this symposium was to stimulate interactions between scientists concerned with the preparation and characterization of inorganic, coordination, and organometallic compounds and scientists concerned with the photochemical, electrical, and other properties of compounds of these types.

Of the 33 papers presented at the symposium 32 are included in this volume. They are of two general types. One type discusses inorganic compounds exhibiting a specific property, most frequently electrical or photochemical. The second type reviews recent developments in the synthetic chemistry of a class of inorganic compounds—most often transition metal or lanthanide complexes—which are of potential interest to persons looking for new systems with unusual properties. Classes of compounds covered by this second type include metal alkyls, metal alkoxides, metal alkylamides, metal chelates, and metal clusters as well as metal complexes with polyboranes, polyphosphines, macrocyclic derivatives, di- and triketones, and polypyrazolylborates as ligands.

I would like to acknowledge the tremendous help of my University of Georgia colleagues, John K. Ruff, Charles R. Kutal, and Robert H. Lane, who served on the Symposium Planning Committee in connection with the planning and arranging of both the scientific and non-scientific programs of the symposium. We thank the Army Research Office (Durham), Petroleum Research Fund of the American Chemical Society, and the Xerox Corp. for sponsorship of the symposium through major financial contributions. Finally, I am pleased to acknowledge assistance from both William Johns of the Georgia Center for Continuing Education and my wife, Mrs. Jane K. King, with the non-scientific program of the symposium.

R. BRUCE KING

Athens, Ga.
February 1975

ix

Electrical Property Studies of Planar Metal Complex Systems

LEONARD V. INTERRANTE

Physical Science Branch, General Electric Corporate Research and
Development, P.O. Box 8, Schenectady, N. Y. 12301

Electrical conductivity and thermopower measurements on single crystals of $Pt(NH_3)_4PtCl_4$ (MGS) and (TTF) $NiS_4C_4H_4$ (TTF = tetrathiafulvalene) illustrate the application of electrical property measurements to the study of intermolecular interactions and electronic structure in planar metal complex systems. MGS is an anisotropic, p-type semiconductor; conductivity along the metal-chain direction $(10^{-5}–10^{-2}$ ohm^{-1} cm^{-1} at $52°C)$ is 18–25-fold that perpendicular to this direction. The importance of impurity effects in determining the conductivity of this compound is demonstrated, and a possible extrinsic band model for the conduction process is presented. Conductivity measurements on the donor–acceptor compound $(TTF)_2NiS_4C_4H_4$ as a function of direction in the crystal are discussed in the context of a recent crystal structure analysis which suggests a two-dimensional interaction among the constituent TTF units.

The strong current interest in inorganic and organic compounds which display pseudo-one-dimensional solid-state properties is noted in several other papers in this volume. This interest has been manifested in the last few years in a large number of special symposia (*1, 2, 3, 4, 5*) and publications, particularly in the physics journals.

In many of these compounds, the one-dimensional solid state properties derive from interactions between planar molecular units which are stacked together to form columns or even metal atom chains along one direction in the solid. The tendency of planar molecules to stack in columns can be understood simply from the standpoint of packing efficiency considerations although, in certain cases, specific bonding interactions are evidenced by unusually short intermolecular separations (*6,*

7). In planar transition metal complexes, the interactions among the metal and ligand π orbitals which occur within these stacks can provide a continuous pathway for electron delocalization along one direction in the solid which is detectable as anisotropic electrical conductivity. Such effectively one-dimensional solid-state interactions can result in unique properties and property combinations.

For example, in addition to metal-like conductivity and optical reflectivity along the stacking direction and effectively insulating character perpendicular to this direction, there is evidence of unusual structural distortions, analogous to Jahn-Teller effects in molecular systems, which result in gradual metal-to-insulator transitions at low temperatures (8). In certain cases, it is believed that these structural distortions give rise to cooperative electron transport *via* charge density waves which presumably enhance the conductivity at low temperatures above that which would be expected on the basis of ordinary one-electron scattering theory (8, 9).

In addition to systems with one-dimensional metallic properties, there are a considerably larger number of planar metal complexes where stacking interactions in the solid state give rise to unusual properties including highly anisotropic conductivity behavior but where electronic or structural factors lead to thermally activated conductivities (10, 11). Such one-dimensional semiconductors constitute an important area of study within the general topic of solids with one-dimensional interactions, and their study has provided much useful information regarding structure–property relationships.

The primary objective of this paper is to illustrate by specific examples from our past and current research how electrical property measurements can be of value in deducing information regarding the solid-state electronic structure and in studying intermolecular orbital interactions in such transition metal complex systems. To facilitate this discussion, a brief description of electrical conductivity and some other electrical properties is included. For a more detailed account as well as for a description of the various experimental techniques which are used to determine these properties, the reader is referred to any of several excellent books on the subject (12, 13).

Measurement of Electrical Properties

Electrical Conductivity. The measurement of electrical conductivity is essentially the determination of the amount of charge transported per second across a unit area of sample as the result of an applied unit electric field. If it is assumed that the conductivity is primarily electronic in nature (as opposed to ionic), as seems to be the case for most transition metal complex solids, the charge carriers are either electrons or their positive counterparts, holes, and the conductivity is determined by the number of such species and their velocity (mobility) in the applied field, *i.e.* $\sigma \alpha n \mu$, where σ is the conductivity, n is the number of carriers, and μ is their mobility in $cm^2/volt$ sec. In very simple terms, the mobility reflects the facility with which the carrier can move through the sample, and thus, in a crystal comprised of molecular units, mobility provides a good measure of the intermolecular orbital interactions.

The reciprocal of conductivity is resistivity, ρ, the resistance per unit cube. Resistivity is measured in ohm cm, and the unit of conductivity is thus ohm^{-1} cm^{-1}. The conductivities at 25°C of a number of representative materials are given in Figure 1 on a logarithmic scale of ohm^{-1} cm^{-1}. The terms metal, semiconductor, and insulator imply a certain level of conductivity, as illustrated by the approximate ranges in Figure 1, but, moreover, they refer to the manner in which the conductivity varies with temperature as well as other characteristic electrical, optical, and magnetic properties. With a metal, conductivity generally decreases with increasing temperature in the manner $\sigma \alpha 1/T$ whereas with semiconductors and insulators, conductivity generally increases with increasing temperature, often exponentially, *i.e.* $\sigma \alpha \exp(-\Delta E/kT)$.

As is indicated in Figure 1, molecular solids, and indeed most organic and metal complex materials fall in this category, usually exhibit rather low electrical conductivities (10^{-10} ohm^{-1} cm^{-1}) that reflect the rather weak van der Waals interactions between the molecular units in the solid. There are a number of molecular solids with substantially higher conductivities, however, that range up to ~ 800 ohm^{-1} cm^{-1} for tetraselenofulvalene · tetracyanoquinodimethane (TSeF · TCNQ) (*14*).

In this compound and several other structurally analogous organic and inorganic compounds, the conductivity is inversely dependent on temperature over a limited temperature region, and other characteristically metal-like properties are observed. Included in this group are polymeric sulfur nitride, $(SN)_x$, and a group of partially oxidized, square planar platinum and iridium complexes that are collectively referred to as "KCP" in Figure 1; both are discussed elsewhere in this volume.

Except for these relatively few materials, most transition metal complexes exhibit properties more typical of semiconductors with conductivities usually less than 10^{-2} ohm^{-1} cm^{-1} and a thermally activated conductivity behavior. The thermal activation energy, which is derived from plots of log σ *vs.* $1/T$ in these cases, may result from either the production of charge carriers (as for example by direct band gap excitation in an intrinsic semiconductor) or from the carrier motion (as in systems where the conduction proceeds by a short range, thermally activated, hopping process).

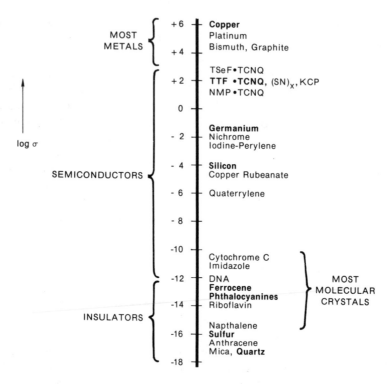

Figure 1. Approximate conductivity scale for some representative solid state materials at 25°C

In either case, extremely small amounts of impurities and crystalline defects can play an important role in determining both the magnitude of the conductivity and the activation energy. The elucidation of such effects is often difficult, and careful control of sample purity and form as well as experiments on samples in various stages of purification are required.

Other Electrical Properties. A characteristic feature of many semiconductors and insulators is an increase in conductivity on absorption of light in a particular frequency range. This photoconductivity usually reflects an increase in the number of free carriers in the solid due to their photoexcitation from filled bands, donor levels, or trapping states. The spectral response and quantum efficiency of the photoconductivity can be used to deduce information regarding the energy band structure of a semiconductor and the carrier lifetimes (or mobilities).

Another electrical property of fundamental importance is the thermoelectric power, or Seebeck effect. Its sign, magnitude, and temperature dependence can provide information about the sign of the charge carriers and their mobility as well as the band structure of the solid. It is essentially the voltage generated between two junctions of a material when the two junctions are held at different temperatures; it is determined simply by measuring the induced voltage and dividing by the temperature difference. The sign of the thermopower is defined as the polarity of the cold junction, and it usually corresponds to the sign of the majority carrier (*i.e.* positive for holes and negative for electrons). Its magnitude is characteristically small ($< 100\ \mu V$/degree) for metals and considerably larger ($> 100\ \mu V$/degree) for semiconductors. The temperature dependence is also usually larger for semiconductors where a relation of the type $S = B/T + K$ (where S is the thermopower and B and K are constants) is often observed. Various theoretical expressions of this general form have been derived on the basis of band theory and have been used to relate the experimental data to such features of the conduction process as the position of the Fermi level in relation to the band edges, the effective mass of the carriers, the electron/hole mobility ratio, and the nature of the carrier scattering mechanism.

A variety of other electrical measurements have been used in the study of semiconductor materials. Hall effect measurements have been used to determine the sign and concentration of charge carriers as well as, in conjunction with conductivity measurements, to deduce mobilities. The mobilities, which are quite sensitive to the intermolecular orbital overlaps in molecular solids, can also be determined directly, with less ambiguity, by injecting a pulse of charge carriers, using light or an electric field, and measuring their transit time over a known distance in a known field. Although such methods have proven quite useful in the study of many inorganic semiconductors, they have not found wide application thus far in work on molecular solids as a result, perhaps in part, of difficulties in obtaining suitable single crystal samples of appropriate dimensions, as well as the somewhat higher degree of sophistication required for their meaningful application. For details regarding these and the various other electrical property measurement methods which are not discussed here, the cited reference sources (*12, 13*) should be consulted.

Application of Electrical Measurements to the Study of Some Metal Complex Systems

Magnus' Green Salt. The crystal structure of Magnus' Green Salt (MGS) (*15*), [Pt(NH$_3$)$_4$PtCl$_4$], is illustrated in Figure 2. Of particular interest are the linear chains of Pt atoms which result from the columnar stacking of the constituent planar complex units. The Pt–Pt separations within these chains (3.25 A) are substantially shorter than the separations between chains (6.39 A) which suggests the possibility of highly direc-

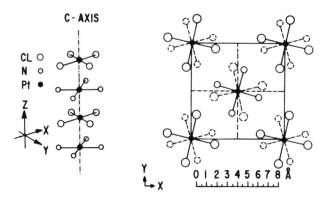

Figure 2. The crystal structure of Magnus' Green Salt
$[Pt(NH_3)_4^{2+}][PtCl_4^{2-}]$ *(MGS). Left: chain structure.*
Right: atomic arrangement of MGS projected on (001);
——: z = 0; - - - -: z = ½.

tional solid-state interactions. The first evidence for the existence of such interactions was the green color of this compound which was not the expected consequence of the combination of colorless $Pt(NH_3)_4^{2+}$ and pink $PtCl_4^{2-}$ ions. Spectral studies of MGS during the past 25 years have led to identification of the source of the green color as a window in the absorption spectrum at $\sim 20,000$ cm^{-1} that arises from a red shift and intensification of the largely intramolecular $d \rightarrow d$ transitions on the $PtCl_4^{2-}$ ion (*16, 17, 18, 19, 20*).

In one of the earlier studies of this absorption spectrum, an apparent broad, strong, absorption band was observed in the near-IR region in the vicinity of 6000 cm^{-1} which was thought to be possibly intermolecular in origin (*21*). This observation led to the postulation of an energy band description for the electronic structure of MGS which has been widely used in subsequent work on this and other metal-chain systems (Figure 3).

The extended interaction of the filled d_{z^2} and empty p_z orbitals on the platinum atoms in the linear chains is viewed here as giving rise to energy bands of appreciable width in the solid, much like the energy bands in covalent inorganic semiconductors, only highly directional in character. The near-IR absorption band was attributed to transitions across the forbidden energy gap between these bands, which then presumably amounts to < 1 eV.

The possibility that thermal or photoexcitation of electrons between these bands could lead to mobile charge carriers in either or both bands, and consequently to anisotropic semiconductivity or photoconductivity, was anticipated. Subsequent observations by Collman and co-workers (*22*) and by Gomm, Thomas, and Underhill (*23*) verified these predictions and, moreover, their data seemed to be in reasonable quantitative agreement with the intrinsic band model description in Figure 3. This agreement was short lived however; later spectral studies on single crystals of MGS (*24, 25*) failed to substantiate the earlier claim of a strong absorption band in the near-IR region which would be expected from the presumed band gap of < 1 eV, and semiquantitative band structure calculations (*26*) led to an estimate of 4.5 eV for the $d_{z2}-p_z$ band gap which would be characteristic of a good insulator.

To elucidate the origin of the conductivity in MGS and the nature of the solid-state electronic interactions, suitable single-crystal samples of MGS were grown, and their electrical properties were studied in detail. The development of a silica gel procedure (*27*) for growing MGS crystals permitted for the first time the preparation of large, high quality crystals of appreciable cross section. Rectangular parallelepiped crystals 1–2 mm long and up to 0.7×0.7 mm in cross section were obtained in this manner, and, after screening for defects by examination under a microscope and study of the crystal morphology by x-ray precession methods, they were mounted for conductivity and thermoelectric power measurements. Conductivities were measured by the standard four-elec-

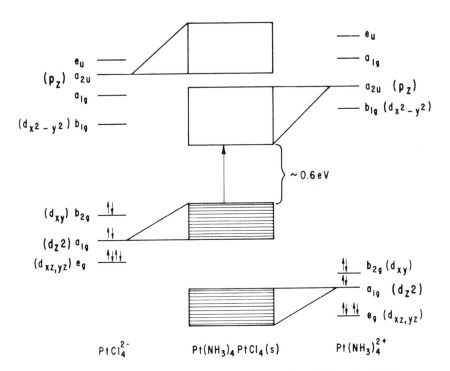

Figure 3.   Energy band model for MGS (adapted from Ref. 21)

trode, voltage probe method (*13*) with two end contacts and two voltage probes attached along one direction in the crystal, and also by the Montgomery method (*28*) with four electrodes at four parallel edges of the crystal, to determine the conductivity anisotropy. Contacts were made using Dupont silver paint No. 7941 and fine gold wire leads.

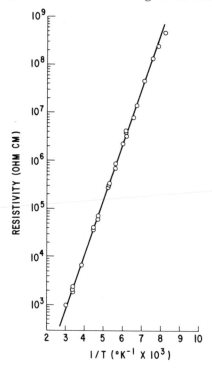

Figure 4.   Representative log ρ vs. 1/T plot for an MGS crystal wired for dc voltage probe measurements along the c axis

A typical log σ *vs.* 1/T plot obtained in the voltage probe studies is presented in Figure 4. The linearity of this plot over the wide temperature range studied (120°–350°K) suggests that we are dealing with one thermally activated conduction process in this temperature regime. Measurement of the conductivity anisotropy both by four-probe measurements on individual samples and by the Montgomery method (*28*) gave values of 18/1–25/1 for the ratio of the *c* (metal chain direction) and *a* axis conductivities. This is somewhat less than the anisotropy exhibited by other one-dimensional systems (*8, 29*), but it is still basically consistent with the suggestion that the predominant solid-state interactions in MGS are along the platinum chains.

Conductivity as a function of frequency was not studied in detail; however, preliminary two-electrode ac measurements at room temperature on a single crystal sample revealed essentially no change from the dc conductivity value with ac frequencies up to at least 1000 Hz. This observation agrees with expectations for band-type semiconductors where the carriers are relatively mobile in an applied field.

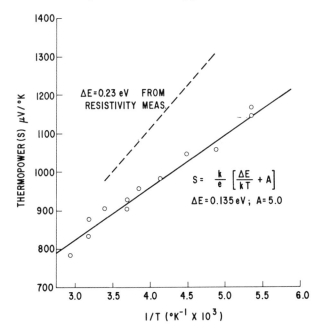

Figure 5.   The thermopower of MGS as a function of 1/T. The slope of the log ρ vs. 1/T curve for this same crystal is included for comparison.

In addition to these conductivity studies, the thermopower of several crystals was measured as a function of temperature by using two small gold-Advance thermocouples attached to the ends of the crystal with silver paint. The crystal and thermocouples were suspended from single-crystal quartz blocks, one of which was electrically heated to provide the required temperature gradient. In each case, positive thermopower values were observed which suggests that holes are the majority carriers. The large magnitude of the thermopower and its essentially linear dependence on 1/T (Figure 5) is consistent with previous observations on semiconductors and with the expression $S = k/e[\Delta E/kT + A]$ derived by Fritzsche for a semiconductor in which only one band is involved in the conduction process (*30*). The ΔE in this expression is basically the activation energy for charge carrier production, and, with a band-type semiconductor, it is equivalent to the thermal activation energy for conductivity. The substantially higher value observed for the conductivity activation energy in this same crystal could arise in part from an activation energy associated with the charge carrier motion, as in a narrow band, hopping type semiconductor.

The data from the voltage probe dc conductivity measurements on 19 different crystals of MGS obtained from four crystal growth preparations are summarized in Table I. These data reveal several interesting features regarding conduction in MGS. In particular, whereas the room temperature conductivities of crystals obtained from the same preparation are in reasonably good agreement, the variation from one preparation to the next is as much as three orders of magnitude. A marked variation in the thermal activation energy for conductivity is also apparent between preparations with a generally lower thermal activation energy for the samples of higher conductivity. Careful examination of these crystals under a microscope and by x-ray precession methods revealed no apparent structural or morphological differences among the samples from different preparations. Microanalysis revealed small amounts (ppm level) of other transition elements (primarily Fe, Cu, and Pd) but no obvious variations in composition from one preparation to the next.

Table I.   Dc Conductivity Data for MGS Crystals[a]

Preparation	Number of Samples	c-Axis Conductivity at 25°C, $ohm^{-1} cm^{-1}$	Thermal Activation Energy, eV
A	6	$4.7(\pm 2.4) \times 10^{-5}$	0.352 ($\pm$0.033)
B	3	$8.5(\pm 4.2) \times 10^{-4}$	0.257 ($\pm$0.025)
C	4	$1.3(\pm 0.3) \times 10^{-3}$	0.255 ($\pm$0.020)
D	6	$7.5(\pm 1.6) \times 10^{-3}$	0.212 ($\pm$0.012)

[a] Average values; average deviations are given in parentheses.

Additional experiments on the same $Pt(NH_3)_4^{2+}$ and $PtCl_4^{2-}$ solutions demonstrated that certain solutions consistently gave crystals of higher conductivity, which suggests that differences in the composition of these solutions were responsible for the observed conductivity variations. On the basis of these findings and the previously mentioned arguments against the intrinsic band model, it was suggested that the conductivity in MGS was impurity dominated (24, 31).

The first definitive information regarding the nature of these impurities was provided by EPR measurements on both single-crystal and powder samples of MGS prepared under different experimental conditions (32, 33). An axially symmetric resonance was observed in the single- crystal samples which was identified, on the basis of the g values and the observed hyperfine pattern, as arising from an electron in a

d_{z^2}-like state that extended over several Pt atoms. Spectrophotometric analysis of the solutions used to prepare MGS and doping experiments suggested further that the source of these unpaired electrons was Pt^{IV} complexes present in the solutions. These Pt^{IV} complexes apparently induce a partial oxidation of the Pt^{II} ions in the solid which is compensated by the addition of negative ions (presumably halides) either at interstitial sites or as a replacement for the neutral NH_3 groups. The presence of such halide ions, particularly at interstitial sites, could facilitate electron transfer between the platinum chains and thus account for the observation of the substantially lower (18–25 ×) but still significant conductivity perpendicular to the metal-chain direction in MGS. The existence of Pt^{IV} species in the nominal Pt^{II} complexes used to prepare MGS was a general occurrence, presumably the result of air oxidation or incomplete reduction of the Pt^{IV} starting material in the preparation of the Pt^{II} complexes.

EPR studies of some of the same crystals used in the conductivity studies revealed a general correspondence between the intensity of the EPR resonance and the conductivity of the crystals, thus establishing a relationship between the impurity-induced conductivity effects and the presence of Pt^{III}-like states in the solid.

Using the available information regarding the nature of the impurity states in MGS and the findings from the semiquantitative band structure calculations, a modified band model for the MGS electronic structure can be constructed which appears to account for much of the available data. This model (Figure 6) is similar to the intrinsic band model proposed by Miller (21) in that d_{z^2} and p_z bands of significant width are postulated; however, consistent with the theoretical calculations (26) and spectral observations (16–20, 24, 25) here the $d_{z^2}–p_z$ band gap is on the order of 4.5 eV, and acceptor levels, corresponding to the Pt^{III}-like states localized near the charge-compensating defects, have been introduced close to the top of the d_{z^2} band.

Intrinsic MGS in this context would be a good insulator, and the observed conductivity is attributed entirely to the presence of the impurity states. Thermal excitation of electrons from the d_{z^2} band into these acceptor levels should lead to hole conduction in the d_{z^2} bands, which is consistent with the observed p-type conductivity behavior. Also the essentially frequency-independent conductivity as well as the weak character of the photoconductivity response observed by Collman et al. (22) in the near-IR region is understandable on this basis, as is the lack of any strong absorption bands in this region, considering the relatively low concentration of impurity states present [200 ppm Pt^{III} sites estimated by EPR (32, 33)]. It is also possible to account for the apparently smaller thermal activation energies observed for the crystals of higher conductivity by

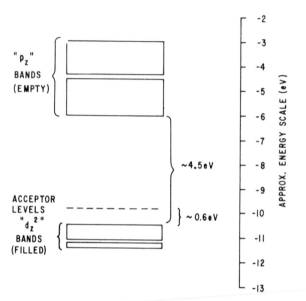

Figure 6.   Extrinsic band model description for the electronic structure of MGS. The acceptor levels correspond to localized Pt^{III}-like states, each containing one unpaired electron.

considering the likelihood of impurity banding in these presumably more highly doped samples.

Although this extrinsic band model can be used to account for much of the available information concerning the electrical properties of MGS, especially in view of the thermopower data obtained, the alternative possibility that conventional band theory is not appropriate to MGS and that a more localized description of the electronic structure is required clearly cannot be discounted at this point. A definite answer to this question must await further studies on this system, including a direct determination of the charge carrier mobility.

(TTF)$_2$NiS$_4$C$_4$H$_4$. Bis(tetrathiafulvalene)-bis(1,2-ethylenedithiolene)-nickel [(TTF)$_2$NiS$_4$C$_4$H$_4$] is a member of a new series of π-donor–acceptor compounds prepared by the interaction of tetrathiafulvalene (TTF) with neutral bisdithiolene metal complexes (*34, 35*). It crystallizes from acetonitrile solution as large (∼ 2 mm on a side) single crystals that are suitable for detailed conductivity measurements. The data from such measurements, when examined in the light of the other solid state properties and information obtained from a complete three-dimensional crystal structure determination, suggest specific interactions of a quite directional character among the constituent planar molecular units in the solid and thus provide a good illustration of the utility of electrical conductivity as a tool for investigating solid-state interactions.

In addition to the conductivity studies, electronic absorption, spectral, magnetic susceptibility, and ESR measurements were made on this compound (*34*). The findings indicate that charge transfer between the neutral diamagnetic TTF and $NiS_4C_4H_4$ units occurs in the solid state to give paramagnetic $NiS_4C_4H_4^-$ ions. These were identified by their characteristic absorption spectrum and by their anisotropic g values. The magnetic measurements indicated essentially Curie-type paramagnetism but with a Curie constant appropriate for only one unpaired electron per $(TTF)_2NiS_4C_4H_4$ formula unit. This was identified as the electron on the $NiS_4C_4H_4^-$ ions on the basis of the ESR and spectral data. The fate of the electron left on the TTF^+ units after charge transfer is suggested by the structural investigation (*35*). This compound crystallizes in the monoclinic space group $C2/m$ with $a = 25.80$ A, $b = 10.67$ A, $c = 9.99$ A, and $\beta = 119.67$ A.

Views normal to the (010) and (100) planes are presented in Figures 7 and 8, respectively. The structure consists of four $(TTF)_2NiS_4C_4H_4$ formula units per unit cell with the TTF and $NiS_4C_4H_4$ molecules arranged in alternate strips parallel to the (100) plane. The $NiS_4C_4H_4$ units are oriented with their molecular planes parallel to the (100) plane and are located at $y = 0$ and $1/2$, well separated from one another and

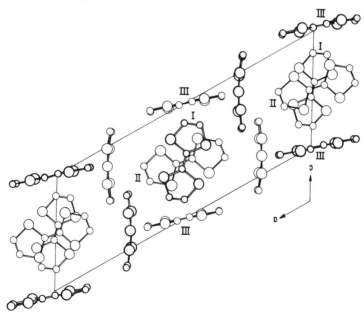

Figure 7.   The unit cell of the $(TTF)_2NiS_4C_4H_4$ crystal structure viewed down [010]. The molecules whose centers are at y $= \frac{1}{2}$ are darkened; except for those labeled II, all other molecules have their centers at y $= 0$.

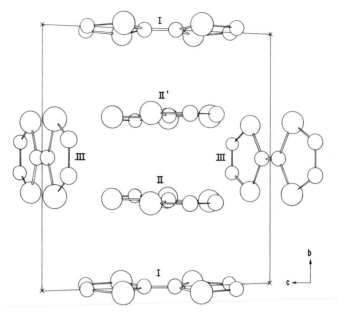

*Figure 8.   A view of the $(TTF)_2NiS_4C_4H_4$ structure normal
to the (100) plane showing the three types of TTF molecules*

the other molecules in the unit cell. Three structurally distinct TTF
molecules occur, labeled I, II, and III in the figures.

Molecules I and II form a columnar stack along [010] with the two
units at II and II′ in Figure 8 rotated by 60° with respect to those at I
and in a fully eclipsed configuration. These two units (II and II′) are
separated by 3.48 A which suggests a significant bonding interaction.
These are identified as the TTF⁺ ions in the structure which are presumably held together in an eclipised configuration by interaction of the
half-filled b_{1u} orbitals (36) on the two units. The electrons left after
charge transfer are thus presumably paired up in the resultant bonding
molecular orbital. These $(TTF^+)_2$ dimer units are effectively connected
to one another in two-dimensional sheets parallel to the (100) plane by
the neutral TTF molecules at I and III. Those at I are separated by
3.60 A from the $(TTF^+)_2$ pairs and could potentially interact through
overlap of the b_{1u} π orbitals on each unit. The ones at III also make
close sulfur–sulfur contacts to the TTF⁺ units in the dimer (3.51 A) thus
bridging these dimers together along the [001] direction (Figure 9). In
this case the interaction, if indeed one occurs, must involve the π orbitals
of the III unit and the σ orbitals of the $(TTF^+)_2$ dimer.

Conductivity of single crystal samples cut into rectangular parallele-
pipeds $\sim 1.0 \times 0.7 \times 0.4$ mm was measured by the Montgomery method
(28) with contacts placed at four parallel edges of the samples. Single

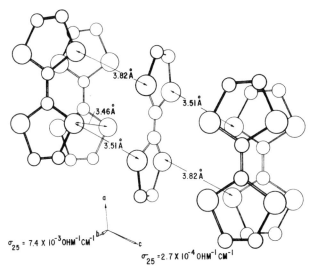

Figure 9. Interaction of TTF units in $(TTF)_2NiS_4C_4H_4$

$\sigma_{25} = 7.4 \times 10^{-3} OHM^{-1}CM^{-1}$

$\sigma_{25} = 2.7 \times 10^{-4} OHM^{-1}CM^{-1}$

crystal x-ray precession measurements were used to determine the relationship to the principal crystallographic directions. In this manner, the conductivities along the [010] and [001] direction and an approximate value for the conductivity perpendicular to the (100) plane were determined as a function of temperature. A thermally activated conductivity behavior much like that in MGS was observed with a thermal activation energy of 0.23 ± 0.02 eV, which was approximately independent of direction. The conductivity, however, showed a strong direction dependence and was highest along the [010] or TTF-stacking direction in the solid. The value observed (7.4×10^{-3} ohm^{-1} cm^{-1}) is relatively high on the scale of molecular semiconductors and indicates an appreciable π orbital interaction among the TTF units in the stack. The conductivity along the [001] axis, although definitely lower than that along the [010] axis, is also appreciable which suggests that the TTF $\sigma-\pi$ orbital interactions in this direction also provide a pathway for electron delocalization in the solid.

The lowest conductivity value determined ($\sim 5.0 \times 10^{-5}$ ohm^{-1} cm^{-1}) is that perpendicular to the (100) plane through the $NiS_4C_4H_4{}^-$ units where the interatomic separations are relatively large and the orbital interactions are weakest.

The origin of the conductivity in this compound is not yet known with certainty although it is tempting to speculate that the electron pair presumably involved in the $(TTF^+)_2$ binding can be thermally activated into conduction band states to produce carriers. Alternative mechanisms, perhaps even involving impurity species, certainly cannot be discounted at this point, especially in view of the findings for MGS. Further investigations of this unusual material are currently in progress.

Acknowledgment

The contributions of H. R. Hart, Jr., I. S. Jacobs, J. S. Kasper, G. D. Watkins, and S. H. Wee to the work described herein are gratefully acknowledged. We are indebted, in particular to H. R. Hart, Jr. and W. R. Giard of General Electric Corporate Research and Development for assistance with the conductivity and thermopower measurements.

Literature Cited

1. *Low Dimensional Cooperative Phenomena and the Possibility of High Temperature Superconductivity, NATO Advanced Study Institute,* Starnberg, Germany, 1974.
2. *German Phys. Soc. Conf., One-Dimensional Conductors,* University of Saarbrücken, July 1974.
3. *Symp. Conducting Org. Transition Metal Salts,* Lake Arrowhead, Calif., May 1974.
4. Interrante, L. V., Ed., "Extended Interactions Between Metal Ions in Transition Metal Complexes," *ACS Symp. Ser.* (1974) **5**.
5. *One-Dimensional Physics Workshop,* Boseman, Mont., July 1973.
6. Kistenmacher, T. J., Phillips, T. E., Cowan, D. O., *Acta Crystallogr. B* (1974) **30**, 763.
7. Krogmann, K., *Angew. Chem. Int. Ed. Engl.* (1969) **8**, 35.
8. Zeller, H. R., *Adv. Solid State Phys.* (1973) **13**, 31.
9. Tanner, D. B., Jacobsen, C. S., Garito, A. F., Heeger, A. J., *Phys. Rev. Lett.* (1974) **32**, 1301.
10. Interrante, L. V., in "Low-Dimensional Cooperative Phenomena: The Possibility of High Temperature Superconductivity," H. J. Keller, Ed., NATO Adv. Study Inst. Ser. B, Vol. 7, pp. 299–314, Plenum, New York, 1975.
11. Underhill, A. E., in "Low-Dimensional Cooperative Phenomena: The Possibility of High Temperature Superconductivity," H. J. Keller, Ed., NATO Adv. Study Inst. Ser. B, Vol. 7, pp. 287–297, Plenum, New York, 1975.
12. Gutmann, F., Lyons, L. E., "Organic Semiconductors," John Wiley and Sons, New York, 1967.
13. Lark-Horovitz, K., Johnson, V. A., Eds., "Methods of Experimental Physics. Solid State Physics," Vol. 6, Part B, Academic, New York, 1959.
14. Engler, E. M., Patel, V. V., *J. Am. Chem. Soc.* (1974) **96**, 7376.
15. Atoji, M., Richardson, J. W., Rundle, R. E., *J. Am. Chem. Soc.* (1957) **79**, 3017.
16. Martin, Jr., D. S., Rush, R. M., Kroening, R. F., Fanwick, P. F., *Inorg. Chem.* (1973) **12**, 301.
17. Day, P., *Inorg. Chim. Acta Rev.* (1969) **3**, 81.
18. Anex, B. G., *ACS Symp. Ser.* (1974) **5**, 276.
19. Day, P., *ACS Symp. Ser.* (1974) **5**, 234.
20. Martin, D. S., *ACS Symp. Ser.* (1974) **5**, 254.
21. Miller, J. R., *J. Chem. Soc.* (1965) 713.
22. Collman, J. P., Ballard, L. F., Monteith, L. K., Pitt, C. G., Slifkin, L., in "International Symposium on Decomposition of Organo-Metallic Compounds to Refractory Ceramics, Metals and Metal Alloys," K. S. Mazdiyasni, Ed., pp. 269–283, Dayton, Ohio, 1968.
23. Gomm, P. S., Thomas, T. W., Underhill, A. E., *J. Chem. Soc. A* (1971) 2154.
24. Fishman, E., Interrante, L. V., *Inorg. Chem.* (1972) **11**, 1722.

25. Day, P., Orchard, A. F., Thomson, A. J., Williams, A. J. P., *J. Chem. Phys.* (1965) **43**, 3763.
26. Interrante, L. V., Messmer, R. P., *Inorg. Chem.* (1971) **10**, 1174.
27. Henisch, H. K., "Crystal Growth in Gels," Pennsylvania State University, University Park, 1970.
28. Montgomery, H. C., *J. Appl. Phys.* (1971) **42**, 2971.
29. Cohen, M. J., Coleman, L. B., Garito, A. F., Heeger, A. J., *Phys. Rev. B* (1974) **10**, 1298.
30. Fritzsche, H., *Solid State Commun.* (1971) **9**, 1813.
31. Interrante, L. V., *J. Chem. Soc., Chem. Commun.* (1972) 302.
32. Mehran, F., Scott, B. A., *Phys. Rev. Lett.* (1973) **31**, 99.
33. Scott, B. A., Mehran, F., Silverman, B. D., Ratner, M. A., *ACS Symp. Ser.* (1974) **5**, 331.
34. Interrante, L. V., Browall, K. W., Hart, Jr., H. R., Jacobs, I. S., Watkins, G. D., Wee, S. H., *J. Am. Chem. Soc.* (1975) **97**, 889.
35. Kasper, J. S., Interrante, L. V., Secaur, C. A., *J. Am. Chem. Soc.* (1975) **97**, 890.
36. Gleiter, R., Schmidt, E., Cowan, D. O., Ferraris, L. P., *Elec. Spectr. Related Phenom.* (1973) **2**, 207.

RECEIVED January 30, 1975. Work supported in part by the Air Force Office of Scientific Research (AFSC), United States Air Force, under contract F-44620-71-C-0129.

2

Unusual Properties Associated with One-Dimensional Inorganic Complexes

JOEL S. MILLER

Physical and Chemical Sciences Laboratory, Webster Research Center, Xerox Corp., 800 Phillips Rd.-114, Webster, N. Y. 14580

In recent years there has been increased interest in inorganic and organic materials which exhibit highly anisotropic properties arising from a one-dimensional or pseudo-one-dimensional structure in the solid state. Besides being pleochroic, such one-dimensional materials may exhibit cooperative magnetic interactions (ferro and antiferromagnetic coupling) as well as unusual electrical properties (e.g. high conductivity and low thermal electric power). The unusual properties of 1-D systems are surveyed while focusing is on the structural and electronic features of the materials to aid in predicting new classes of materials which may exhibit unusual electrical, magnetic, and optical properties. These properties are discussed in terms of transition metal (row and group), oxidation state, geometry, ligand field strength, stoichiometry, etc.

One-dimensional(1-D) systems (*1, 2, 3, 4, 5, 6*) are best defined in terms of their anisotropic electrical, magnetic, and optical properties. All noncubic crystals will exhibit such anisotropic properties by virtue of the symmetry, but for this discussion a 1-D system is defined by an intensive property which is pronounced in a single direction with respect to the remaining orthogonal directions—*e.g.* the electrical conductivity (*7*) and optical (*8*) and reflectance (*9*) spectra of $K_2Pt(CN)_4Br_{0.300} \cdot 3H_2O$ (*10, 11*). These pronounced anisotropic properties arise from stronger intermolecular interactions in one dimension in the crystal than in the remaining directions. In the context of the broad subject of inorganic chemistry, the complexes which exhibit strong anisotropic properties are relatively few in number, and, having been studied in detail only recently, they are classified as unusual. The most extensively studied 1-D inorganic complex $K_2Pt(CN)_4Br_{0.300} \cdot 3H_2O$, is depicted in Figure 1.

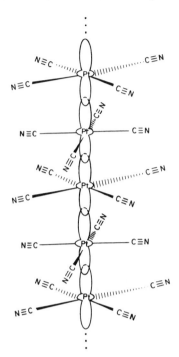

Figure 1. Structure of a conducting chain of the 1-D complex $K_2Pt(CN)_4Br_{0.300} \cdot 3H_2O$

This paper focuses on the unusual properties associated with 1-D inorganic complexes which makes them worthy of special study. Although a variety of properties are unusual because of intermolecular interactions in the solid, only those most familiar to inorganic chemists are discussed —namely, the optical, magnetic, and electrical properties. Topics more familiar to physicists (2, 3, 6)—e.g., thermoelectric power, neutron scattering, piezoresistance, and specific heat—are not discussed. In addition, special features of structure, stoichiometry, ligand field strength, and oxidation state are discussed since they frequently yield clues that a complex may exhibit unusual properties. The survey is not intended to be exhaustive but rather to give an indication of which properties are most useful in the characterization of new and, in particular, highly conducting 1-D complexes.

The following is a case history of the one-dimensional inorganic complex, $K_{\sim0.5}Ir(CO)_2Cl_2$, which exhibits high conductivity at room temperature. This complex exhibits a variety of unusual properties which indicate a columnar structure and a partially oxidized character in spite of the original proposals for its structure. This material serves as an instructive example to point out several features which make the complex worthy of detailed study. To date, the detailed study of its electrical, magnetic, and optical properties has not been reported.

In 1961 Malatesta and Canziani (12) reported the reaction of potassium hexachloroiridate(IV) with carbon monoxide and a copper catalyst above $100°C$ at high-pressure to yield $KIr_2(CO)_4Cl_4$ and $K_2Ir(CO)_4Cl_{4.8}$, which suggests a mixed valent character for the materials. These complexes had a metallic luster and were reported to be diamagnetic in the solid and in acetone. The bromo analogs, $KIr_2(CO)_4Br_4$ and $K_2Ir_2(CO)_4$-Br_5, were also prepared as were a variety of salts with different cations. The former bromo complex exhibited ν_{co} at 2092 and 2053 cm^{-1} which agrees with the ν_{co} absorptions of various halocarbonyl Ir^I and Ir^{II} complexes. An almost complete and continuous IR absorption was also noted which presumably arises from the high reflectivity of the material. The authors formulated dinuclear structures to account for the stoichiometry and the terminal carbonyl groups. However, these structures do not account for the two carbonyl absorptions in the IR, the solid-state and solution diamagnetism, or the metallic color of these complexes which is in contrast with the colors of other iridium complexes.

A decade later Cleare and Griffith (13) reported that the reaction of concentrated hydrohalic and formic acids with hexachloroiridate(IV) results in the formation of diamagnetic needle crystals of $[Ir_2(CO)_4X_4]^{1-}$ stoichiometry which exhibit metallic reflections. In accord with the work of Malatesta and Canziani (12), they prepared the bromo complexes as well as complexes containing different cations, but they did not prepare complexes containing more than four halides. To account for the observed diamagnetism, they proposed a planar tetranuclear structure based on the structure of the isoelectronic $Re_4(CO)_{16}^{2+}$ ion (14).

In 1972 Buravov et al. (15) reported the temperature-dependent four-probe polycrystalline dc conductivity of $KIr_2(CO)_4Cl_4$ and K_2Ir_2-$(CO)_4Cl_{4.8}$ prepared by the method of Malatesta and Canziani (12). Both complexes had high conductivity, which implies intermolecular interactions and a partially oxidized character in the solid. The previously proposed dinuclear and tetranuclear structures lack intermolecular interactions and thus must be discounted.

Recently Krogmann et al. postulated a cation-deficient, partially oxidized formulation for this material, $K_{\sim 0.5}Ir(CO)_2Cl_2$, and a columnar structure in the solid (16). This one-dimensional structure is consistent with the observed diamagnetism, stoichiometry, IR spectra, dc conductivity, and visual appearance of the complexes. Preliminary powder x-ray data support this formulation with a short 2.86 A Ir–Ir distance. This information sets a basis for the interpretation of detailed optical, electrical, and magnetic measurements that will help in further understanding of the chemistry and physics of one-dimensional inorganic complexes.

In addition to a discussion of the physical properties, a short description of the features important in the understanding of highly conducting materials is presented. These include crystallographic, stoichiometric, transition metal, oxidation state, and ligand field strength artifacts associated with partially oxidized inorganic systems.

Optical Properties

Solution. Inorganic complexes typically exhibit electronic absorption spectra in the UV–visible region which are assigned to $\pi \rightarrow \pi$, d–d, M $\rightarrow$ L charge transfer and L $\rightarrow$ M charge transfer absorptions. Near-IR absorptions have been reported for fewer complexes, and they are generally assigned to d–d and charge transfer absorptions [*e.g.* of $Ni(S_2C_2(CN)_2)_2{}^{z-}$ (*17, 18*)] or to intravalence transfer (IVT) absorptions (*19, 20, 21, 22, 23, 24*) arising from the Franck–Condon barrier to electron transfer between metal sites of a mixed valent dimer [*e.g.* $(H_3N)_5RuN$⬡NRu-$(NH_3)_5{}^{5+}$ (Complex I), Equation 1 (*25, 26*)]. In terms of the Robin and Day classification, this is a Class II material in which the metal ions are distinguishable (*21, 24*).

$$(H_3N)_5Ru^{II}N\text{⬡}NRu^{III}(NH_3)_5{}^{5+} \xrightarrow{\quad h\nu \quad}$$

$$(H_3N)_5Ru^{III}N\text{⬡}NRu^{II}(NH_3)_5{}^{5+} \quad (1)$$

The properties of the mixed valent complexes typified by I are unusual because of the nature of the low energy absorption, and unusual magnetic and electrical properties would be expected if the discrete dimer molecules interacted in the solid. Unusual optical, magnetic, and electrical properties are expected for mixed valent complexes of infinite length, *i.e.* going from Robin and Day Class III-A to Class III-B (*21, 24*).

Another unusual optical property is solvatochromism (*27, 28*). Most inorganic complexes have small solvent-dependent absorption spectra. Pronounced solvatochromism has not been extensively characterized for inorganic complexes in solution with the exception of a recent report on the electronic spectra of a variety of dipolar dithiolene α-diimine nickel complexes (*29*).

Solid. Most inorganic complexes have solid-state electronic spectra that are virtually superimposable with the solution spectra. For these complexes it can be stated that the same species (same electronic structures) exist in both states. Deviations from the identity of solution and solid-state data may result from (a) equilibria arising from association, *e.g.*

$$Ru[S_2C_2(CF_3)_2]_2PPh_3 + PPh_3 \rightleftharpoons Ru[S_2C_2(CF_3)_2]_2(PPh_3)_2 \quad (30)$$

or from different conformations in the solid than in solutions [*e.g.* $Ru(S_2C_2(CF_3)_2)(CO)(PPh_3)_2$ (*30, 31*)] or (b) intermolecular interactions arising from metal–metal bonding [*e.g.* $M(HDPG)_2$ ($H_2DPG =$ diphenylglyoxime; M = Ni (*32, 33*), Pd (*32, 33*), Pt (*34*))], from dimerization [*e.g.* $Ni(S_2C_2(CN)_2)_2^{1-}$ (*17, 18, 35*)], from infinite metal–halo interactions [*e.g.* $[Pt^{II}(en)X_2]$ $[Pt^{IV}(en)X_4]$ (*36*)], or from the delocalization of free electrons resulting in a Drudé behavior and a plasma absorption characteristic of a metal [*e.g.* $K_2Pt(CN)_4Br_{0.300} \cdot 3H_2O$ (*2, 6, 9*)]. In aqueous solution $K_2Pt(CN)_4Br_{0.300} \cdot 3H_2O$ has an absorption spectrum which can be decomposed into the spectra associated with Pt^{II}-$(CN)_4^{2-}$ and *trans*-$Pt^{IV}(CN)_4Br_2^{2-}$. Neither of these latter species in solution or in the solid exhibits the absorptions observed for $K_2Pt(CN)_4$-$Br_{0.300} \cdot 3H_2O$ in the solid (*10*).

Single crystals which do not have cubic symmetry should exhibit dichroism or pleochroism. Pronounced dichroism is generally associated with large anisotropic properties and is characteristic of one-dimensional materials, *e.g.* $M(CO)_2(acac)$ (M = Rh, Ir (*37*) and $K_2Pt(CN)_4Br_{0.300} \cdot 3H_2O$ (*8, 9*). Band semiconductors should have an absorption that corresponds to the band gap, but it may be difficult to observe. Highly conducting one-dimensional materials have a plasma edge which is the signature of a metallic state.

Color. In addition to white, inorganic complexes exhibit colors reminiscent of components of the rainbow. Atypically some complexes have high reflectivity in the visible region and a metallic luster. A metallic luster need not indicate a metallic state since a plasma absorption is necessary for the existence of a metallic state. For example, whereas crystals of $K_2Pt(CN)_4$ are light yellow and crystals of $K_2Pt(CN)_4 \cdot 3H_2O$ are white, the crystal faces of $K_2Pt(CN)_4Br_{0.300} \cdot 3H_2O$ are copper colored and have a plasma absorption (and metallic state) above 620 nm. Surfaces of semiconducting (*38*) crystals of $Ir(CO)_2(acac)$ are also metallic gold in appearance and have a strong reflectance at 568 nm (*16, 37*). The eye cannot distinguish between the sources of strong reflectances, and a metallic luster is a necessary but not sufficient condition for a metallic state; thus "All is not gold that glistens" (*39*).

Magnetic Properties

Coordination complexes typically exhibit identical magnetic behavior (diamagnetism or paramagnetism) in both the solid state and in solution. Deviations from solution and solid-state magnetic data suggest equilibria [*e.g.* T_d Ni(II) $\rightleftharpoons$ D_{4h} Ni(II)] or association with ferromagnetic or anti-ferromagnetic coupling. Temperature independent (Pauli) paramagnetism arising from the magnetic moment of conduction electrons usually is not observed for inorganic complexes in the solid state but is characteristic of a metallic system (*40*). Thus the molar magnetic susceptibility should be positive in sign and lower in value than would be expected for one unpaired electron per metal site, *i.e.* $\mu_{eff} < 1.73$ BM. A one-dimensional metallic system should exhibit temperature-independent paramagnetism. A variety of first-row paramagnetic metal complexes, *e.g.* Me_4NMnCl_3, forms columnar structures in the solid state which exhibit antiferromagnetic coupling (*6, 41, 42*). Examples of systems with ferromagnetic coupling have also been reported (*6, 41*).

Electrical Properties

Most inorganic and organic materials have low conductivity in either single-crystal or polycrystalline form. Values typically less than 10^{-10} ohm^{-1}cm^{-1} are observed, and the complexes are labeled as to insulators. These materials are probably large gap semiconductors with large impurity contributions, and only a careful, detailed study of the temperature dependence of the electrical conductivity (including electric field dependence) on single crystals can adequately characterize them. A plot of log conductivity *vs.* inverse temperature should result in a straight line for a semiconductor (assuming that the mobility is temperature independent). Some inorganic complexes atypically have a much larger conductivity at room temperature which may arise from a metallic state. In contrast to a semiconductor, the conductivity of a metal increases with decreasing temperature and exhibits a more complex temperature dependence. Thus a detailed study of the temperature dependence is necessary to distinguish between a small gap semiconductor and a metallic state (*6*). $K_2Pt(CN)_4$-$Br_{0.300} \cdot 3H_2O$ (and possibly the chloro analogue) is the only 1-D transition metal complex which has been characterized so far as a metal (*2, 6*). At room temperature, $K_2Pt(CN)_4Br_{0.300} \cdot 3H_2O$ has a corrected conductivity (*43*) of $\sim5.5 \times 10^3$ ohm^{-1}cm^{-1} (*7*) [compared with values of $\sim9 \times 10^{-6}$ for $K_2Pt(CN)_4 \cdot xH_2O$ (*44*) and 9.4×10^4 for platinum metal (*45*)] and an anisotropic conductivity ratio ($\sigma_{||}/\sigma_{\perp}$) of 10^5 (*7*). In addition to high conductivity, a low value for the thermoelectric power also characterizes a metallic state (*6*).

Crystallography

By definition, 1-D complexes have a columnar structure. The properties associated with these complexes arise from the electronic and steric properties of the molecules in the chain and the inter- and intrachain interactions. For 1-D square planar complexes which have collinear metal atoms, the spacings characterize the resultant properties. The spacings may be equivalent (Structure II) or inequivalent (Structure III) throughout the chain.

II   III

With equivalent spacing the shorter the spacings, the stronger the anisotropic properties. Complexes which have short equivalent spacings, *i.e.* $\lesssim 3.0$ A, have been characterized as partially oxidized as they have high electrical conductivity as well as strongly anisotropic properties, *e.g.* $K_2Pt(CN)_4Br_{0.300} \cdot 3H_2O$ with ~ 2.88 A spacings (46) (*see* Table I). For moderate spacings, $\sim 3.0 \simeq 3.5$ A, the observed anisotropic properties are not as dramatic and low conductivity is observed, *e.g.* $K_2Pt(CN)_4 \cdot xH_2O$ [~ 3.5A (49)]. For crystals with spacings which exceed the van der Waal radii, the complexes do not exhibit enhanced properties. Uniform short spacings are not sufficient to warrant enhanced properties. Strong metal–metal interactions are important. Thus, square planar third row transition metal complexes with the greater spatial extent and interaction with neighboring complexes of the $5d_{z^2}$ orbital enhance the anisotropic properties with respect to the first and second row congeners (4, 6).

A variety of divalent 1-D inorganic complexes are comprised of alternating cations and anions. The most extensively studied member of this class is Magnus' Green Salt, $[Pt(NH_3)_4][PtCl_4]$ (4, 6). Complexes of this type, as well as complexes comprised of chains of identical neutral molecules (6) [*e.g.* $Pt^{II}(ethylenediamine)Cl_2$ or $M(HDPG)_2$ (M = Ni, Pd, Pt)] containing metals with filled valence shells exhibit moderate ($\lesssim 3.5$ A) to long ($\gtrsim 3.5$ A) spacings, dichroic optical properties, diamagnetism, and low conductivity.

Several 1-D inorganic complexes are comprised of alternating mixed valent complexes with long equivalent metal–metal spacings. Complexes of this type have been observed for Au^I–Au^{III}, Pd^{II}–Pd^{IV}, and Pt^{II}–Pt^{IV} (4, 6). The complexes of $[M(amine)_2X_3]$ (X = halide) stoichiometry have halide atoms bridging the M^{II} and M^{IV} atoms, but they are clearly associated with the tetravalent metal (6). The resultant properties are best described in terms of the sum of the individual molecules and are placed in Class II by Robin and Day (21, 24). Low conductivity has been observed for compounds of this type.

Some collinear complexes do not have collinear metal–metal atoms, *e.g.* $P(C_6H_5)_3CH_3^+Ni[S_2C_2(CN)_2]_2^{1-}$ (Complex IV) (*50*). Complex IV does not have equivalent interplanar spacings. This complex exhibits a single ground state with a low lying triplet excited state (*51*). Its electrical properties characterize Complex IV as an intrinsic semiconductor along the chain axis with an anisotropic conductivity (*35*).

Structures of 1-D antiferromagnetically coupled paramagnetic first row transition metals exhibit ligand bridged polymers with equivalent long metal–metal spacings (*6, 41, 42*). For example, NMe_4MnCl_3 has a chain with three bridging halide atoms between the d^5 Mn^{II} atoms that provides a pathway for magnetic coupling.

In summary, crystal structure directly relates to the type and extent of intermolecular interactions which characterize the properties of the system. In general, planar molecules of the third row transition metal with short equivalent spacings exhibit the strongest anisotropic properties. These 1-D crystals are dark, dichroic needles which may appear metallic. Chain complexes of the first row paramagnetic ions frequently exhibit 1-D antiferromagnetic coupling.

Stoichiometry

An important artifact of highly conducting 1-D materials is the partially oxidized character of the metal which results from nonstoichiometry of ions in the unit cell (Table I). For example, $K_2Pt(CN)_4Br_{0.300} \cdot 3H_2O$ may be made reproducibly with a 0.300:1::Br:Pt (*11*) ratio: thus the system is nonstoichiometric. Historically this was explained by the invocation of mixed valence states for the metal. Recent work has revealed unequivocally that all the metal atoms are equivalent (*1, 2, 3, 4, 5, 6*) and they are classified by Robin and Day as Class III-B (*21, 24*). The unusual conductivity of these systems arises directly from the electronic properties associated with the partially oxidized (nonstoichiometric) system.

Nonstoichiometry can be assessed by precise elemental analysis. Extreme care must be taken to ensure that reproducible nonstoichiometry is characterized. For example, absorption spectra, x-ray fluorescence, neutron activation, and mass spectral analysis were used to determine the 0.300 ± 0.006:1::Br:Pt ratio for $K_2Pt(CN)_4Br_{0.300} \cdot 3H_2O$ (*10*). The errors associated with routine elemental analysis allow significant differences in the stoichiometry of nonstoichiometric materials. In order to understand fully the physics of 1-D systems, it is imperative that the exact stoichiometry be known as this relates directly to the band filling which, in general, cannot be obtained by alternative techniques.

A pair of computer programs was written to aid in the interpretation of chemical analysis (*52, 53*). For example, oxidation of bis(diphenyl-

Table I. Classes of Highly Conducting

Complex	*M–M, A*	*Cations/Anions*
$(Cation)_2Pt(CN)_4X_{\sim 0.3} \cdot yH_2O$	2.88–2.95	K^+, Rb^+, $C(NH_2)_3^+$ Cl^-, Br^-
$K_{1.74}Pt(CN)_4 \cdot yH_2O$	~2.95	
$(Cation)_xPt(O_2C_2O_2)_2 \cdot yH_2O$	2.88–2.9	alkali$^+$ alkaline earth^{2+}
$Ir(CO)_{2.93}Cl_{1.07}$ or $Ir(CO)_3Cl_{1.1}$	2.85	
$(Cation)_xIr(CO)_2X_2$	~2.86	H^+, K^+, Cs^+, NMe^+_4, $AsPh^+_4$, Li^+ Mg^{2+}, Ba^{2+} Cl^-, (Br^-?, I^-?)
$CaPt_2O_4$ [b]	2.8–3.0	
$Hg_{2.86}AsF_6$ [b]	2.64	
$(Cation)_xPt_3O_4$ [d]	~2.85	Na^+, Mg^{2+}, Cd^{2+}, Ni^{2+}
$(SN)_x$		

[a] Poorly characterized.
[b] 1-D in two dimensions (*see* footnote *f*).
[c] From Ref. *47*.
[d] 1-D in three dimensions (*see* footnote *f*).

```
                              NI(HDPG)2(BR)X


#    1 :  THF MOLECULAR WEIGHT IS   600.352 FOR:
 1.00  NI(HDPG)2,  .79  BR          ,  .00            ,  .0C           ,  .00                ,  .00             ,  .00          ,
                C      H      N     NI    BR      O                             C      H      N     NI     BR     O
PERCENT      56.02   3.69   9.33   9.78  10.52  10.66    .00    .00    .00     .26    .08   -.03   9.78   -.06  10.66    .00    .00
EMP. FORM.   35.44  27.85   5.06    .00   1.00    .00    .00    .00    .00     .38    .80    .02    .00    .00    .00    .00    .00
FORMULA      28.00  22.00   4.00   1.00    .79   4.00    .00    .00    .00

#    2 :  THF MOLECULAR WEIGHT IS   601.151 FOR:
 1.00  NI(HDPG)2,  .80  BR          ,  .00            ,  .00           ,  .00                ,  .00             ,  .00          ,
                C      H      N     NI    BR      O                             C      H      N     NI     BR     O
PERCENT      55.94   3.69   9.32   9.77  10.63  10.65    .00    .00    .00     .18    .08   -.04   9.77    .05  10.65    .00    .00
EMP. FORM.   35.00  27.50   5.00    .00   1.00    .00    .00    .00    .00    -.06    .45   -.05    .00    .00    .00    .00    .00
FORMULA      28.00  22.00   4.00   1.00    .80   4.00    .00    .00    .00

#    3 :  THF MOLECULAR WEIGHT IS   601.950 FOR:
 1.00  NI(HDPG)2,  .81  BR          ,  .00            ,  .00           ,  .00                ,  .00             ,  .00          ,
                C      H      N     NI    BR      O                             C      H      N     NI     BR     O
PERCENT      55.87   3.68   9.31   9.75  10.75  10.63    .00    .00    .00     .11    .07   -.05   9.75    .17  10.63    .00    .00
EMP. FORM.   34.57  27.16   4.94   -00   1.00    .00   -.00    .00    .00    -.49    .11   -.11    .00    .00    .00    .00    .00
FORMULA      28.00  22.00   4.00   1.00    .81   4.00    .00    .00    .00

#    4 :  THF MOLECULAR WEIGHT IS   602.749 FOR:
 1.00  NI(HDPG)2,  .82  BR          ,  .00            ,  .00           ,  .00                ,  .00             ,  .00          ,
                C      H      N     NI    BR      O                             C      H      N     NI     BR     O
PERCENT      55.80   3.68   9.30   9.74  10.87  10.62    .00    .00    .00     .04    .07   -.06   9.74    .29  10.62    .00    .00
EMP. FORM.   34.15  26.83   4.88    .00   1.00    .00    .00    .00    .00    -.92   -.22   -.17    .00    .00    .00    .00    .00
FORMULA      28.00  22.00   4.00   1.00    .82   4.00    .00    .00    .00


                              NI(HDPG)2(BR)X


THERE ARE    4 POSSIBILITIES.    3 OUT OF 4 AT   .30 PER-CENT, WHILE ALL ARE LESS THAN OR EQUAL TO   .30 PER-CENT.
THE TOLERANCE FOR THE EMPIRICAL FORMULA COMPARISON IS 999.00.
   .00  IS THE TOLERANCE FOR THE MOLECULAR WEIGHT COMPARISON WITH          .0


                            C      H      N      NI     BR     O
PERCENT                   55.76   3.61   9.36    .00  10.58    .00    .00    .00
EMPIRICAL FORMULA         35.06  27.05   5.05    .00   1.00    .00    .00    .00
                                                         UNITS  VARIED
         NAMF     CHARGE  M.W.    C  H  N NI BR  O          FROM   TO    BY
UNIT#1  NI(HDPG)2   0    537.22  28 22  4  1  0  4  0  0   1.00  1.00  1.00
UNIT#2  BR          0     79.91   0  0  0  0  1  0  0  0    .00  1.50   .01
 *STOP* 0
```

Figure 2. Sample computer output characterizing $Ni(HDPG)_2Br_x$ ($0.79 < x < 0.82$). The numbers on the right side indicate deviations from the observed microanalytical data (%: C 55.76, H 3.61, N 9.36, Br 10.58). Only possibilities in which all four analyses fall within 0.30% of the calculated values (left side) are printed. Note that in formulations #2 and #3 all four analyses fall within 0.20% of the calculated values.

Inorganic Complexes (6)

x	y	Conductivity $\sigma 273°$, $ohm^{-1}cm^{-1}$
	≲3.2	<300
	~1.8	a
~1.65	≲6	a
~0.83		
		~0.2 a
~0.5		>1 a
?		
~0.5		
		>10 a
		8 × 10³ c
<1, >0		>10³ a
		>10³ e,g

e From Ref. *48*.
f Anisotropy is suggested by crystallographic data, but it has not been confirmed experimentally.
g A superconducting transition has been observed at ~0.3°K, Ref. *61*.

glyoximato)nickel(II), Ni(HDPG)$_2$, with halogen (*54, 55, 56*) yields a series of materials of M(HDPG)$_2$X$_y$ (X = Br, I; $0 < y < 1.15$) stoichiometry in contrast to previous literature reports. A preliminary report suggests similar properties for bis(1,2-benzoquinonedioximato) complexes of the nickel triad (*57*). Figure 2 depicts a sample computer output which characterizes a set of analysis as Ni(HDPG)$_2$Br$_{0.79-0.82}$. Microscopy and crystallography demonstrated that this material is not a mixture of Ni(HDPG)$_2$ and Ni(HDPG)$_2$Br as is suggested by the data that appeared previously in the literature (*58*).

A new material may or may not be nonstoichiometric. Most complexes are stoichiometric. If an inorganic complex exhibits high anisotropic conductivity, the complex may be nonstoichiometric.

Upon reviewing the properties of inorganic complexes which exhibit high conductivity, one notices several trends which may be instructive in the prediction of new materials. The inorganic 1-D materials that are characterized by high conductivity (Table I) are comprised of partially oxidized square planar Ir^I and Pt^{II} complexes which contain rigorously planar ligands. The overall geometry allows for short intermolecular spacings and strong intermolecular interactions (provided for by the greater spatial extent of the $5d_{z^2}$ orbital with respect to the $4d_{z^2}$ and $3d_{z^2}$ orbitals). Partial oxidation seems to be necessary for high conductivity and seems to require a pair of stable oxidation states, but partial oxidation may stabilize a previously uncharacterized oxidation state of a metal. To date no partially oxidized Au^{III} complexes have been reported. This may be attributable to the instability of Au^{IV} or to the higher charge which will decrease the spatial extent of the $5d_{z^2}$ orbital. The ligands $(1, 4, 6)$ which have been successful in stabilizing partially oxidized materials range from monodentate (CO, Cl, CN) to bidentate (oxalate) and from a strong field ligand (CO) to a weak field ligand (oxalate). Thus no conclusion can be made about the ligands except for their planarity. Complexes containing macrocylic ligands are poor candidates for forming a highly conducting 1-D materials as the van der Waal radii for the intermolecular π interactions (*i.e.* > 3.3 A) is greater than the spacings required for strong metal-metal overlap (*i.e.* < 3.0 A). Cationic square planar complexes are also poor candidates for the formation of a highly conducting complex as the plus charge will contract the $5d_{z^2}$ orbital diminishing the overlap between adjacent molecules in the chain. Thus a likely candidate for the formation of a new highly conducting material would be based on $Ir^I(CN)_4{}^{3-}$. The 3− charge on this d^8 complex should increase the spatial extension of the $5d_{z^2}$ orbital enabling stronger intermolecular overlap along the chain. Initial attempts at preparing $Ir(CN)_4{}^{3-}$ has led to the isolation of $Ir^{III}(CN)_5H^{3-}$ from alcohol and water solutions $(1, 59)$ although the Rh^I analog has recently been isolated as the salt of a bulky cation (60).

New materials are necessary to characterize the inorganic parameters associated with the highly conducting materials as well as to understand the physics of one dimension $(2, 3, 5, 6)$.

Literature Cited

1. Krogmann, K., *Angew. Chem. Int. Ed. Engl.* (1969) **8**, 35.
2. Zeller, H. R., *Advan. Solid State Phys.* (1973) **13**, 31.
3. Shchegolev, I. F., *Phys. Status Solidi A* (1972) **12**, 9.
4. Thomas, T. W., Underhill, A. E., *Chem. Soc. Rev.* (1972) **1**, 99.
5. Garito, A. F., Heeger, A. J., *Acc. Chem. Res.* (1974) **7**, 232.
6. Miller, J. S., Epstein, A. J., *Prog. Inorg. Chem.* (1975) **20**, 1 and references therein.
7. Zeller, H. R., Beck, A., *J. Phys. Chem. Solids* (1974) **35**, 77.
8. Bernasconi, J., Brüesch, P., Kuse, D., Zeller, H. R., *J. Phys. Chem. Solids* (1974) **35**, 145.
9. Geserich, H. P., Hausen, H. D., Krogmann, K., Stampel, P., *Phys. Status Solidi A* (1972) **9**, 187.
10. Saillant, R. B., Jaklevic, R. C., *ACS Symp. Ser.* (1974) **5**, 376.
11. Saillant, R. B., Jaklevic, R. C., Bedford, C. D., *Mater. Res. Bull.* (1974) **9**, 289.
12. Malatesta, L., Canziani, F., *J. Inorg. Nucl. Chem.* (1961) **19**, 81.
13. Cleare, M. J., Griffith, W. P., *J. Chem. Soc. A* (1970) 2788.
14. Churchill, M. R., Bau, R., *Inorg. Chem.* (1968) **7**, 2606.
15. Buravov, L. N., Stepanova, K. N., Khidekel', M. L., Shchegolev, I. F., *Dokl. Chem.* (1972) **203**, 283.

16. Krogmann, K., Geserich, H. P., Wagner, H., Zielke, H. J., "Abstracts of Papers," 167th National Meeting, ACS, March 1974, INOR 221.
17. Shupack, S. I., Billig, E., Clark, R. J. H., Williams, R., Gray, H. B., *J. Amer. Chem. Soc.* (1964) **86**, 4594.
18. Maki, A. H., Edelstein, N., Davison, A., Holm, R. H., *J. Amer. Chem. Soc.* (1964) **86**, 4586.
19. Allen, G. C., Hush, N. S., *Prog. Inorg. Chem.* (1967) **8**, 357.
20. Hush, N. S., *Prog. Inorg. Chem.* (1967) **8**, 391.
21. Robin, M. B., Day, P., *Adv. Inorg. Chem. Radiochem.* (1967) **10**, 247.
22. Cowan, D. O., LeVanda, C., Park, J., Kaufman, F., *Acc. Chem. Res.* (1973) **6**, 1.
23. Day, P., *ACS Symp. Ser.* (1974) **5**, 234.
24. Day, P., *NATO Adv. Study Inst.* (1975) **7B**, 191.
25. Creutz, C., Taube, H., *J. Amer. Chem. Soc.* (1969) **91**, 3988.
26. *Ibid.* (1973) **95**, 1086.
27. Liptay, W., *Angew. Chem. Int. Ed. Engl.* (1969) **8**, 177.
28. Reichardt, C., *Angew. Chem. Int. Ed. Engl.* (1965) **4**, 29.
29. Dance, I. G., Miller, T. R., *J. Chem. Soc. Chem. Commun.* (1973) 433.
30. Miller, J., Balch, A. L., *Inorg. Chem.* (1971) **10**, 1410.
31. Bernal, I., Clearfield, A., Epstein, E. F., Ricci, J. S., Jr., Balch, A., Miller, J. S., *J. Chem. Soc. Chem. Commun.* (1973) 39.
32. Banks, C. V., Barnum, D. W., *J. Amer. Chem. Soc.* (1958) **80**, 3579.
33. *Ibid.* (1958) **80**, 4767.
34. Miller, J. S., Goldberg, S. Z., accepted for publication.
35. Miller, J. S., Epstein, A. J., "Abstracts of Papers," 167th National Meeting, ACS, March 1974, INOR 110.
36. Watt, G. W., McCarley, R. E., *J. Amer. Chem. Soc.* (1957) **79**, 4585.
37. Dessent, T. A., Palmer, R. A., Horner, S. M., *ACS Symp. Ser.* (1974) **5**, 301.
38. Pitt, C. G., Monteith, L. K., Ballard, L. F., Collman, J. P., Morrow, J. C., Roper, W. R., Ulkü, D., *J. Amer. Chem. Soc.* (1966) **88**, 4286.
39. Cervantes, M., "Don Quixote," Part II, Book III, Chap. 33, p. 1615.
40. Kittel, C., "Introduction to Solid State Physics," 4th ed., p. 518, John Wiley & Sons, New York, 1971.
41. Ackerman, J. F., Cole, G. M., Holt, S. L., *Inorg. Chim. Acta* (1974) **8**, 323.
42. Hone, D. W., Richards, P. M., *Ann. Rev. Mater. Sci.* (1974) **4**, 337.
43. Miller, J. S., *J. Amer. Chem. Soc.* (1974) **96**, 7131.
44. Minot, M. J., Perlstein, J. H., *Phys. Rev. Lett.* (1971) **26**, 371.
45. "Handbook of Chemistry and Physics," 53rd ed., p. F145, Chemical Rubber, Cleveland, 1972.
46. Krogmann, K., Hausen, H. D., *Z. Anorg. Allg. Chem.* (1968) **358**, 67.
47. Cutforth, B. D., Datars, W. R., Gillespie, R. J., van Schyndel, A., ADVAN. CHEM. SER. (1975) **150**, 56.
48. MacDiarmid, A. G., MiKulski, C. M., Saran, M. S., Russo, P. J., Cohen, M. J., Bright, A. F., Garito, A. J., Heeger, A. J., ADVAN. CHEM. SER. (1975) **150**, 63.
49. Moreau-Colin, M. L., *Bull. Soc. R. Sci. Liege* (1965) **34**, 778.
50. Fritchie, C. J., Jr., *Acta Crystallogr.* (1966) **20**, 107.
51. Weiher, J. F., Melby, L. R., Benson, R. E., *J. Amer. Chem. Soc.* (1964) **86**, 4329.
52. Miller, J. S., Goedde, A. O., *J. Chem. Educ.* (1973) **50**, 431.
53. Miller, J. S., Kirschner, S. Kravitz, S. H, Ostrowski, P., manuscript in preparation.
54. Edelman, L. E., *J. Amer. Chem. Soc.* (1950) **72**, 5765.
55. Foust, A. S., Soderberg, R. H., *J. Amer. Chem. Soc.* (1967) **89**, 5507.
56. Keller, H. J., Seibold, K., *J. Amer. Chem. Soc.* (1971) **93**, 1309.

57. Endres, H., Keller, H. J., Mégnamisi-Bélombé, Moroni, W., Nörte, D., *Inorg. Nucl. Chem. Lett.* (1974) **10**, 467.
58. Miller, J. S., Griffiths, C. H., manuscript in preparation.
59. Krogmann, K., Binder, W., *Angew. Chem., Int. Ed. Engl.* (1967) **6**, 881.
60. Halpern, J., Cozens, R., Goh, L-Y., *Inorg. Chim. Acta.* (1975) **12**, L35.
61. Greene, R. L., Street, G. B., Suter, L. J., *Phys. Rev. Lett.* (1975) **34**, 577.

RECEIVED January 24, 1975.

Binuclear Transition Metal Complex Systems

ULRICH T. MUELLER-WESTERHOFF

Physical Sciences Dept., IBM Research Laboratory, San Jose, Calif. 95193

*The search for organic materials with high electrical con-
ductivity has led to the highly anisotropic charge transfer
complexes of the TTF–TCNQ type (one-dimensional metals)
as an interim goal. Intermolecular exchange interactions between π-electron systems are limited but may be increased
by use of planar complexes of transition metals in either the
donor or the acceptor part of the charge transfer system.
There is ample precedent for extended interaction in square
planar inorganic complexes, but simple mononuclear organometallic compounds of this type cannot be conductive since
they form only dimers. However, binuclear planar complexes offer all the needed advantages to synthesize stable
anisotropic conductors: their particular combination of intra-
and intermolecular interactions may prevent the distortion
(Peierls instability) that leads to metal-to-insulator transition
in other one-dimensional materials.*

Some recent work of our group has concerned the synthesis of binuclear, completely planar transition metal complexes and their reaction with donor or acceptor molecules to give charge transfer complex systems with strong intramolecular as well as intermolecular interactions. We originally became interested in such systems several years ago because of the assumption that they would have unusual physical properties. This assumption is based on our—certainly not complete—understanding of the structural requirements on the molecular level needed to produce the optical and electronic phenomena that are associated with highly aniso-tropic molecular crystals of organic charge transfer complexes (including also salts of the tetracyanoquinodimethane (TCNQ) radical anion) (1, 2, 3) of inorganic mixed valence systems (such as KCP, $K_2Pt(CN)_40.3Br \cdot 3H_2O$, commonly called Krogmann's salt) (4, 5) and of our own mixed valence ferrocenophane complexes (6). An article on the design of organic metals appeared recently (7). In order to present a logical reasoning for our assumptions, we preface this discussion with a brief and general introduction to the peculiarities of organic conductors.

Conductivity

Electrical conductivity σ (expressed in $\Omega^{-1}cm^{-1}$ and defined as the reciprocal of the experimentally determinable resistivity ρ of a sample) is principally a function of the product of two main variables: the concentration of charge carriers n and their mobility μ. This statement holds for all normal conductors, excluding superconductors (σ infinite). With the latter, all electrical resistivity vanishes below a certain critical temperature T_c where a different conductivity mechanism takes over. The generally accepted Bardeen–Cooper–Schrieffer theory (8) of superconductivity is based on the existence below T_c of Cooper Pairs of electrons which travel with a common momentum and which therefore are not subject to the same scattering processes as single electrons, which limit the conductivity of normal metals. Free electrons are the carriers in metals like Cu and Ag, and, since their probability for collisions with scatterers increases with temperature, for metals $\sigma \simeq 1/T$. This means that even if the carrier concentration remains approximately constant, their net (forward) mobility and therefore also $n \cdot \mu$ decrease with increasing T. The opposite is true for other conductive materials like semiconductors. It is obvious that when a certain activation energy is required for the hopping of an electron from one site to the next, the mobility is proportional to T; therefore, when $n = $ const, $\sigma \simeq T$. When both n and μ are variable, a maximum in σ may result at some intermediate temperature.

Most organic and organometallic materials are closed shell systems and as such are insulators because there are no carriers unless impurities or injected carriers are present. Even then, mobilities are usually low, since the individual molecules in the solid are spaced relatively far apart. It therefore seems necessary to fulfill both of two conditions in order to create conductive organics or organometallics. These are (a) carriers must be generated by transforming the usual closed shell (all electrons paired) materials into open shell (unpaired spin) free radical or mixed valence systems; and (b) the overlap between these species (and with it the carrier mobility) must be increased by dense packing in the crystal (since parallel overlap of π systems is most efficient, dense packing in linear stacks is the most advantageous arrangement for creating high mobilities).

The intermolecular spacing, in addition to being close, must also be uniform along the entire stack within the crystal. Any deviation from equidistance will lead to the opening of a gap at the Fermi surface so that a semiconductor or insulator rather than a metal will be formed. In 1955,

Peierls (9) predicted that no strictly one-dimensional metal could exist because there would always be a distorted state of lower energy with pairs, (or, more general, multiples) of the basic unit (Figure 1).

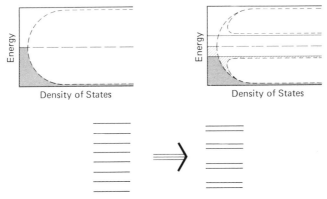

Figure 1.   The Peierls instability. A lattice distortion opens a gap, and a lower energy state is reached which has a completely filled valence band; the distorted state is insulating.

A crystal composed of such densely packed independent stacks of planar molecules must necessarily have unique anisotropic properties. There are indeed a number of purely organic and also organometallic materials which fulfill the above requirements and which seem to be one-dimensional metals. At present, these materials provide excitement and stimulus to physicists since they allow the study of physics in one dimension (anisotropic electrical, optical, and magnetic properties) and also offer the only promising possibility of obtaining high temperature superconductors. For technical applications, anisotropic properties may be a problem as well as a particularly useful asset.

In TTF–TCNQ, an organic charge transfer salt which at present is the most prominent example of an organic metallic system (2, 3), because of the suggestion (3) that it shows superconducting fluctuations, an electron is transferred from the strong donor TTF to the strong acceptor TCNQ thereby creating a radical cation–radical anion salt: both partners become open shell systems. In addition, both the donor and the acceptor moieties form independent stacks (10) allowing the transport of electrons and holes in the respective columns. The exact amount of charge transfer, although it is one of the most crucial quantities, is still disputed (11, 12) TTF–TCNQ has a maximum conductivity at about 60°K which is more pronounced in some samples than in others (3). All samples, however, become insulating below 50°K. The concentration of free spins as determined by ESR does not have such a maximum: it declines monotonically as the temperature is lowered (*12*). This would indicate that either the carrier mobility has a very sharp and so far unexplainable maximum, or that indeed there are superconductive fluctuation at 60°K. That TTF–TCNQ and related one-dimensional conductors become insulating at lower temperatures has been attributed to the Peierls instability, but other mechanisms were also proposed. An apparent exception to this rule is the TCNQ complex of bis-trimethylene-tetraseleno-fulvalene that has been prepared and studied by Cowan and co-workers (*13*); it maintains considerable conductivity even at 1.8°K.

The conduction mechanism and the nature of the metal–insulator transitions in organic charge transfer compounds with high anisotropies have been widely discussed. Numerous publications on TTF–TCNQ and related salts appeared during the past three years; it would be inappropriate to discuss them in this context. The unambiguous quantitative determination of conductivity in anisotropic media is at present a task frought with many pitfalls. An added difficulty in assessing exactly the absolute maximum of the conductivity (*e.g.* in single crystals of TTF–TCNQ) is the variation in sample purity and crystal perfection. Each crystal defect amounts to an interruption in the conductive one-dimensional strand, and, since each jump by an electron to an adjacent strand of these highly anisotropic materials (anisotropy ratios of the conductivity as high as 10^5 have been found) requires considerable energy, the conductive properties of these materials are sensitive to even the smallest amounts of impurities.

Observations were similar for inorganic mixed valence materials with linear stacks of transition metal complex ions such as KCP (*14, 15, 16, 17, 18, 19*). KCP is a Pt(II–IV) mixed valence system that is prepared by the partial oxidation of $K_2Pt(CN)_4$ with bromine. The most interesting consequence of this oxidation is the shrinking of the interplanar spacings of the square planar $Pt(CN)_4^{2-}$ units from 3.35 to 2.88 A, so that there is a sharply increased overlap of the platinum $5d_z^2$ AO's. The initial d_z^2 band widens considerably and, because of the removal of electrons from the d_z^2 level upon oxidation, this band is only partially filled, thus giving rise to metallic properties along the stacking axis (Figure 2). Although on an absolute scale the value of σ (about 500 $\Omega^{-1}cm^{-1}$) is not extraordinary when compared with that of typical metals like Cu or Ag ($\sigma \sim 10^6$ $\Omega^{-1}cm^{-1}$) and is even less than the conductivity of TTF–TCNQ ($\sigma \sim 2000$ $\Omega^{-1}cm^{-1}$ at room temperature), this type of linear interaction is what we want to exploit.

The principle of creating linear stacks of organometallic complexes and depleting the valence band of electrons by transforming the complexes into mixed valence cationic species seems tempting, especially if the counterions can also form independent stacks. To realize this, we

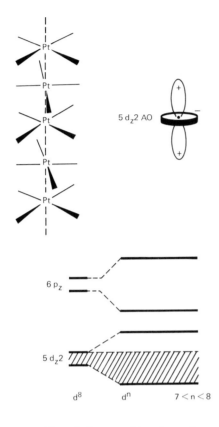

5 d$_{z^2}$ AO

6 p$_z$

5 d$_z$2

d^8 d^n 7 $<$ n $<$ 8

Figure 2.   KCP and related complexes. Overlap of 5d$_{z^2}$ AO's creates linear stacks. Removal of electrons from the 5d$_{z^2}$ band shortens the intermetallic distance from 3.2 to 2.9 A, the band widens, and, since it is not completely filled, this gives rise to anisotropic metallic properties.

considered the combination of organometallic mixed valence systems as open shell cations with the TCNQ radical anion as the counterion. (We shall not discuss here the related series of charge transfer salts derived from organic donors and organometallic acceptors.) The first aim then is to prepare organometallic planar and sterically not demanding donor molecules, which, either by themselves or upon oxidation by TCNQ or another strong acceptor that is known to be able to form segregated stacks, would arrange themselves into columns within which the overlap of the d_{z^2} atomic orbitals would produce the desired extended interactions.

Mononuclear Transition Metal Charge Transfer Systems

For the sake of simplicity in the following discussion, we shall consider only the very popular TCNQ as the example of an acceptor, and we shall also limit ourselves to the square planar Pt complexes as organometallic donors. The simplest example of a CT complex of the type we would want to have would then be a direct combination of the two.

According to the equation

$$nPt^{2+} + 2mTCNQ \rightarrow (n - m)Pt^{2+} + mPt^{4+} + 2mTCNQ^-$$

we would create Pt(II–IV) mixed valence stacks in which the ratio of Pt(II) to Pt(IV) would depend on the amount of TCNQ that could be accomodated in the parallel stack (the ratio a/b in Figure 3 determines the ratio n/m).

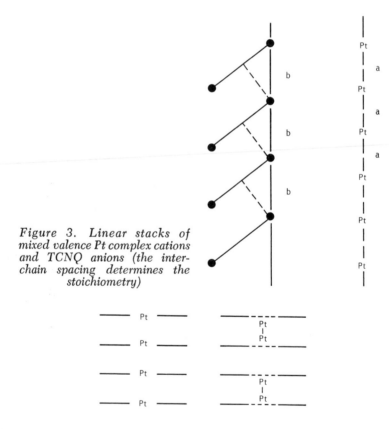

Figure 3. Linear stacks of mixed valence Pt complex cations and TCNQ anions (the interchain spacing determines the stoichiometry)

Figure 4. Linear stacks of transition metal square planar complexes with organic ligands. Left: uniformly spaced; no significant interactions. Right: strong pairwise interactions in the distorted state.

Unfortunately, such a simple system has to be insulating. In order to create sufficient overlap between the Pt atoms (Pt–Pt distance 2.7–2.9 A), the molecules would have to distort so that the ligands can maintain their minimum distance of 3.1 A (Figure 4). This dimer structure is reminiscent of the distorted Peierls state discussed above. It could possibly also induce a distortion in adjacent TCNQ stacks so that the conductivity would drop to the level of poor semiconductors or insulators. Such a distortion, through which pairwise metal–metal interactions are maximized, was observed (20) with the unsubstituted dithiene complexes $Pd(C_2H_2S_2)_2$ and $Pt(C_2H_2S_2)_2$. Both compounds form distinct dimers in the solid state and maybe in solution also.

One way to avoid this dilemma is to use some solid state aggregates like TlC_5H_5 (CpTl). As a gas, CpTl is monomeric, but it forms an extended, somewhat irregular polysandwich chain in the solid phase. The two stable oxidation states for Tl are +1 and +3 so the formation of mixed valence systems by the oxidation of CpTl is conceivable (Figure 5). Experiments at our laboratory to produce such species were unsuccessful.

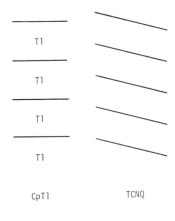

CpTl   TCNQ

Figure 5. Thallium cyclopentadienide as its TCNQ complex—a possible alternative to distortion

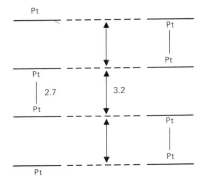

Figure 6. The ladder-type arrangement of binuclear complexes. While the ligands retain their minimum distance, transition metals can interact strongly in a pairwise fashion. Strong intramolecular coupling will lead to significant extended interactions.

Binuclear Charge Transfer Systems

We consider the use of completely planar, binuclear transition metal complexes as donors (with TCNQ as acceptor) to be the most promising approach to synthesizing organometallic metals. The main reason is that the distortion just described for mononuclear complexes is no longer a detriment, but rather it is a requirement for extended interactions. Of the several ways in which such binuclear complexes can stack in the crystal (21), let us consider just the ladder-type arrangement (Figure 6). Metal–metal bonds can form without creating a gap in the band structure—the distortion on one side of the system has its symmetric counterpart on the other half. If there is significant intramolecular interaction between the transition metals, then this arrangement (in the partially oxidized mixed valence state) must lead to pseudo one-dimensional behavior. This arrangement may also preclude a further distortion of the Peierls kind, thus avoiding the metal–insulator transition; these systems would be metals in the sense of the definition given in the introduction.

Even without extended intermolecular interactions, the intramolecular exchange processes of mixed valence systems appear to have a significant effect on the conductivity of adjacent TCNQ radical anion stacks. In the complex $(TCNQ)_2^-$ salt of bisfulvalene-diiron, BFD (1) (6), the polarizability associated with the strong metal–metal interactions apparently modulates the Coulomb repulsion within the TCNQ radical anion stack as an electron passes along. In this sense, $BFD(TCNQ)_2$ is a model for a Little superconductor (22). At 10 $\Omega^{-1}cm^{-1}$, the room temperature conductivity of $BDF(TCNQ)_2$ pellets is comparable to that of compressed disks of TTF–TCNQ. Unfortunately, it is exceedingly difficult to obtain reasonably sized, single crystals of this material so that a complete study has long been impossible. Only recently were we able to obtain a few sufficiently large crystals of reasonable quality to begin a structural study.

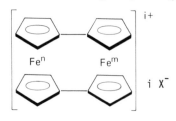

1

1a: $n = m = 2, i = 0$

1b: $n = 2, m = 3; i = 1; X = BF_4$, picrate, $(TCNQ)_2$

1c: $n = m = 3, i = 2, X = BF_4$

The conductivity at 300°K along the long axis of one such crystal (only one sample has been investigated so far) was 40 $\Omega^{-1}cm^{-1}$. This value seems low for the main axis conductivity of this material if one considers the high value found for the compressed pellet measurement. It may be that the axis along which this measurement was made is not the b axis but one of the other crystallographic axes. X-ray work to elucidate the structure is under way.

In the mixed valence BFD salts (in addition to the TCNQ compound, the picrate and tetrafluoroborate were also studied), the metal–metal interactions are so strong [Class III in the classification of Robin and Day (23)] that the metals are equivalent in every measurable respect. The doubly oxidized (24) Fe(III)–Fe(III) species is diamagnetic which is further evidence for strong interactions. Since the irons in BFD are 3.98 A apart (25), direct metal–metal interactions are not likely. We must then assume that in this case there exist metal–ligand–metal interactions of the magnitude in which we are interested. We propose that the BFD system be considered a delocalized aromatic species.

Equally strong interactions do not necessarily have to exist in square planar binuclear complexes, a problem that is of great concern to us. There is very little information in the literature about the synthesis or properties of such materials. Possibly planar binuclear systems are the ethanetetraaldehyde-bridged nickel aminotroponeiminato complexes, 2, reported by Trofimenko (26). The preparation of the bisimidazole complexes 3 was described (27) of which the tetracarbonyl derivatives might well be the very first example of completely planar binuclear transition metal compounds that are capable of extended interactions.

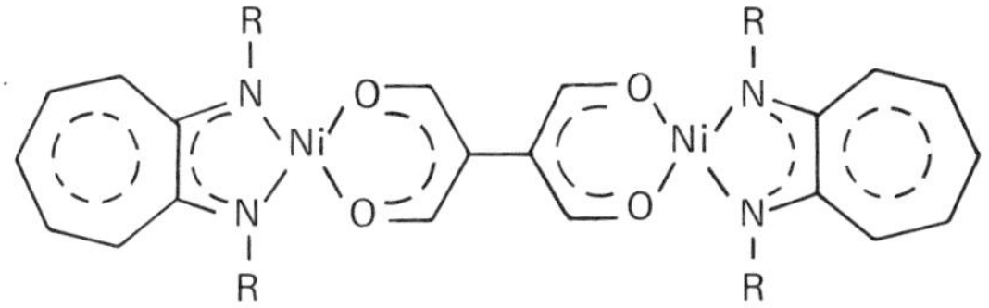

2

R = Me, Et, *n*-Pr

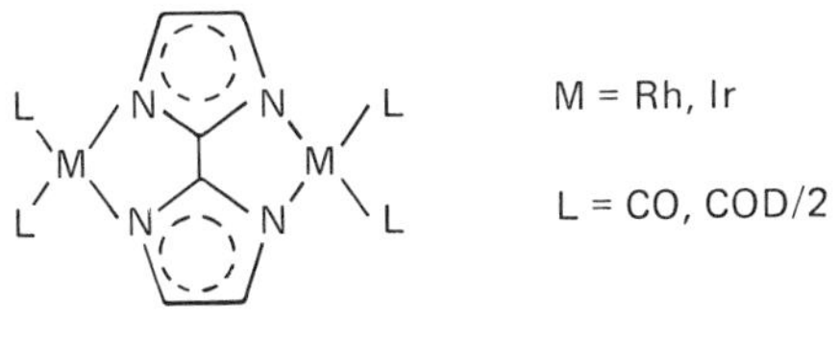

M = Rh, Ir

L = CO, COD/2

3

In our work, we considered three approaches to the synthesis of binuclear square planar systems: (a) preforming fully conjugated bifunctional macrocyclic ligands and then introducing the transition metal (this is a completely new field); (b) coupling appropriate mononuclear species [a binuclear nonplanar complex was prepared this way recently (28)]; and (c) using bridging ligands as in 2 and 3. A considerable part of our efforts has gone into the first approach.

Knowing the difficulties in the synthesis of porphine-like ligands, we expected that the synthesis of bisporphines would be no simple task but that it would also offer some synthetic rewards. The results exceeded our expectations in many respects except for the utility of the products in complex-forming reactions. We could synthesize compounds like 4 and 5

4   5

and thus demonstrate the feasibility of our proposal, but the complexes were mostly too bulky and insoluble to be of much use. In the area of one-dimensional conductors, product purity is of crucial importance since even minute amounts of impurities will cause interruptions in the conductive strands. Therefore, compounds which are difficult to purify are of little value. Because of these considerations, we have—at least temporarily—placed less emphasis on this approach. The idea of generating binuclear complexes with fully conjugated bifunctional rigid ligands still has its merits since, in complexes of this type, strong metal–ligand–metal interactions are guaranteed.

This is certainly not the case for ligand-bridged binuclear species. It is well known from the work of Taube (29) and others that, in systems

$$\left[(NH_3)_5 Ru - N \bigcirc N - Ru(NH_3)_5 \right]^{n+}$$

6

like the pyridazine-bridged ruthenium pentammine complex **6,** there is only very little interaction between the two metals. One of the reasons for this might be the π–MO symmetry properties of this bridging ligand. The important MO's have nodal planes through the nitrogens so there is no direct overlap between the ligand π system and the metal d orbitals; metal–metal interactions are thus restricted to the σ framework of the ligand.

For our own binuclear complexes, we prefer bridging ligands with two coordination sites for each metal. Neutral bridging ligands such as 2,2'-bipyrimidyl (**7**), TTF (**8**), and the ethylene tetrathioethers (**9**), all require dinegative outer ligands in order to form neutral Pt(II) species. Although calculations to predict the extent of the metal–metal interaction in such systems are still incomplete, we explored some of the possibilities for using these ligands. One additional consideration is that our final products should have electron donor properties. Although it is known

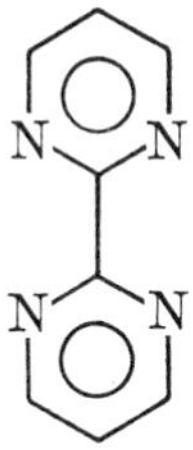

$$R = CH_3, \ C_2H_5, \ CH_2C_6H_5$$

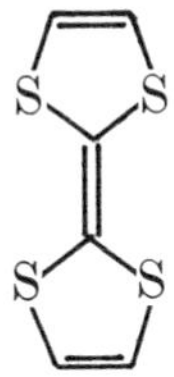

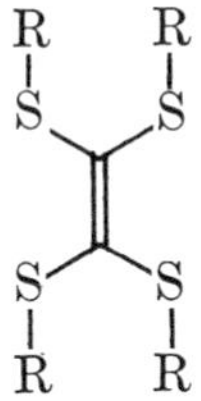

7   8   9

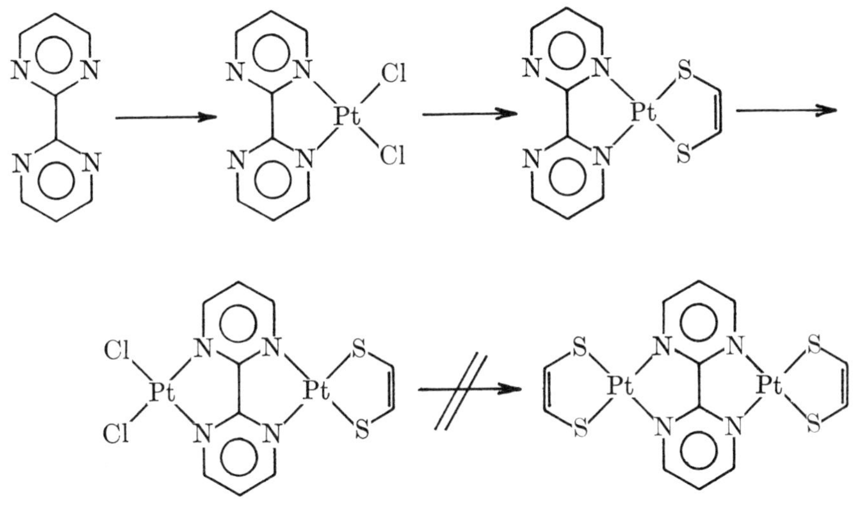

10

(*30*) that the dithiene complexes are reasonably strong acceptors instead, one of our first attempts was to synthesize the bipyrimidine-bridged dithiolate platinum complex **10**. The reaction of $PtCl_2$ or K_2PtCl_4 with bipyrimidine (bipyi) produces only the monobipyi–$PtCl_2$ complex. Apparently, the presence of one transition metal reduces the donor properties at the other two coordination sites sufficiently to prohibit complexation with a second metal. Reaction with ethylene dithiolate produces the mixed bipyi–dithiolate complex in which there is enough backbonding from the dithiolate ligand to allow the bipyi to coordinate to a second metal: $PtCl_2$ adds to form Cl_2Pt–bipyi–Pt–dithiolate. So far we have been unable to convert this material to **10**; all attempts lead to the isolation of the mononuclear bipyi–Pt–dithiolate. Although we found that $PdCl_2$ coordinates to both sides of the bridging ligand, the chemistry of this system has not yet been explored.

In general, the preparation of binuclear complexes with the features in which we are interested is hampered by the poor solubility of the intermediates. Furthermore, there are only a few ligands which will enhance the donor properties of, *e.g.*, Pt(II). This led us to investigate a new ligand system, propene–thione–thiol (PTT) or dithiomalonaldehyde, **11**. The PTT complexes are related to the well known SacSac compounds **12**, they but offer a better possibility for intermolecular interactions because of the absence of sterically unfavorable methyl groups. Schrauzer predicted (*31, 32*) that complexes like Pt(PTT)$_2$ would not be as stable as the dithienes since high-lying filled ligand MO's would considerably increase their reactivity. We believe that in PTT complexes the ligand HOMO is bonding and is further stabilized by complex formation. However, its energy is still high enough to provide the PTT complexes with

11   **12**

reasonably strong donor properties. In the last section of this paper, we describe this new class of compounds.

The Ni, Pd, and Pt PTT complexes are highly colored, crystalline, stable compounds which can be obtained by reacting tetraethoxypropane with gaseous HCl–H_2S in the presence of the appropriate transition metal chloride (*33*). Their NMR spectra indicate extensive delocalization. As can be seen from the chemical shift data for the acac, SacSac, malonaldehyde, and PTT complexes of Pd(II) (Table I), both the strong downfield shift and the large coupling constant in the PTT complex are indicative of an aromatic, delocalized, electronic ground state. The difference between the spectra of Pd(PTT)$_2$ and of its malonaldehyde congener 13 is particularly striking. We prepared 13 from tetraethoxypropane with

13

PdCl$_2$/HCl, followed by treatment with triethylamine; the yellow needles (mp 137°C) are considerably less stable than the PTT analog. Additional evidence for the aromaticity of the PTT complexes is the magnitude of the ^{195}Pt–H coupling constant in Pt(PTT)$_2$: the α-hydrogens couple with J = 104 Hz, and even a long range coupling to the β-hydrogen of 8 Hz is observed. For the unsubstituted dithiene Pt(C$_2$H$_2$S$_2$)$_2$, a ^{195}Pt–H coupling of 103 Hz was reported (20).

Table I.   Ring Proton NMR Data for Pd Compounds

Compound	Chemical Shift, τ Units
Pd(acac)$_2$	S 4.62
Pd(SacSac)$_2$	S 3.01
Pd(C$_3$H$_3$O$_2$)$_2$	D 2.90, T 4.44, J = 4 Hz
Pd(PTT)$_2$	D 0.98, T 2.52, J = 9 Hz

The donor properties of Pt(PTT)$_2$ are sufficiently pronounced to allow the direct formation of a TCNQ complex of 1:1 stoichiometry by reaction with an acetonitrile solution of the acceptor. It was noted with initial satisfaction that, in accord with our earlier prediction concerning the distortion of mononuclear species, the TCNQ complex is insulating: its main crystal axis conductivity at 300°K is 5 · 10^{-8} Ω^{-1}cm^{-1}. However, preliminary x-ray diffraction data for Pt(PTT)$_2$–TCNQ indicate that the crystal consists not of independent but of alternating D–A–D–A stacks. Therefore, there is still no proof for the distortion that we postulated in charge transfer salts of monouclear complexes.

The obvious extension of the PTT synthesis to include the tetrathio derivative of ethanetetraaldehyde as the bridging ligand in order to obtain the binuclear species 14 has not yet been successful. We are currently investigating complexes of PTT with other bridging ligands and also the charge transfer complex formation of Pt(PTT)$_2$ with the dithienes as

14

acceptors. We expect these materials to be better conductors than the purely organic charge transfer species because they have an uninterrupted conduction path along the stacks of the binuclear complexes. However, even if the absolute value of σ at room temperature should be no higher than for example that in TTF–TCNQ, these compounds will offer the possibility of achieving anisotropic conduction without the transition to an insulating state at low temperatures since the conductive state is already fixed into a distortion. Therefore, these materials are metals in the exact sense, and, regardless of the magnitude of their room temperature conductivity, they may even lead to a superconducting state.

Acknowledgment

Most of the work presented in this paper was carried out by the very able postdoctoral fellows P. Eilbracht (Darmstadt), F. Heinrich (Gottingen), and A. Alscher (Cologne), and by Adel Nazzal. We thank R. L. Greene of this laboratory for the conductivity measurements.

Literature Cited

1. LeBlanc, Jr., O. H., in "Physics and Chemistry of the Organic Solid State," Vol. 3, p. 182, Interscience, New York, 1963.
2. Ferraris, J. P., Cowan, D. O., Walatka, Jr., V., Perlstein, J. H., *J. Am. Chem. Soc.* (1973) **95**, 948.
3. Coleman, L. B., Cohen, M. F., Sandman, D. J., F. G., Yamagishi, Garito, A. F., Heeger, A. J., *Solid State Commun.* (1973) **12**, 1125.
4. Minot, M. J., Perlstein, J. H., *Phys. Rev. Lett.* (1971 **26**, 371.
5. Krogmann, K., *Angew. Chem. Int. Ed. Engl.* (1969) **8**, 35.
6. Mueller-Westerhoff, U. T., Eilbracht, P., *J. Am. Chem. Soc.* (1972) **94**, 9272.
7. Carito, A. F., Heeger, A. J., *Acc. Chem. Res.* (1974) **7**, 218.
8. Bardeen, J., Cooper, L. N., Schrieffer, J. R., *Phys. Rev.* (1957) **108**, 1175.
9. Peierls, R. E., "Quantum Theory of Solids," Oxford University, Oxford, 1955.
10. Kistenmacher, T. J., Phillips, T. E., Cowan, D. O., *Acta Crystallogr. Cryst Sect. B* (1974) **30**, 763.
11. Butler, M. A., Ferraris, J. P., Bloch, A. N., Cowan, D. O., *Chem. Phys. Lett.* (1974) **24**, 600.
12. Tomkiewicz, Y., Scott, B. A., Tao, L. J., Title, R. S., *Phys. Rev. Lett.* (1973) **32**, 1363.
13. Cowan, D. O., private communication.
14. Scott, B. A., *et al.*, *ACS Symp. Ser.* (1974) **5**, 331.
15. Krogman, K., Geserich, H. P., *ACS Symp. Ser.* (1974) **5**, 350.
16. Bloch, A. N., Weisman, R. B., *ACS Symp. Ser.* (1974) **5**, 356.

17. Zeller, H. R., Bruesch, P., *ACS Symp. Ser.* (1974) **5**, 372.
18. Saillant, R. B., Jaklevic, R. C., *ACS Symp. Ser.* (1974) **5**, 376.
19. Interrante, L. V., Messmer, R. P., *ACS Symp. Ser.* (1974) **5**, 382.
20. Browall, K. W., Interrante, L. V., *J. Coord. Chem.* (1973) **3**, 27.
21. Mueller-Westerhoff, U. T., Heinrich, F., *ACS Symp. Ser.* (1974) **5**, 403.
22. Little, W. A., *Phys. Rev. A.* (1964) **134**, 1416.
23. Robin, M. B., Day, P., *Adv. Inorg. Chem. Radiochem.* (1967) **10**, 247.
24. Mueller-Westerhoff, U. T., Eilbracht, P., *Tetrahedron Lett.* (1973) 1855.
25. Churchill, M. R., Wormwald, J., *Inorg. Chem.* (1969) **8**, 1970.
26. Trofimenko, S., *J. Org. Chem.* (1964) **29**, 3046.
27. Kaiser, S. W., Saillant R. B., Rasmussen, P. G., *J. Am. Chem. Soc.* (1975) **97**, 425.
28. Cunningham, J. A., Sievers, R. E., *J. Am. Chem. Soc.* (1973) **95**, 7183.
29. Taube, H., *Acc. Chem. Res.* (1969) **2**, 321.
30. McCleverty, J. A., *Prog. Inorg. Chem.* (1968) **10**, 49.
31. Schrauzer, G. N., *Acc. Chem. Res.* (1969) **2**, 72.
32. Schrauzer, G. N., *Transition Met. Chem.* (1969) **4**, 299.
33. Mueller-Westerhoff, U. T., Alscher, A., unpublished data.

RECEIVED February 26, 1975.

4

Mixed Valence, Semiconducting Ferrocene-Containing Polymers

CHARLES U. PITTMAN, JR., B. SURYNARAYANAN, and
YUKIHIKO SASAKI

The University of Alabama, University, Ala. 35486

A series of mixed valence, ferrocene-containing polymers of varying structural types was prepared. Their bulk conductivity was examined as a function of the Fe^{II}/Fe^{III} ratio. Polymers prepared included polyvinylferrocene, polyferrocenylene, polyethynylferrocene, and poly(3-vinylbisfulvalenediiron). Each was oxidized with electron acceptors such as dichlorodicyanoquinone, iodine, and TCNQ. The Fe^{II}/Fe^{III} ratio was controlled by varying the stoichiometry. The unoxidized polyvinylferrocenes, polyferrocenylenes, and polyethynylferrocenes were insulators ($\sigma = 10^{-14}\Omega^{-1}cm^{-1}$). Conductivities increased rapidly upon partial oxidation. Maximum conductivities, always obtained at 35–65% Fe^{III}, were 10^{-8}–$10^{-6}\Omega^{-1}cm^{-1}$ and were largely independent of the anion. Treatment of poly(3-vinylbisfulvalenediiron) with TCNQ gave $Fe^{II}Fe^{III}$ $(TCNQ)_2^-$ salts. Increasing the percentage of monooxidized bisfulvalenediiron moieties increased the conductivity. At 71% oxidation, $\sigma = 6 \times 10^{-3}\Omega^{-1}cm^{-1}$.

Organic polymers are usually insulators, but highly conjugated polymers (1) and organic charge-transfer complexes (2) can be semiconducting. Indeed, the poly(N-vinylcarbazole)–trinitrofluorenone complex is an excellent photoconductor. It was even demonstrated recently that several organic salts have metallic properties (3). For example, the salt formed by tetrathiafulvalene and tetracyanoquinodimethane (TCNQ), **1**, has a room temperature conductivity of $10^3\Omega^{-1}$-cm^{-1}, a negative temperature coefficient, and a negative thermoelectric power which is linear with temperature above 100°K (4, 5, 6, 7). Similarly, the (bisfulvalenediiron)$^+$ (TCNQ)$_2^-$ salt, **2**, has a high conductivity

46

$(\sigma = \simeq 10\Omega^{-1}cm^{-1})$ (8, 9). Thus, mixed valence cations in systems where the anion TCNQ⁻ can form ordered stacks, improves the conductivity. The conductivity of **2** was measured on a powder, and it would be expected to increase if a single crystal could be obtained.

The introduction of mixed valence states can also induce semiconductivity. Cowan and Kaufman (10, 11) demonstrated that the conductivity of biferrocene [Fe(II)Fe(III)] picrate, **4**, was six orders of magnitude greater than that of biferrocene, **3**, itself. Could this be extended to polymeric systems? We previously reported (12) the synthesis of mixed valence [Fe(II)Fe(III)] polyvinylferrocene and polyferrocenylene systems with significantly enhanced conductivities (10^{-7}–$10^{-8}\Omega^{-1}cm^{-1}$) as compared with the [Fe(II)Fe(II)] analogs ($10^{-14}\Omega^{-1}cm^{-1}$).

Thus, it was of interest to prepare, polymerize, and copolymerize vinylbisfulvalenediiron and then to convert the pendant bisfulvalenediiron (BFD) groups to mixed valence TCNQ salts in order to examine the conductivity of the resulting polymers. In addition, it was considered important to see if the completely conjugated system, polyethynylferrocene, would exhibit greater conductivities than polyvinylferrocene in mixed valence systems.

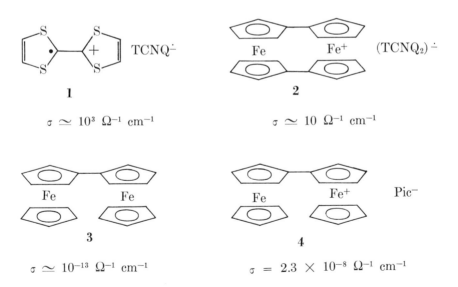

1

$\sigma \simeq 10^3 \ \Omega^{-1} \ cm^{-1}$

2

$\sigma \simeq 10 \ \Omega^{-1} \ cm^{-1}$

3

$\sigma \simeq 10^{-13} \ \Omega^{-1} \ cm^{-1}$

4

$\sigma = 2.3 \times 10^{-8} \ \Omega^{-1} \ cm^{-1}$

Results and Discussion

The synthesis, purification, and characterization of polyvinylferrocene (13), polyferrocenylene (14), and polyethynylferrocene (15) have been thoroughly described. Polyvinylferrocene was made by the azo-bis(isobutyronitrile) (AIBN)-initiated radical polymerization of vinyl-ferrocene, and various molecular weight distributions were obtained by fractionation (*13*). Polyferrocenylene was prepared by the polyrecom-bination technique using di-*tert*-butyl peroxide in molten ferrocene (*14*). Polyethynylferrocene was obtained by radical (AIBN-initiated) polym-erization of ethynylferrocene (*15*).

Poly(3-vinylbisfulvalenediiron), **10**, never previously reported, pre-sented an especially difficult synthesis because of logistical problems (*see* Figure 1). BFD, **6**, was prepared by the improved Mueller-Wester-hoff method (*11, 16*) in yields as high as 14%. We were unable to attain the 18–22% yields previously reported. The NMR spectrum of **6** had two triplets at τ 4.78 and 6.27 as reported, and its IR spectrum was identical to that of an authentic sample. Numerous attempts to mono-acylate **6**, using a variety of conditions, failed in dilute CH_2Cl_2 solutions.

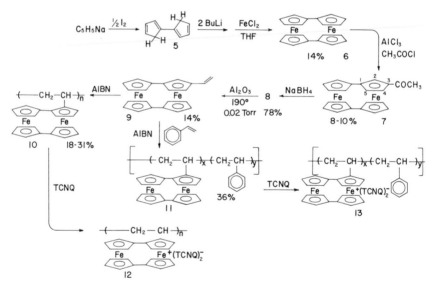

Figure 1.   Synthesis of poly(3-vinylbisfulvalenediiron)

BFD is only slightly soluble in benzene, CS_2, tetrahydrofuran (THF), CCl_4, acetone, CH_2Cl_2, and other organic solvents. This contributed to the difficulty in carrying out standard reactions. The use of acetic anhydride and $BF_3 \cdot OEt_2$, a method used successfully by Hauser on ferrocene (*16*), failed as did the use of acetic anhydride–H_3PO_4 as per Graham (*17*). Similarly, $SnCl_4$ and CH_3COCl in CH_2Cl_2 or benzene failed to give 3-acetylbisfulvalenediiron, **7**. The heterogeneous reaction of **6**, $AlCl_3$, and CH_3COCl (2.5 mmoles of each) in CH_2Cl_2 (30 ml) at

0°C followed by 24 hrs at 22°C gave at least seven products from which a mixture of four diacetylferrocenes was isolated by dry column chromatography with a 22% yield; however, 7 was not detected.

Acylation of 6 (7.5 mmoles) with $AlCl_3$ (7.5 mmoles) and CH_3COCl (7.5 mmoles) at 0°C in CH_2Cl_2 (350 ml) followed by 4 hrs at 22°C was successful and gave 0.41 g of crude, CH_2Cl_2-soluble material. Evaporation followed by acetone extraction yielded an insoluble product that was purified by dry column and high pressure liquid chromatography. The monoacetyl derivative, 7, was isolated with an 8–10% yield from a mixture of three diacetyl derivatives. The yields of 7 in six experiments were consistently within this range. A 16–22% yield of diacetylated derivatives was obtained, and 35–45% of unreacted 6 were recovered. Purified 7 had NMR bands ($CDCl_3$) at τ 4.5 (7H, multiplet, ring H's at 3 and 4 positions), 6.02 (8H, m, ring H's at 2 and 3 positions) and 7.62 (3H, singlet, acetyl methyl). A parent ion at $m/e = 410$, its IR spectrum, and the elemental analysis agree with the structure.

Reduction of 7 to 1-hydroxyethylbisfulvalenediiron, 8, in 78% isolated yield was effected with excess $NaBH_4$ in 10:10:1 $CHCl_3$:CH_3-OH:H_2O at 22°C for 4 hrs followed by dry column chromatography (deactivated silica, $CHCl_3$). The dehydration of 8 was effected by heating a thoroughly mixed sample of 8 (0.08 g) with alumina (1 g) in a vacuum sublimator (this technique was pioneered by Rausch and Siegal (17) for the synthesis of vinyl- and 1,1'-divinylmetallocenes). At At 210°–215°C and 0.7 torr, yield of the vinyl derivative 9 was only 4%. Reducing the pressure and temperature to 0.002 torr and 190°C gave 4 in 14% yield. Thus 9 was prepared with a 0.02% overall yield from NaC_5H_5. Compound 9 was identified by IR spectrum (strong C=C stretch at 1630 cm^{-1}), a parent ion at $m/e = 394$, and satisfactory analysis.

Radical-initiated polymerization of 9 in dilute degassed benzene solutions (40 mg 9, 25 ml benzene) using AIBN (10 mole % of 9) at 70°C for 24 hrs gave an 18% yield of low molecular weight (5000 *via* gel permeation chromatographic analysis) polymer, 10. Yields up to 31% were achieved by successive additions of 9 and initiator in more concentrated solutions. Polymer 10 was only slightly soluble in benzene and other solvents. This precluded kinetic studies of the polymerization and frustrated attempts to form higher molecular weight polymers. Solution copolymerizations of 9 with styrene (1:4 mole ratio) gave a 36% yield of copolymer 11 containing 28 mole % of 9. Polymers 10 and 11 were purified by repeated precipitations into petroleum ether and methanol.

Polymers 10 and 11 were then oxidized to their $(TCNQ)_2^-$ complexes. Both 10 and 11 were dissolved in excess benzene, $CHCl_3$, or benzonitrile and treated with excess TCNQ to give the mixed valence [Fe(II)Fe(III)] polysalt precipitates, 12 and 13. Analysis revealed that 71% of the BFD units in 12 were oxidized to the mixed valence form while 88% of those in 13 were oxidized in the most highly oxidized samples.

Studies were then conducted to establish that 13 had a mixture of BFD and BFD$^+$ (TCNQ)$_2^-$ moieties and not BFD plus BFD^{+2} or BFD$^+$ TCNQ$^-$.

Nitrogen-to-iron weight ratios of 0.71 and 0.88 in 6 and 8, respectively, established the 71% and 88% oxidation ratios. The near-IR spectra of 12 and 13 had a broad absorption at 1000–2000 nm with maximum intensity at 1400–1700 nm. This band in BFD$^+$ (TCNQ)$_2^-$ had previously been assigned to a photon-assisted intramolecular intervalence exchange (8, 9) and it confirmed that monooxidation of BFD units to [Fe(II)Fe(III)] occurred (BFD was dioxidized to its [Fe(III)Fe(III)] salts by Mueller-Westerhoff and Eilbracht (19); these salts had no absorption at 1400–1700 nm). Absorption at 600 nm was also pronounced. The Mössbauer spectrum of 12 was dominated by a single symmetrical absorption with a quadrupole splitting of 1.73 mm/sec which further confirms the BFD$^+$ structure. Neutral BFD's doublet (2.40 mm/sec) was also observed.

The suggestion that BFD$^+$ could be a delocalized system with each iron atom formally Fe(2½) was based on the Mössbauer spectrum, which shows only one type of iron present at 77°C, and on the ESCA spectrum (20).

Polyvinylferrocene, 14, polyferrocenylene, 15, and polyethynylferrocene, 16, were each oxidized by dichlorodicyanoquinone (DDQ), iodine, and TCNQ to give a polymer series (e.g. Equations 1, 2, and 3) with a range of Fe(II)/Fe(III) ratios. The amount of oxidation was controlled by the reactant stoichiometry, and Fe(II)/Fe(III) ratios were firmly established by both elemental analyses and by Mössbauer spectroscopy (21). In each case, the isomer shifts and quadrupole splittings were about 0.78 and 0.0–0.2 for ferricenium units. Each DDQ molecule incorporated into the polysalts was reduced to DDQ$^-$ as was indicated by the absence of the 1680 cm^{-1} ν_{CO} of DDQ and the presence of the 1590 cm^{-1} ν_{CO} of DDQ$^-$ (21). The Fe(II)/Fe(III) ratios from elemental analyses agreed within ±5% with those determined by Mössbauer spectroscopy. All mixed valence polymers were blue-black or black because of the ferricenium 620 nm $^2E_{2g} \rightarrow \, ^2E_{1u}$ transition. Polysalts of 15 had a broad electron-transfer band (12, 22) at 1100–1900 nm.

Oxidation of polyethynylferrocene with TCNQ was successfully effected in nitrobenzene or benzonitrile (but not in benzene, THF, or CH$_2$Cl$_2$) at 50°–90°C. Only 1:1 complexes could be obtained (each

$$(1)$$

$$(2)$$

$$(3)$$

ferricenium unit generated was associated with $TCNQ^-$ but not with $(TCNQ)^{2-}$. However, under these conditions, polyvinylferrocene gave only loose charge-transfer complexes with TCNQ, which exhibited no ferricenium absorption in the Mössbauer spectrum. TCNQ could be continously extracted from these polyvinylferrocene charge-transfer complexes with benzene. We were unable to prepare polysalts of **14, 15,** or **16** with $(TCNQ)_2^-$ groups incorporated for each ferricenium unit. This problem represents a challenge which is so far unsolved. There probably exists a delicate balance going from ferricenium–$(TCNQ)_2^-$ complexes in the presence of excess ferrocene groups to further oxidation of more ferrocene giving $TCNQ^-$ complexes.

Conductivity Measurements

The room temperature bulk conductivity of compressed disks of **12** was $6 \times 10^{-3} \Omega^{-1} cm^{-1}$ whereas that of **13** was $2.5 \times 10^{-5} \Omega^{-1} cm^{-1}$. With only 71% of the BFD units in **12** oxidized, and since **12** cannot pack in as orderly a manner as $BFD^+(TCNQ)_2^-$, the conductivity of **12** is remarkably high. Preliminary microwave conductivity measurements on **12** appear to give results similar to bulk studies with disks. The increase in conductivity as the $BFD^+/BFD + BFD^+$ ratio increases in polymer **12** is plotted in Figure 2. From the shape of the curve, one might predict a conductivity of 10^{-1}–$10^{0}\Omega^{-1}cm^{-1}$ when the polymer is completely oxidized.

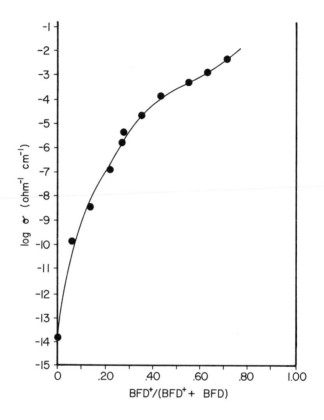

Figure 2. Variation of conductivity with increasing oxidation of polymer 12

Currently, we are attempting to monooxidize **10** completely and to prepare block copolymers of **9** with 1,3-butadiene, to be followed by oxidation with TCNQ.

Compressed pellets of the mixed valence polysalts of **14**, **15**, and **16** (0.6–1.0 mm thick) were made, and their conductivities were measured (*12*) at 200 V. Conductivity *vs.* Fe(III) content is plotted in Figures 3 and 4. Several conclusions are apparent.

  (a) Polymers **14**, **15**, and **16** are insulators with $\sigma = 8 \times 10^{-15}$, 1.2×10^{-14}, and $4.5 \times 10^{-14}\Omega^{-1}$ cm^{-1} respectively. A previous report (*23,*

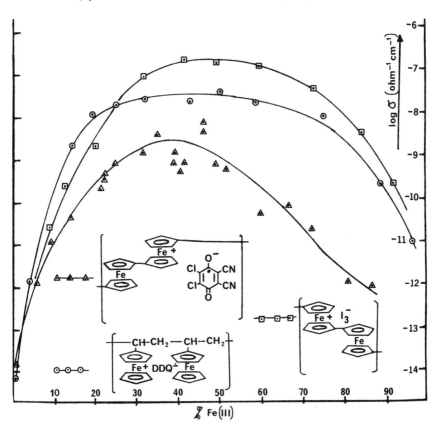

Figure 3. Conductivity of mixed valence DDQ polymers

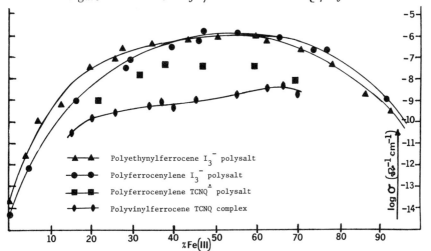

Figure 4. Conductivity of mixed valence ferrocene polymers

24) that the conductivity of **16**, prepared from acetylferrocene in $ZnCl_2$, was $10^{-10}\Omega^{-1}cm^{-1}$ must have been based on a polymer that was partially oxidized or that contained metallic impurities. To get accurate conductivity values, polymers **14–16** must be reduced before compaction in order to avoid incorporation of small amounts of ferricenium.

(b) Upon partial oxidation, all three polymers become semiconductors. Small percentages of oxidation cause large increases in conductivity.

(c) For all samples, conductivity maximized at 35–65% Fe(III), and in this region the curves are not steep.

(d) Conductivity increases 6–8 powers of 10 were achieved, but to a first approximation the conductivities were independent of the anion present.

(e) Increasing conjugation (going from complexes **14** to those of **16**) did not markedly affect the conductivity. Thus, no synergism existed between extended conjugation and mixed valence states in these samples.

These observations are consistent with an electron-hopping model where ferrocene is surrounded by a maximum number of nearest neighbor ferricenium moities. None of the polysalts exhibited photoconductivity. Under 200 and 500 W tungsten lamp irradiation, the conductivities were essentially unchanged. Conjugation in **15** or its polysalts could be interrupted by occasional serious tilting of adjacent ferrocene units with respect to each other, although the preferred conformation is the staggered structure shown in Equation 2 where conjugation is possible. It is thought from Mössbauer (*11*), UV (*6*) and magnetic susceptibility (*14*) studies that the positive charge in ferricenium and biferrocene [Fe(II)-Fe(III)] is somewhat localized on iron. However, Hartree–Fock calculations suggest that extensive charge delocalization onto the ligands has taken place. Thus, conjugative assistance of electron transfer through the π-system, in polysalts, might have been expected to have an appreciable effect on bulk conductivity of **15** and **16** (*vs.* **14**). Extended conjugation, if present in **15** or **16**, did not affect bulk conductivity. It is too early to suggest the mechanism of conduction in **12**.

Acknowledgment

T. K. Mukherjee of the Air Force Cambridge Research Laboratory is thanked for making a large number of conductivity measurements.

Literature Cited

1. Little, W. A., "Superconductivity of Organic Polymers," in "Electrical Conduction Properties of Polymers," *J. Polym. Sci. Part C* (1967) **17**.
2. Foster, R., "Organic Charge-Transfer Complexes," Academic, New York, 1969.
3. Garito, A. F., Heeger, A. J., *Acc. Chem. Res.* (1974) **7**, 232.

4. Ferraris, J., Cowan, D. O., Walatka, V., Perlstein, J. H., *J. Amer. Chem. Soc.* (1973) **95**, 948.
5. Coleman, L. B., Cohen, J. J., Sandman, D. J., Yamagishi, F. G., Garito, A. G., Heeger, A. J., *Solid State Commun.* (1973) **12**, 1125.
6. Cohen, M. J., Coleman, L. B., Garito, A. F., Heeger, A. J., *Phys. Rev. B,* in press.
7. Bloch, A. N., Ferraris, J. P., Cowan, D. O., Poehler, T. O., *Solid State Commun.* (1973) **13**, 753.
8. Cowan, D. O., LeVanda, C., *J. Amer. Chem. Soc.* (1972) **94**, 9271.
9. Mueller-Westerhoff, U. T., Eilbracht, P., *J. Amer. Chem. Soc.* (1972) **94**, 9272.
10. Cowan, D. O., Kaufman, F., *J. Amer. Chem. Soc.* (1970) **92**, 219.
11. *Ibid.* (1970) **92**, 6198.
12. Cowan, D. O., Park, J., Pittman, Jr., C. U., Sasaki, Y., Mukherjee, T. K., Diamond, N. A., *J. Amer. Chem. Soc.* (1972) **94**, 5110.
13. Sasaki, Y., Walker, L. L., Hurst, E. L., Pittman, Jr., C. U., *J. Polym. Sci. Part A-1* (1973) **11**, 1213.
14. Bilow, N., Landis, A. L., Rosenberg, H., *J. Polym. Sci. Part A-1* (1969) **7**, 2719.
15. Pittman, Jr., C. U., Sasaki, Y., Grube, P. L., *J. Macromol. Sci. Chem.* (1974) **A8** (5), 923.
16. Hauser, C. R., Lindsay, J. K., *J. Org. Chem.* (1957) **22**, 482.
17. Graham, P. J., Lindsay, R. V., Parshall, G. W., Peterson, M. J., Whitman, G. M., *J. Amer. Chem. Soc.* (1957) **79**, 3416.
18. Rausch, M. D., Siegal, A., *J. Organomet. Chem.* (1968) **11**, 317.
19. Mueller-Westerhoff, U. T., Eilbracht, P., *Tetrahedron Lett.* (1973) 1855.
20. Cowan, D. O., LeVanda, C., Collins, R. L., Candela, G. A., Mueller-Westerhoff, U. T., Eilbracht, P., *J. Chem. Soc. Chem. Commun.* (1973) 329.
21. Pittman, Jr., C. U., Lai, J. C., Vanderpool, D. P., Good, M., Prados, R., *Macromolecules* (1970) **3**, 746.
22. Kaufman, F., Cowan, D. O., *Macromolecules* (1970) **92**, 6198.
23. Paushkin, Ya. M., Vishnyakova, T. P., Polak, L. S., Patalah, I. I., Macius, F. F., Sokolinskaja, T. A., *Visokomol. Soed.* (1964) **6**, 545.
24. Paushkin, Ya. M., Vishnyakova, T. P., Polak, L. S., Patalah, I. I., Macius, F. F., Sokolinskaja, T. A., *J. Polym. Sci. Part C* (1964) **4**, 1481.

RECEIVED January 24, 1975. Work supported by the Office of Naval Research, the Air Force Cambridge Research Laboratory (contract F-1968-71-C-0107), and the National Science Foundation.

5

Hg$_{2.86}$ AsF$_6$—A Novel Structure with Unusual Electrical Properties

B. D. CUTFORTH, W. R. DATARS, R. J. GILLESPIE, and
A. VAN SCHYNDEL

Departments of Chemistry and Physics, McMaster University,
Hamilton, Ontario, Canada

The structure of Hg$_{2.86}$AsF$_6$ can be described as octahedral AsF$_6^-$ ions arranged on a lattice that contains linear non-intersecting channels in two mutually perpendicular directions. Within these channels are infinite chains of mercury atoms, each with a fractional formal charge of +0.35 and a mercury–mercury distance of 2.64 (1) Å. Conductivity experiments confirm the metallic nature of the compound. The resistivity, measured by a four-probe ac technique, decreases by a factor of 10^3 between room temperature and 4.2°K. No metal–insulator transition was detected.

R ecent studies on compounds containing mercury in oxidation states lower than +1 demonstrated that Hg$_3$(AsF$_6$)$_2$ (1) and Hg$_4$(AsF$_6$)$_2$ (2) can be prepared by treating mercury with a stoichiometric amount of AsF$_5$ in liquid SO$_2$. Both compounds contain linear, discrete, mercury polycations. Hg$_3$(AlCl$_4$)$_2$ has also been prepared (3) by the reaction of mercury with a molten HgCl$_2$–AlCl$_3$ mixture, and subsequent structural determination (4) revealed that the compound contained a very nearly linear Hg$_3^{2+}$ cation.

The initial product obtained in the reaction of mercury with AsF$_5$ was a crystalline solid with a distinct golden metallic lustre. X-ray crystallography (5) demonstrated that the compound could be formulated as Hg$_{2.86}$AsF$_6$, and it revealed the presence of infinite linear chains of mercury atoms in two mutually perpendicular directions. It was realized that the electrical properties of such a compound would be of considerable interest because the compound might be a highly anisotropic metallic conductor.

In recent years, the study of one-dimensional metallic systems has attracted a great deal of interest. A number of excellent reviews have been published (6, 7, 8, 9, 10) which were mostly concerned with two classes of one-dimensional compounds: the partially oxidized tetracyanoplatinate complexes such as $K_2[Pt(CN)_4]Br_{0.30} \cdot 3H_2O$, and the charge-transfer organic complexes involving tetracyanoquinodimethane (TCNQ) of which the best example to date is tetrathiofulvalinium tetracyanoquinodimethane [(TTF)(TCNQ)]. Structural studies of the partially oxidized teracyanoplatinate complexes (11) revealed that the platinum–platinum distance in the bromide is an unusually short 2.88 A. In such a partially oxidized compound, electrons have been removed from the highest occupied band arising from overlap of the platinum (d_{z^2}) orbitals, thus giving a partially filled one-dimensional conduction band. The anisotropic metallic nature of this compound was confirmed by extensive conductivity studies (12). Measurements of (TTF)(TCNQ) (13, 14) also demonstrated that the conductivity is metallic at room temperature with a large anisotropy ratio. We demonstrate in this paper that $Hg_{2.86}AsF_6$ is highly conducting although the one-dimensional nature of the conductivity has not yet been confirmed.

Experimental

The compound was prepared by a method which differed from that described previously (5). Because of the metallic nature of the compound and its resultant insolubility in available solvents, crystals suitable for conductivity studies were prepared by low temperature disproportionation of an SO_2 solution of the $Hg_4(AsF_6)_2$ cation. As the temperature was lowered from room temperature to $-30°C$ over a few days, the red solution of $Hg_4(AsF_6)_2$ disproportionated to yield soluble $Hg_3(AsF_6)_2$ and $Hg_{2.86}AsF_6$ which deposited as large (1 mm³) crystals on the side of the borosilicate glass tube. The crystals were washed with SO_2 to remove traces of $Hg_3(AsF_6)_2$ and $Hg_4(AsF_6)_2$, and they were then stored under vacuum in a sealed borosilicate glass tube.

Resistance was measured with four platinum wire probes in spring contact with the sample. Alternating current with a frequency of approximately 25 Hz was passed through the sample between two current probes. A phase-sensitive detector was used to amplify the voltage between the two potential probes and to obtain the in-phase component of the signal. The sample was mounted in a holder which was sealed by a rubber O-ring onto a borosilicate glass tube. (These manipulations were performed in a very tight dry box because of the extremely hygroscopic nature of the compound.) All contacts were checked before removing the sealed sample from the dry box. The sample holder was transferred to a stainless-steel helium dewar for resistance measurements. The temperature was monitord by a copper–constantan thermocouple mounted next to the sample. Bismuth and antimony samples were also measured to determine the resistivity for our contact configuration from the voltage–current relationship. The error in the absolute value of the resistivity was ~50% because of uncertainty in the probe position and in the effective sample thickness.

Results and Discussion

The structure of $Hg_{2.86}AsF_6$ (Figure 1) is an array of octahedral AsF_6^- anions conforming to the space group symmetry (tetragonal, $I4_1/amd$). There is an array of non-intersecting channels within the lattice running along directions a and b. Infinite chains of mercury atoms lie in these channels with a mercury–mercury distance of 2.64(1) A, a distance not commensurate with the lattice dimensions. The mercury–mercury bond length is considerably shorter than that found in metallic mercury (3.005 A). The chains are electron deficient (each mercury may be assigned a formal positive charge of +0.35). These facts suggest a metallic state. It seems reasonable to suppose that the crystals may display anisotropic conductivity, i.e. they might be highly conducting along the a and b axes and less so along the c axis.

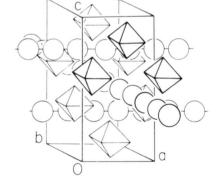

Figure 1. An isometric view of $Hg_{2.86}AsF_6$ showing the chains of mercury atoms (circles) running through the lattice composed of AsF_6^- ions (octahedra)

Resistivity as a function of temperature is plotted in Figure 2 for a typical sample. Measurements were made without regard for crystal orientation because it was not possible to obtain crystals with sufficiently well defined morphology to allow orientation of the crystals along the principal axes. The compound is highly conducting, the resistivity at room temperature being of the order of 130 $\mu\Omega$cm. This corresponds to a conductivity of 8×10^3 $(\Omega cm)^{-1}$, a value that is considerably larger than that found for the partially oxidized $K_2[Pt(CN)_4]Br_{0.30} \cdot 3H_2O$ which is typically about a few hundred $(\Omega cm)^{-1}$ along the highly conducting axis (12). It is even more extraordinary that the resistivity decreases so rapidly with decreasing temperature. The resistivity at 4.2°K was ~0.1 $\mu\Omega$cm which was the lower limit of detection.

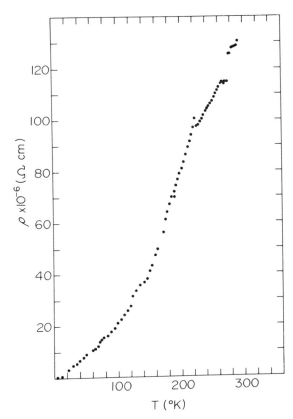

Figure 2. The resistivity of $Hg_{2.86}AsF_6$ as a function of temperature

Low temperature ($< 77°K$) data represent the average for two different samples; they are plotted in Figure 3 together with the measured temperature-dependent resistivity of a bismuth sample used for calibration. The decrease in resistivity by a factor of almost 10^3 over the temperature range studied is considerably larger than that observed in metals that are not highly purified, and it is consistent with a metallic state arising from the interaction of mercury atoms in a chain forming a partially filled band. The increase in conductivity with decreasing temperature normalized to the conductivity at room temperature is plotted in Figure 4.

Peierls (15) showed that a partially filled one-dimensional band can always lower its energy by splitting into filled and empty bands. If the compound under study is in fact one-dimensional, the failure to observe the transition is somewhat puzzling. $Hg_{2.86}AsF_6$ seems to fulfill the necessary conditions for a one-dimensional metallic system. These conditions may be summarized briefly as follows: (a) the structure must contain linear parallel rows of atoms; (b) the interaction between these atoms must be sufficiently strong to make band formation possible; and (c) the atoms composing the chain must have an odd or fractional formal valency, which would lead to a partially filled conduction band.

It is not yet clear to what extent the rather short interchain separation of 3.085 A affects conduction along the c axis. It may well be that although this distance is considerably longer than the intrachain mercury–mercury distance of 2.64 A, is is still short enough to provide a conduction path between the chains along the c axis. If this is true, it may mean that, although the conductivity is anisotropic, the anisotropy may be too small for detection. The fact that the conductivity was consistently high for a large number of samples mounted randomly with respect to orientation does not necessarily mean that the conductivity is not anisotropic. It can be shown that if the orientation is not within a few degrees of the crystal axes, then the component along the highly conducting direction will dominate the conductivity.

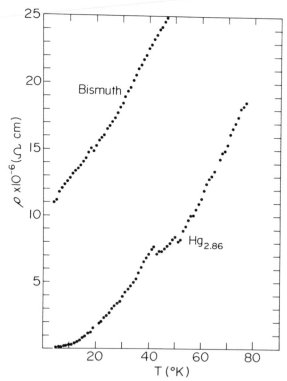

Figure 3.   The low temperature region of the resistivity of $Hg_{2.86}AsF_6$ and bismuth as a function of temperature

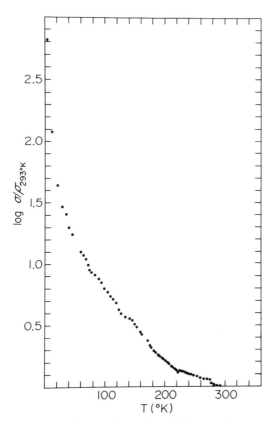

Figure 4. The conductivity of $Hg_{2.86}AsF_6$ as a function of temperature normalized to the room temperature conductivity

It is hoped that in the near future sufficiently well defined crystals will be obtained which will permit us to investigate the possible electrical anisotropy of this most interesting, highly conducting material.

Literature Cited

1. Cutforth, B. D., Davies, C. G., Dean, P. A. W., Gillespie, R. J., Ireland, P. R., Ummat, P. K., *Inorg. Chem.* (1973) **12**, 1343.
2. Cutforth, B. D., Gillespie, R. J., Ireland, P. R., *J. Chem. Soc. Chem. Commun.* (1973) 723.
3. Torsi, G., Fung, K. W., Begun, G. H., Mamantov, G., *Inorg. Chem.* (1971) **10**, 2285.
4. Ellison, R. D., Levy, H. A., Fung, K. W., *Inorg. Chem.* (1972) **11**, 283.
5. Brown, I. D., Cutforth, B. D., Davies, C. G., Gillespie, R. J., Ireland, P. R., Vekris, J. E., *Can. J. Chem.* (1974) **52**, 791.
6. Krogmann, K., *Angew. Chem. Int. Ed. Engl.* (1969) **8**, 35.
7. Thomas, T. W., Underhill, A. E., *Chem. Soc. Rev.* (1972) **1**, 99.

8. Shchegolev, I. F., *Phys. Status Solidi A* (1972) **12**, 9.
9. Zeller, H. R., *Festkoerperprobleme (Adv. Solid State Phys.)* (1973) **13**, 31.
10. Miller, J. S., Epstein, A. J., *Prog. Inorg. Chem.* (1975) **20**.
11. Krogmann, K., Hansen, H. D., *Z. Anorg. Allg. Chem.* (1968) **67**, 358.
12. Zeller, H. R., Beck, A., *J. Phys. Chem. Solids* (1974) **35**, 77.
13. Ferraris, J., Cowan, D. O., Walatka, V., Jr., Perlstein, J. H., *J. Amer. Chem. Soc.* (1973) **95**, 948.
14. Coleman, L. B., Cohen, M. J., Sandman, D. J., Yamagishi, F. G., Garito, A. F., Heeger, A. J., *Solid State Commun.* (1973) **12**, 1125.
15. Peierls, R. E., "Quantum Theory of Solids," p. 108, Oxford University, London, 1955.

RECEIVED January 24, 1975.

6

Synthesis and Selected Properties of Polymeric Sulfur Nitride, (Polythiazyl), $(SN)_x$

A. G. MacDIARMID, C. M. MIKULSKI, M. S. SARAN, P. J. RUSSO, M. J. COHEN, A. A. BRIGHT, A. F. GARITO and A. J. HEEGER

Departments of Chemistry and Physics, University of Pennsylvania, Philadelphia, Penna. 19174

Analytically pure single crystals of polymeric sulfur nitride (polythiazyl), $(SN)_x$, suitable for solid state studies were prepared by the spontaneous room temperature polymerization of single crystals of pure S_2N_2. The polymer is relatively inert to oxygen and water. A single-crystal x-ray study revealed that an $(SN)_x$ polymer molecule consists of an almost planar chain of alternating sulfur and nitrogen atoms in which the sulfur–nitrogen bond lengths are all very similar. The optical reflectance of crystalline films of $(SN)_x$ from the near uv through the visible region is metal-like in character for light polarized parallel to the polymer chain axis. The covalent polymer has the optical and electrical properties of a low-dimensional metal.

Metals, by definition, are those elements which possess certain characteristic chemical properties such as ease in forming positive ions by chemical processes and certain characteristic solid state physical properties such as high electrical conductivity (which increases as the temperature is lowered), high reflectivity of light, good thermal conductivity, ductility, and malleability. Until very recently, it was believed that these collective properties were unique to metallic elements, but it now appears that many of these properties may be found in a simple inorganic polymer that contains no metal atom. Studies of the compound, polymeric sulfur nitride (polythiazyl), $(SN)_x$, reveal that it possesses many of the above properties and may therefore be termed a metal even though its metallic characteristics are strongly anisotropic. It is believed that this polymer may be the forerunner of a whole new class of polymeric metals.

Space does not permit a complete discussion of the background information available on $(SN)_x$, but key landmarks in its study are summarized very briefly. Polymeric sulfur nitride was first prepared in 1910 by Burt (1) by passing S_4N_4 vapor over silver gauze or quartz wool at 100–300°C. It was deposited outside the hot zone as very small crystals or as a film—when thin, the film was blue by transmitted light, but, as it became thicker, it became opaque and took on a bronze metallic luster by reflected light.

In 1971 Boudeulle et al. (2, 3, 4, 5, 6) reported that S_2N_2, which was formed as a primary product when S_4N_4 vapor was passed over heated silver wool, would polymerize at low temperatures to give pseudo single crystals composed of layers of parallel fiberlike crystals of $(SN)_x$. The crystals were too small for single-crystal x-ray studies, but they were used in a preliminary electron diffraction investigation. Boudeulle reported recently that the S–N bond lengths alternated between short (1.58 A) and long (1.72 A) bonds which are close to double and single S–N bond lengths, respectively. Also bond angles of 113.5° for S–N–S and 111.5° for N–S–N were reported. By using more reliable single-crystal x-ray diffraction methods, we recently demonstrated (7) that these data are not correct (the S_2N_2 and $(SN)_x$ structures were determined by the Molecular Structure Corp., College Stations, Texas).

In 1973 Labes et al. (8, 9) synthesized crystalline bundles of impure $(SN)_x$ fibers. Although the S:N atomic ratio was 1:1, the material contained 5.48% impurity (4.93% O, 0.42% H, and 0.13% C). However, metallic-like conductivity was observed in directions parallel to the $(SN)_x$ fibers, and this increased sharply with decrease in temperature. Six different samples had conductivities at room temperature of 10, 89, 230, 640, 1470, and 1730 ohm^{-1} cm^{-1}. Since the electrical conductivity of an anisotropic substance can be affected enormously by even traces of impurities, we decided that it was most important to attempt to synthesize analytically pure crystals of $(SN)_x$ and to examine the physical and chemical properties of the material. Only in this way would it be possible to determine whether the metallic-like properties reported for $(SN)_x$ (8, 9) were characteristic of the pure material.

The Labes (8, 9) method of preparing $(SN)_x$ usually involved passing S_4N_4 vapor over heated silver wool and then immediately condensing the issuing vapors on a cold finger at 0–8°C. However, we found that, in order to synthesize reproducibly large single crystals of analytically pure $(SN)_x$ it is first necessary to prepare absolutely pure S_2N_2 (free of traces of S_4N_2) by pumping S_4N_4 vapor formed from solid S_4N_4 at 85°C through silver wool at 220°C. After being condensed on a liquid nitrogen-cooled cold finger, the S_2N_2 is slowly sublimed from the cold finger (by warming it to room temperature) into a trap with rectangular walls that is immersed in a 0°C bath. When S_2N_2 crystals of the correct shape and dimensions have formed, the rectangular trap is raised to room temperature. The initially colorless, tabular, monoclinic crystals of S_2N_2 rapidly turn intense blue-black and become paramagnetic ($g = 2.005$). After several hours, these crystals change spontaneously to the bright, lustrous golden color of $(SN)_x$; however, chemical and x-ray studies (7, 10) reveal that they still contain large amounts of unpolymerized S_2N_2. Hence, polymerization may possibly start at the surface of an S_2N_2 crystal and proceed inward. The crystals are left at room temperature for approximately two additional days, and they are finally pumped on at 75°C to remove all traces of unpolymerized S_2N_2. The material does not undergo any change in appearance during this two-day period at room temperature and the final heating to 75°C. The $(SN)_x$ so obtained gives no EPR signal; the absence of S_4N_4 and S_2N_2 is demonstrated by an x-ray powder diffraction study (10, 11). Two typical analyses for the obtained $(SN)_x$ are presented in Table I.

Table I. $(SN)_x$ Crystals[a]

Analysis	S, %	N, %	C, %	H, %	O,[b] %	Total, %
Calcd.	69.59	30.41	0.00	0.00	0	100.00
Found	69.29[c]	30.56	0.00	0.00	0[d]	99.85
	70.18[e]	30.20				100.38

[a] Crystals were from two different preparations. Semiquantitative emission spectrographic analyses did not detect the presence of any metal impurities at the parts-per-million level.
[b] Traces of oxygen are difficult to determine accurately experimentally.
[c] First set of data obtained by Galbraith Laboratories, Inc., Knoxville, Tenn.
[d] Reported as "none or trace."
[e] Second set of data obtained by Schwarzkopf Microanalytical Laboratory, Woodside, N. Y. Analyses were for sulfur and nitrogen only.

The $(SN)_x$ crystals formed in this study have an extremely high lustrous golden metallic appearance on all faces, but the ends are dull, dark blue-black. This is expected (*see* below). The lustrous surfaces of less well formed crystals are highly striated; this causes considerable scattering of light from these surfaces at certain incident angles and consequently they appear dark blue-black under these particular lighting conditions. Electron micrograph studies (magnification 100–1300) of a specially chosen, imperfect, single crystal of $(SN)_x$ revealed that the crystals are composed of layers of fibers stacked parallel to each other along the long axis (*b* axis) of the crystal. At places where the crystal was separated mechanically, long fibrous strands of $(SN)_x$ are apparent. The crystals are highly anisotropic, and they may be mechanically cleaved very easily along a plane parallel to the $(SN)_x$ fibers. The crystals are also soft and malleable; they can be flattened readily by mild pressure applied perpendicular to the fibers to give thin, lustrous, golden sheets.

The crystalline material also exhibits very high optical anisotropy in the visible portion of the spectrum: only that component of incident light which is polarized parallel to the $(SN)_x$ fibers is reflected ($R_{||}$) with high metallic reflectance. The component of incidental light which is polarized perpendicular to the fibers is reflected ($R_\perp$) only very weakly. Thus, the face of an $(SN)_x$ crystal may first be examined with nonpolarized light at an incidence angle of 90° with the analyzer adjusted to give maximum intensity of the reflected light. When the analyzer is then rotated through 90°, the intensity of the reflected light is reduced enormously, and its color changes to a dark blue-black. It is therefore apparent why the ends of the $(SN)_x$ crystals, which consist only of ends of $(SN)_x$ fibers, appear blue-black—at these surfaces all incident radiation is perpendicular to the $(SN)_x$ fibers; furthermore, light scattering is extensive at these less smooth surfaces.

Although $(SN)_x$ crystals slowly become covered with a whitish-gray powder after they stand in air for several months, no immediate reaction with air is apparent; however, the possibility of instantaneous formation of a nonvisible film of oxidized or hydrolyzed material coating the $(SN)_x$ surface cannot yet be ruled out. During seven days at room temperature, crystalline $(SN)_x$ is not attacked by (a) oxygen at 1-atm pressure, (b) water vapor at a pressure of 4.6 torr, or (c) oxygen saturated with water vapor. It is not attacked by degassed distilled water when completely immersed in it for 24 hr at room temperature; however, after six days, although there is no change in appearance of the $(SN)_x$, a very small amount of white solid material may be obtained by evaporating the water. When it is heated in a sealed tube *in vacuo* at approximately 140°C, $(SN)_x$ decomposes to sulfur, nitrogen, and possibly other as yet unidentified materials.

When $(SN)_x$ is heated with constant pumping at ∼ 140°C, golden, lustrous, polycrystalline cohesive films of $(SN)_x$ can be condensed on a variety of substrates held at temperatures of 0°–50°C. X-ray powder patterns of scrapings of these sublimed films demonstrate that the $(SN)_x$ has the same crystal structure as the $(SN)_x$ crystals from which they were sublimed and that the films are completely free of S_4N_4 and S_2N_2. Optical reflectance and x-ray studies of the films formed on a glass surface revealed that the $(SN)_x$ fibers always lie parallel to the glass surface (11); no $(SN)_x$ fibers are observed perpendicular to the surface. Many interesting possibilities of epitaxial growth of $(SN)_x$ films therefore appear likely since fully oriented epitaxial films of $(SN)_x$ can be deposited on various substrates including Mylar, Teflon, and polyethylene by this method (12).

We prepared single crystals of analytically pure $(SN)_x$ that were sufficiently large for single-crystal x-ray studies (10) by solid state poly-

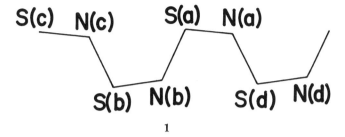

1

merization of S_2N_2 crystals. The $(SN)_x$ crystals have the same space group $(P2_1/c)$ as, and are pseudomorphs of, the S_2N_2 crystals from which they are derived (*10*). The studies reveal that $(SN)_x$ consists of an almost planar chain of alternating sulfur and nitrogen atoms as in 1. There are four SN units per unit cell which has a $= 4.153(6)$, b $= 4.439(5)$, and c $= 7.637(12)$ A, $\beta = 109.7(1)°$, and $D_c = 2.30$ g/cm³. The refined structure $(R = 0.11)$ has as its major feature $(SN)_x$ chains with intrachain distances of $S(a)-N(a) = 1.593(5)$, $S(a)-N(b) = 1.628(7)$, $S(a)-S(b) = 2.789(2)$, $N(a)-N(b) = 2.576(7)$, and $S(a)-N(c) = 2.864(5)$ A and bond angles of $S-N-S = 119.9(4)°$ and $N-S-N = 106.2(2)°$. The S–N bond lengths are all very similar, and they correspond to a sulfur–nitrogen bond order that is intermediate between those expected for a single and a double bond.

Certain of the sulfur–sulfur interchain distances (3.48 A) between $(SN)_x$ chains lying in the same ($\overline{1}02$) plane are less than the sum of the van der Waals radii of two sulfur atoms (3.70 A). This suggests the presence of a weak but important interchain interaction (shown as dotted

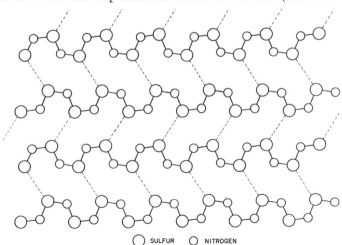

○ SULFUR ○ NITROGEN

Figure 1. Diagrammatic representation of $(SN)_x$ chains in the $\overline{1}02$ plane of an $(SN)_x$ crystal

lines in Figure 1) which is probably largely responsible for causing $(SN)_x$ to be, at the very least, an anisotropic two-dimensional metal rather than a one-dimensional metal.

The optical reflectance of thin films of $(SN)_x$ was measured from the near UV (30,000 cm^{-1}) to the far IR (500 cm^{-1}) regions (Figures 2 and 3). Reflectance characteristic of metals is observed in the IR through the visible regions, with a well defined plasma minimum at 22,000 cm^{-1} that corresponds to light polarized parallel ($R_{||}$) to the polymer chain axis (11). Furthermore, $R_{||}$ and $R_\perp$ were recently measured separately using a face of a carefully polished single crystal (12). The peaks in the IR at 995 and 685 cm^{-1} (Figure 2) are probably caused by S–N vibrational stretching modes. The expanded plasma edge portion of this curve is presented in Figure 3. In both curves, the dashed line is the least squares computer Drude fit (11) to the measured reflectance. Drude reflectance is concerned only with the metallic-like reflectance of light, i.e. that component of light that is parallel to the $(SN)_x$ fibers (11). It was demonstrated that the difference between the curve for measured reflectance and that calculated by the Drude equation is attributable to substantial interchain coupling (12). The maximum metallic reflectance observed is about 45%. This value is actually very much greater than it first appears. Thus, it may be recalled that the maximum metallic reflectance of unpolarized light possible in this system is 50% since each crystal domain in the film will reflect in a metallic fashion only half the incident light which falls upon it, i.e. only that component polarized parallel to the $(SN)_x$ fibers. The real metallic reflectance is therefore on the order of 90% which is characteristic of a metal.

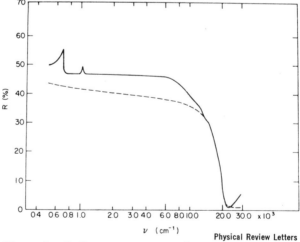

Physical Review Letters

Figure 2. Reflectance spectrum (500–30,000 cm^{-1}) of a thin film of (SN)$_x$. Dashed line: Drude fit to the measured reflectance (11).

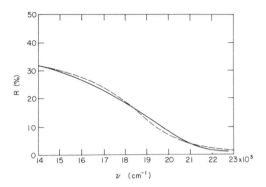

Physical Review Letters

Figure 3. Expanded view of the plasma edge portion of the reflectance spectrum of $(SN)_x$ in Figure 2. Solid line: experimental spectrum; dashed line: Drude fit to the experimental spectrum (11).

From the plasma minimum (plasma frequency, ω_p), one can readily calculate τ the electronic relaxation time by the classical Drude equation which is well established for metal films (*11*). This gives a value of 1.9 × 10^{-15} sec for $(SN)_x$. From τ and ω_p, one can calculate the dc conductivity of a metal by the well-known equation (*11*):

$$\text{dc conductivity} = \omega_p^2 \tau / 4\pi$$

Using this relationship, a dc conductivity of 3 × 10^3 ohm^{-1} cm^{-1} is calculated for $(SN)_x$ along the fiber axis. This value agrees well with the preliminary dc conductivity (four-probe method) of a single crystal of $(SN)_x$ along the direction of the fiber axis. We obtained values of 1200–3700 ohm^{-1} cm^{-1} at room temperature; the conductivity increases markedly when the temperature is lowered to 4.2 K, increasing approximately 50–205-fold. Thus, $(SN)_x$ has a conductivity at room temperature on the same order of magnitude as that of a metal such as mercury (*see* Table II). It was recently reported (*13*) that $(SN)_x$ becomes superconducting at 0.26K.

Table II. Electrical Conductivity (16)

Substance	Conductivity at $20°C$, ohm^{-1} cm^{-1}
$(SN)_x$	3.7 × 10^3
Bi	8.33 × 10^3
Nichrome	10.0 × 10^3
Hg	10.4 × 10^3
Sb	23.9 × 10^3
Fe	1.0 × 10^5
Cu	5.80 × 10^5

Since the length of a sulfur–nitrogen single bond is expected to be approximately 1.74 A and that of a sulfur–nitrogen double bond ~ 1.54 A (*14, 15*), for a sulfur–nitrogen bond order of 1.5, an experimental length of approximately 1.64 A would be expected. From simple valence bond concepts, the bonding between sulfur and nitrogen atoms in an $(SN)_x$ chain may therefore be considered as derived, to a first approximation, from the two extreme resonance forms (**2** and **3**) (in which all atoms

exhibit normal oxidation states and valences) to give a resonance hybrid species (**4**) in which all bonds are intermediate between double and

single bonds. It should be stressed that this description is obviously an over-simplification since, although all S–N bonds are intermediate in length between single and double bonds, they are not all exactly equal experimentally.

Alternatively, since the NS monomer is the sulfur analog of NO (which has one unpaired electron in a π^* orbital), $(SN)_x$ may be regarded as made up from the polymerization of

monomer units, which like NO, might be expected to have a nitrogen–sulfur bond order of 2.5. Polymerization then gives **5**; each nitrogen atom

5

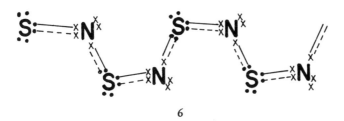

6

is depicted as being associated with eight electrons, each sulfur with nine electrons. The ninth electron can be regarded as present in one of the π^* orbitals of each $-S \cdots N-$ unit. These π^* orbitals on adjacent $-S \cdots N-$ units can then overlap to give some multiple-bond character to the other S–N linkages, viz. **6**. It should be noted that this qualitative treatment differs from the previous resonance hybrid argument—it does not suggest that the amount of multiple bond character in all S–N linkages is necessarily exactly the same. The half-filled overlapping π^* orbitals of each NS unit are expected to produce a half-filled conduction band in the polymer molecule in a manner that is somewhat similar, for example, to the production of a half-filled conduction band in metallic Li_x "polymer" by the overlapping of half-filled $2s$ atomic orbitals.

From these observations, it appears that $(SN)_x$ exhibits strongly all the major collective properties characteristic of a metal in a direction parallel to the $(SN)_x$ fibers. It is, therefore, the first member of a new class of covalent polymeric materials—lower dimensional, polymeric metals. It is highly likely that a whole broad new field of polymeric metals based on diatomic neutral, cationic, or anionic monomeric units that contain one unpaired electron could indeed exist. A prime requisite appears to be that the real or hypothetical monomer unit have one unpaired electron. Neutral species of this type might be sought from binary compounds that contain one element from an odd-numbered group in the Periodic Table and one from an even-numbered group, e.g. AsS. Cationic or anionic units could, in principle, be obtained by removing or adding an electron, respectively, to binary compounds in which both atoms were from either even-numbered or odd-numbered groups in the Periodic Table, e.g. $(CS)^-$ and $(PN)^+$. In this respect, it may be noted that a species such as $(CS)_x^{-x}$ is isoelectronic with $(NS)_x$. Species isoelectronic with $(NS)_x$ are, of course, not limited to ions. For example, HCS is isoelectronic with NS and consequently $(HCS)_x$—and its derivatives, $(RCS)_x$, if they could be synthesized—might be expected to be metals.

It appears that there is much new challenging synthetic chemistry—both inorganic and organic—to be carried out in order to ascertain the extent and importance of this new area of chemistry that is concerned with covalent polymers which are metals.

Acknowledgment

The authors wish to thank J. M. Troup and B. A. Frenz of the Molecular Structure Corp. for many helpful discussions concerning the single-crystal x-ray studies.

Literature Cited

1. Burt, F. P., *J. Chem. Soc.* (1910) 1171.
2. Boudeulle, M., Douillard, A., Michel, P., Vallet, G., *C. R. Acad. Sci Ser. C* (1971) **272**, 2137.
3. Boudeulle, M., Michel, P., *Acta Crystallogr. A* (1972) **28**, S199.
4. Boudeulle, M., Douillard, A., *J. Microsc. Paris* (1971) **11**, 3.
5. Boudeulle, M., Ph.D. thesis, University of Lyon, 1974.
6. Boudeulle, M., *Cryst. Struct. Commun.* (1975). **4**, 9.
7. MacDiarmid, A. G., Mikulski, C. M., Russo, P. J., Saran, M. S., Garito, A. F., Heeger, A. J., *Chem. Commun.* (1975) 476.
8. Walatka, Jr., V. V., Labes, M. M., Perlstein, J. H., *Phys. Rev. Lett.* (1973) **31**, 1139.
9. Hsu, C. H., Labes, M. M., *J. Chem. Phys.* (1974) **61**, 4640.
10. MacDiarmid, A. G., Mikulski, C. M., Russo, P. J., Saran, M. S., Garito, A. F., Heeger, A. J., *J. Am. Chem. Soc.* (1975) **97**, 6358.
11. Bright A. A., Cohen, M. J., Garito, A. F., Heeger, A. J., Mikulski, C. M., Russo, P. J., MacDiarmid, A. G., *Phys. Rev. Lett.* (1975) **34**, 206.
12. Bright, A. A., Cohen, M. J., Garito, A. F., Heeger, A. J., Mikulski, C. M., MacDiarmid, A. G., *Appl. Phys. Lett.* (1975) **26**, 612.
13. Greene, R. L., Street, G. B., Suter, L. J., *Phys. Rev. Lett.* (1975) **34**, 577.
14. Lu, C., Donohue, J., *J. Am. Chem. Soc.* (1944) **66**, 818.
15. Sharma, B., Donohue, J., *Acta Crystallogr.* (1963) **16**, 891.
16. "Handbook of Chemistry and Physics," 52nd Ed., p. E-72, The Chemical Rubber Co., Cleveland (1971–1972).

RECEIVED January 24, 1975. Work supported in part by the National Science Foundation through the Laboratory for Research on the Structure of Matter and grants GH-39303 and GP-41766X, and by the Advanced Research Projects Agency through grant DAHC 15-72-C-0174.

Redox Properties of Polymetallic Systems

THOMAS J. MEYER

University of North Carolina, Chapel Hill, N. C. 27514

The redox properties of three classes of polymetallic systems were studied. In compounds with strong metal–metal bonds, multiple oxidation state properties are found for metal clusters and in compounds where bridging ligands reinforce the metal–metal bond. Compounds with weak interactions between metal ion sites have electronic and chemical properties that are essentially those of isolated monomeric complexes. Electronic interactions between metal centers, electrostatic effects, and statistical effects affect reduction potential values. In systems more complicated than dimers, there are ambiguities about the site of oxidation. In mixed-valence ions, intervalence transfer bands appear; their energies and intensities are functions of both bridging and non-bridging ligand effects. When metal–metal interactions across a bridging ligand are sufficiently strong, the system is delocalized and chemical and electronic properties are significantly modified.

R ecent work has led to the synthesis of a variety of compounds in which metal atoms or ions are held in close proximity by chemical linkages. These polymetallic compounds represent a new class of materials that have distinctive chemical and physical properties, and in some systems the properties can be varied systematically by chemical synthesis. The compounds are of interest because of possible cooperative chemical and electronic interactions between the chemically linked metal centers. In the future it may prove possible: (a) to create solid state materials that have controllable, and perhaps unusual, electrical conductivity properties; (b) to prepare polymeric complexes which in solution have properties that are intermediate between those of solid state materials and those of simple monomeric complexes; and (c) to devise chemical systems in which cooperative chemical interactions lead to net, multiple-electron redox processes, or to simultaneous, two- or more site reactions.

My intention is to develop, as systematically as possible, the redox properties of polymetallic systems in solution. An understanding of redox properties and of metal–metal interactions is essential in order to exploit polymetallic systems. Most of the examples are based on the findings of my own research group with three different classes of compounds which differ in the nature and/or the extent of the metal–metal interaction.

Strong, Direct Metal–Metal Bonding

A metal–metal bond has a profound effect on the properties of the linked metal centers (1). With a metal–metal bond: (a) absorption bands are present which can be assigned to transitions between bonding and antibonding metal–metal orbitals (2, 3, 4); (b) multiple oxidation state properties can appear that are based on the metal–metal bond or bonds; and (c) chemical properties are strongly modified. As an example of the latter, Hughey and Bock demonstrated by flash photolysis (5) that $[(\pi\text{-}C_5H_5)Mo(CO)_3]_2$ and related compounds undergo light-induced homolytic fission (Reaction 1); the monomeric fragments that are produced react rapidly with a variety of substrates under conditions in which the parent compound is unreactive (4, 6).

$$[(\pi\text{-}C_5H_5)Mo(CO)_3]_2 \underset{10^9\text{–}10^{10}\ M^{-1}\ sec^{-1}}{\overset{h\nu}{\rightleftharpoons}} 2(\pi\text{-}C_5H_5)Mo(CO)_3 \qquad (1)$$

Redox processes in relatively simple metal–metal bonds lead to a breakdown in primary structure [see Reaction 2a (7, 8) where S = solvent and Reaction 2b (9)]. However, reversible electron transfer

$$[(\pi\text{-}C_5H_5)Fe(CO)_2]_2 \begin{array}{c} \xrightarrow{\substack{-2e^- \\ +2S}} 2(\pi\text{-}C_5H_5)Fe(CO)_2S^+ \quad (2a) \\ \xrightarrow{+2e^-} 2(\pi\text{-}C_5H_5)Fe(CO)_2^- \quad (2b) \end{array}$$

can occur in metal clusters and in compounds where ligands reinforce the metal–metal bond by bridging (1). Voltammetric experiments in nonaqueous solvents revealed that the cluster systems $[(\pi\text{-}C_5H_5)Fe(CO)]_4$ (10, 11) (Figure 1) and $[(\pi\text{-}C_5H_5)FeS]_4$ (10) remain intact in several different molecular oxidation states: $[(\pi\text{-}C_5H_5)Fe(CO)]_4{}^{2+/+/0/-}$ and $[(\pi\text{-}C_5H_5)FeS]_4{}^{3+/2+/+/0/-}$. The structural details of the compounds $[(\pi\text{-}C_5H_5)Fe(CO)]_4$–$[(\pi\text{-}C_5H_5)Fe(CO)]_4(PF_6)$ (12) and $[(\pi\text{-}C_5H_5)FeS]_4$–$[(\pi\text{-}C_5H_5)FeS]_4(PF_6)$–$[(\pi\text{-}C_5H_5)FeS]_4(PF_6)_2$ (13) and magnetic field Mössbauer data for $[(\pi\text{-}C_5H_5)Fe(CO)]_4(PF_6)$ (14) are consistent with a model in which redox properties are carried, at least in part,

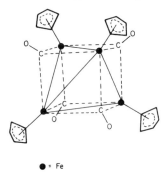

● = Fe

Inorganic Chemistry
Figure 1. Structure of the
$[(\pi\text{-}C_5H_5)Fe(CO)]_4$ *cluster*
unit (14)

by delocalized metal–metal bonding and antibonding orbitals. The clusters appear to undergo facile electron transfer (15).

Chemically reversible electron transfer processes have also been reported for ligand-bridged complexes of Cr, Mo, W, Mn, Fe, Ru, Co, and Ni by Dessy and co-workers who used electrochemical techniques in 1,2-dimethoxyethane (16, 17, 18, 19). Similar behavior was reported for di-*tert*-phosphine-bridged derivatives of $[(\pi\text{-}C_5H_5)Fe(CO)_2]_2$; $[(\pi\text{-}C_5H_5)Fe(CO)]_2(Ph_2P(CH_2)_3PPh_2)^{2+/+/0}$ (20, 21). The multiple oxidation state behavior in ligand-bridged systems also seems to arise because of bonding or antibonding metal–metal orbitals. For example, ESCA and frozen solution EPR data for the once-oxidized form of $[(\pi\text{-}C_5H_5)Fe(CO)]_2(cis\text{-}Ph_2PCH=CHPPh_2)$ (Figure 2) indicate that oxidation

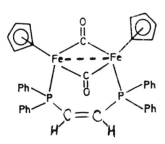

Figure 2. Proposed
structure of $[(\pi\text{-}C_5H_5)Fe(CO)]_2(cis\text{-}Ph_2PCH=CHPPh_2)^+$

occurs at iron and that the unpaired electron resides in a metal–metal orbital (22, 23). From structural studies of the system $[(\pi\text{-}C_5H_5)Fe(CO)SR]_2^{+/0}$, Connelly and Dahl concluded that one-electron oxidation occurs from an antibonding Fe–Fe orbital that also gives a partial metal–metal bond (24). However, in many of these systems, detailed information is needed about the electronic structure. It is conceivable that in some cases the observed redox behavior is carried by orbitals which are largely ligand based, and in related systems it is not always clear whether the metal–metal interaction occurs primarily through space or through a bridging ligand (*see* below).

The ligand-bridged systems are attractive in terms of redox properties because of chemical versatility and the possibility of preparing
polymeric compounds. The compounds are also of interest electronically
since, for a given metal, the orbital character of the metal–metal interaction can be varied. For example, the ions $[(\pi\text{-}C_5H_5)Fe(CO)]_2(cis\text{-}$
$Ph_2PCH{=}CHPPh_2)^+$ and $[(\pi\text{-}C_5H_5)Fe(CO)SCH_3]_2^+$ are formally $d^6\text{-}d^7$
and $d^5\text{-}d^6$ cases, respectively, but in both ions there is a partial metal–
metal bond.

Weak Interactions between Metals through a Bridging Ligand

When metal–metal interactions through a connecting ligand bridge
are weak, the metal centers have the electronic and chemical properties
of isolated, monomeric complexes except for certain special effects. If
appropriate monomeric complexes undergo reversible electron transfer,
a related ligand-bridged system will undergo a series of electron transfer
steps in which each of the metal sites in turn undergoes oxidation or

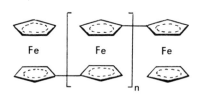

Figure 3. Structure of
the 1,1'-polyferrocenes.
n = 0: Biferrocene (Fc–
Fc), n = 1: 1,1'-terfer-
rocene (Fc–Fc–Fc), and
n = 2: 1,1'-quatrefer-
rocene (Fc–Fc–Fc–Fc)

reduction. Examples are known for the 1,1'-polyferrocenes (Figure
3), $(C_5H_5)Fe(C_5H_4\text{-}C_5H_4)Fe(C_5H_4\text{-}C_5H_4)Fe(C_5H_5)^{3+/2+/+/0}$ [Fc–Fc–
Fc)$^{3+/2+/+/0}$] (25), and for ligand-bridged complexes of ruthenium,
$(bipy)_2ClRu(pyz)Ru(bipy)_2(pyz)RuCl(bipy)_2^{7+/6+/5+/4+}$ (pyz = pyrazine,
bipy = 2,2'-bipyridine) (26, 27, 28). Reduction potentials will reflect
electronic effects (resonance and inductive) if they are large enough,
and they will be affected by simple electrostatic effects. As an example
of the latter, in the complex:

$$(NH_3)_5RuN\text{—}\bigcirc\text{—}CH_2\text{—}CH_2\text{—}\bigcirc\text{—}NRuCl(bipy)_2^{3+},$$

the metal centers are electronically isolated, and yet they are oxidized at
slightly higher potentials than related monomeric complexes because of
the higher charge on the dimer (29, 30).

If there is symmetry, statistical effects appear. Tom, Creutz, and
Taube noted that in the equilibrium in Reaction 3 (where 4,4'-bipy =
4,4'-bipyridine), the mixed-valence ion is favored by a statistical factor
of 4 even in the absence of other effects (31). When one compares the

$$(NH_3)_5Ru(4,4'\text{-bipy})Ru(NH_3)_5{}^{6+} + (NH_3)_5Ru(4,4'\text{-bipy})Ru(NH_3)_5{}^{4+}$$

$$\rightleftharpoons 2(NH_3)_5Ru(4,4'\text{-bipy})Ru(NH_3)_5{}^{5+} \tag{3}$$

reduction potentials for the half-reactions 4 and 5 (where Fc represents ferrocene and a ferrocenyl group), the biferrocene couple $(Fc\text{-}Fc)^{+/0}$ is disfavored by a statistical factor of 2 or 0.018 V $[(RT/nF)\ln 2 = 0.018]$ since there are two ways of forming $(Fc\text{-}Fc)^+$—$Fc^+\text{-}Fc$ and $Fc\text{-}Fc^+$ (25).

$$Fc^+ + e \rightarrow Fc \tag{4}$$

$$(Fc\text{-}Fc)^+ + e \rightarrow Fc\text{-}Fc \tag{5}$$

For systems more complicated than dimers, there are ambiguities regarding the site of oxidation. For example, in the 1,1'-polyferrocenes, Fc_n ($n = 3, 4$) (Figure 3), and in ligand-bridged ruthenium complexes, $(bipy)_2ClRu(pyz)[Ru(bipy)_2pyz]nRuCl(bipy)_2{}^{(2n+2)+}$ ($n = 1$–4) (25, 26, 27), there are chemically different sites. Oxidation gives a series of oxidation state isomers which differ in the site of oxidation (25). For example, for $(1,1'\text{-terferrocene})^+$ there are two energetically equivalent isomers ($Fc^+\text{-}Fc\text{-}Fc$ and $Fc\text{-}Fc\text{-}Fc^+$) and one non-equivalent isomer ($Fc\text{-}Fc^+\text{-}Fc$). Depending on differences in ligand environments, energy differences between isomers may be large and a single isomer may be dominant. It was estimated that in solution the free energy difference between $Fc^+\text{-}Fc\text{-}Fc^+$ and $Fc^+\text{-}Fc^+\text{-}Fc$ is ~ 0.12 V (25) and that between $(NH_3)_5Ru^{III}(pyz)Ru^{II}Cl(bipy)_2{}^{4+}$ and $(NH_3)_5Ru^{II}(pyz)Ru^{III}Cl(bipy)_2{}^{4+}$ is ~ 0.30 V (29, 30). If an orbital pathway exists between metal centers, the different isomers are accessible by thermal- and light-induced intra-molecular electron transfer processes.

It is important to realize that the assignment of oxidation states based on solution information may not apply to the solid state. Although a mixed valence ion like $(bipy)_2ClRu(pyz)RuCl(bipy)_2{}^{3+}$ may be favored in solution, there is no guarantee that it is favored in the solid state over a stoichiometric mixture of the two adjacent ions $(bipy)_2ClRu(pyz)$-$RuCl(bipy)_2{}^{2+}$ and $(bipy)_2ClRu(pyz)RuCl(bipy)_2{}^{4+}$.

In mixed-valence complexes, weak metal–metal interactions lead to intervalence transfer (IT) bands, usually in the visible or near IR regions (25). In an IT transition, light-induced electron transfer occurs

$$(NH_3)_5Ru^{III}N \bigcirc NRu^{II}Cl(bipy)_2{}^{4+} \xrightarrow{\ h\nu\ }$$

$$(NH_3)_5Ru^{II}N \bigcirc NRu^{III}Cl(bipy)_2{}^{4+\ *} \tag{6}$$

between metal sites in different oxidation states (Reaction 6). In the immediate product of light-induced electron transfer, the metal sites are in non-equilibrium vibration and solvation states since nuclear motion is slow compared with electron motion (Franck–Condon principle).

The potential energy–configurational coordinate diagrams used by Hush to describe IT transitions (32) are presented in Figure 4. Diagrams showing IT transitions for both symmetrical [Reactions 7 (31, 34) and 8 (33)] and unsymmetrical (Reaction 6) mixed-valence ions are

$$(NH_3)_5Ru(4,4'\text{-bipy})Ru(NH_3)_5^{5+} \xrightarrow{h\nu} (NH_3)_5Ru(4,4'\text{-bipy})Ru(NH_3)_5^{5+} {}^* \tag{7}$$

$$Fc\text{–}Fc^+ \rightarrow Fc^+\text{–}Fc^* \tag{8}$$

given. The transition in Reaction 6 is at higher energy than that for a symmetrical system because the electron transfer process is energetically unsymmetrical. The product, $(NH_3)_5Ru^{II}(pyz)Ru^{III}Cl(bipy)_2^{4+}$, which is a thermally equilibrated mixed-valence excited state, is disfavored because the oxidation state configuration is reversed from the ground state (29, 30). Similar effects occur for certain oxidation state configurations of the 1,1'-polyferrocenes. The IT band observed for (Fc–Fc–Fc)$^{2+}$ (in 1:1 v:v CH_2Cl_2–CH_3CN) is at a higher energy (λ_{max} 5.99 kK) than the IT band for Fc–Fc$^+$ (λ_{max} 5.26 kK) because the transition is also energetically unsymmetrical (Reaction 9) (25).

$$Fc^+\text{–}Fc\text{–}Fc^+ \xrightarrow{h\nu} Fc^+\text{–}Fc^+\text{–}Fc \quad {}^* \tag{9}$$

Hush developed a theoretical treatment for IT transitions (32). The work of Hush and that of Robin and Day (35) is important because they relate the properties of IT bands to the extent of metal–metal interaction and to the rate of thermal electron transfer between metal sites. The relationships are depicted diagrammatically in Figure 4 where E_{op} and E_{th} are the energies for the optical and the thermal electron transfer processes, respectively. Orbital overlap between metal sites is the origin of the splitting between surfaces and of the intensity of IT bands.

Recent work on dimeric ruthenium complexes demonstrated that there is reasonable agreement between experimental data and the band width and solvent dependence predictd by Hush for IT bands (29, 30, 31). The work with ruthenium complexes also revealed that the energies and intensities of IT bands vary systematically as a function of bridging and nonbridging ligand effects (29, 30, 31, 34). No IT band was observed for the ion:

$$[(NH_3)_5Ru^{III}N\langle\bigcirc\rangle-CH_2CH_2-\langle\bigcirc\rangle NRu^{II}Cl(bipy)_2]^{4+}$$

because the orbital pathway between metal centers is blocked by the saturated $-CH_2-CH_2-$ linkage. IT bands appear for pyrazine, 4,4-bipyridine, and *trans*-1,2-bis(4-pyridyl)ethylene as the bridging ligands where there is an intact π system. The bands are at higher energies for the longer bridging ligands (in acetonitrile: 10.4 kK for pyrazine, $\sim$ 14.4 kK for 4,4'-bipyridine, and $\sim$ 14.7 kK for *trans*-1,2-bis(4-pyridyl)ethylene (*29, 30*) because the intersection region between the $Ru^{II}-Ru^{III}$ and $Ru^{III}-Ru^{II}$ surfaces (Figure 4) and therefore the energy of the IT transition are functions of the distance between the metal centers. The intensity of the band for the pyrazine-bridged dimer is considerably greater than that for the other dimers; this indicates a stronger metal–metal interaction.

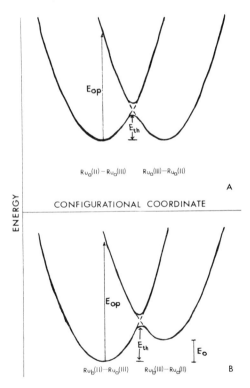

Figure 4. Potential energy–configurational coordinate diagrams for symmetrical (A) and unsymmetrical (B) cases. IT transitions are indicated by the arrows.

Nonbridging ligands can affect the energy and apparently also the intensity of IT bands. In the dimers $(NH_3)_5Ru^{III}(pyz)Ru^{II}L(NH_3)_4^{5+}$ (34) and $(NH_3)_5Ru^{III}(pyz)Ru^{II}X(bipy)_2^{4+}$ (29, 30), the energy of the IT band increases as variations in L or X increase the energy asymmetry between the two ends. For the mixed-valence dimer $(bipy)_2ClRu(pyz)$-$RuCl(bipy)_2^{3+}$, ESCA studies demonstrated that there are discrete $Ru(II)$ and $Ru(III)$ sites (36). In recent work, Callahan and Meyer (37) found an IT band for the ion (λ_{max} 1300 nm, $\epsilon = 450$ in acetonitrile) which has the approximate band width and solvent dependence that were predicted by Hush. IR data are consistent with localized valences. The properties of this ion differ markedly from those of the Creutz and Taube ion $(NH_3)_5Ru(pyz)Ru(NH_3)_5^{5+}$ (33), and the differences are consistent with a much stronger metal–metal interaction in the pentaammine system (37).

Strong Interactions between Metals through a Bridging Ligand

With strong metal–metal interactions across a bridging ligand, the valence redox orbitals are delocalized molecular orbitals both metal and ligand in character. In mixed-valence compounds, different, discrete oxidation states do not exist since the site of oxidation is delocalized. Strongly coupled systems are like metal–metal bonds in that their electronic and chemical properties are significantly modified from those of related monomeric complexes. As with metal–metal bonds, such compounds can have an extensive multiple oxidation state chemistry based on delocalized molecular orbitals.

Strong coupling is expected when the bridging distance is short, especially for second- and third-row metals and for metals in low oxidation states where d orbital extension is great and strong π overlap can occur. Several oxo-bridged complexes of ruthenium(III), $[(AA)_2XRuO$-$RuX(AA)_2]^{2+}$ (where AA is 2,2'-bipyridine or 1,10-phenanthroline and X is Cl or NO_2), have unusual spectral, redox, and chemical properties (38). In the salt $[(bipy)_2(NO_2)RuORu(NO_2)(bipy)_2](ClO_4)_2$, the Ru–O–Ru angle is $\sim 157°$ and the nitro groups are cis to the bridging oxide ion (39). The properties of the complexes are consistent with an MO scheme where the valence orbitals are antibonding with respect to the Ru–O–Ru group (38). On the cyclic voltammetry time scale, in acetonitrile, the complexes exist in a series of molecular oxidation states, $(bipy)_2ClRuORuCl(bipy)_2^{4+/3+/2+/+}$. An ESCA study of the mixed-valence +3 ion suggests that the ruthenium sites are equivalent. The potential for reduction to the +1 ion is low compared with that of monomeric complexes (Table I); this is consistent with a lowest $\pi^*(RuORu)$ orbital for the dimer which has been strongly destabilized compared with the valence $d\pi$ orbitals of monomeric complexes. A similar conclusion was reached for the osmium system [(terpy)(bipy)OsOOs(bipy)-(terpy)]$^{5+/4+/3+}$ (40) whereas magnetic coupling occurs in oxo-bridged complexes of iron(III) but the valence electrons are localized on the iron(III) ions (41, 42, 43).

Table I. $E_{1/2}$ Values for Ru(III)/Ru(II) Couples in 0.1M N(n–C$_4$H$_{94}$+PF$^-_6$–Acetonitrile at 22° ± 2°C $vs.$ the Saturated Sodium Chloride Calomel Electrode[a]

Couple	$E_{1/2}$, V
(bipy)$_2$ClRuORuCl(bipy)$_2$$^{2+/+}$	−0.32
Ru(bipy)$_2$Cl$_2$$^{+/0}$	0.32
Ru(bipy)$_2$(NH$_2$CH(CH$_3$)$_2$)Cl$^{2+/+}$	0.65

[a] The couples are electrochemically reversible so that $E_{1/2}$ is essentially a reduction potential in the medium described.

Strong metal–metal interactions can occur through a bridging ligand or by direct metal–metal bonding, and the electronic origin of the metal–metal interaction is not always clear. With the compound [Ir(NO)-(PPh$_3$)$_2$]$_2$O · C$_6$H$_6$, for example, an x-ray study revealed that there are both an oxide bridge and a short Ir–Ir separation (2.56 A) (43).

In contrast to the oxo-bridged ions the d^5 Ru(III) sites in the pyrazine bridged complexes [(bipy)$_2$ClRu(pyz)[Ru(bipy)$_2$pyz]$_n$RuCl-(bipy)$_2$](ClO$_4$)$_{3n+4}$ ($n = 0, 1, 2$) are strongly localized. No evidence for magnetic interactions was found down to 4°K (44). However, it may be possible to prepare strongly coupled, polymeric systems. Johnson recently devised routes to mercapto-bridged complexes, $e.g.$ (bipy)$_2$ClRu(SPh)-Ru(bipy)$_2$(SPh)RuCl(bipy)$_2$$^{2+}$, which have multiple oxidation state properties (28). Magnuson and Taube suggested that the dinitrogen-bridged osmium complexes, [cis-Cl(NH$_3$)$_4$OsNNOs(NH$_3$)$_5$]$^{5+/4+/3+}$ are delocalized systems (45). This finding is important since it shows that strong coupling can occur through bridging groups that are longer than a single atom or ion.

A series of triangular, oxo-bridged cluster compounds, Ru$_3$O(CH$_3$-CO$_2$)$_6$L$_3$ (L is a neutral ligand), was prepared by Spencer and Wilkinson

1

(*46, 47*) (in **1**, the arcs represent bridging acetate groups). They found from chemical and electrochemical studies that the cluster remains intact as neutral and $+1$ units, $Ru_3O(CH_3CO_2)_6(py)_3^{+/0}$ (*46, 47*). The basic structure was determined by Cotton and Norman ($L = PPh_3$) who developed a qualitative MO scheme that assumes a delocalized system (*48*). Wilson and Salmon (*49*) found that the basic cluster unit remains intact in five different oxidation states, $Ru_3O(CH_3CO_2)_6L_3^{3+/2+/+/0/-}$, on the cyclic voltammetry time scale in acetonitrile. Ruthenium ESCA data for mixed-valence compounds like $[Ru_3O(CH_3CO_2)_6(py)_3](ClO_4)_2$ and $Ru_3O(CH_3CO_2)_6(py)_3$ suggest that the Ru sites are equivalent and that the clusters are delocalized (*49*). We recently succeeded in linking triangular cluster units by ligand bridges (Scheme 1; S = solvent).

$$Ru_3O(CH_3CO_2)_6(py)_2(CO) \underset{+e^-}{\overset{-e^-}{\rightleftharpoons}} Ru_3O(CH_3CO_2)_6(py)_2(CO)^+$$

$$\downarrow \;-CO$$

$$Ru_3O(CH_3CO_2)_6(py)_2S^+$$
$$\downarrow \;+pyz$$
$$Ru_3O(CH_3CO_2)_6(py)_2pyz^+ + S$$

$$(py)_2(CH_3CO_2)_6ORu_3(pyz)Ru_3O(CH_3CO_6)_2(py)_2^{2+}$$

Scheme 1

Cyclic voltammetry demonstrated that the pyrazine-bridged dimer $Ru_3(pyz)Ru_3^{2+}$ has an extensive oxidation state chemistry, and that the cluster–cluster mixed-valence ions $[Ru_3(pyz)Ru_3]^+$ and $[Ru_3(pyz)Ru_3]^-$ are discrete species in solution. Ruthenium ESCA data for the perchlorate salt of the $+1$ ion reveal the presence of distinct cluster units, $Ru_3^+(pyz)Ru_3^0$, and therefore indicate that intercluster interactions across pyrazine are weak. However, the electrochemical data reveal that the extent of cluster–cluster interaction increases as the electron content of the system increases. The synthetic chemistry here is very promising, and we should be able to prepare complex two-dimensional systems including polymers and to investigate cluster–cluster and metal ion–cluster interactions.

Intermediate Cases

Perhaps the most interesting cases are complexes which appear to lie in the transition region between localized and delocalized systems. Much evidence points to trapped valences in the Creutz and Taube ion

$(NH_3)_5Ru(pyz)Ru(NH_3)_5^{+5}$, and yet the $\delta(NH_3)(sym)$ and $\rho(NH_3)$ IR bands for the $+5$ ion are intermediate between the $+4$ and $+6$ ions and the $+5$ IT band does not have the expected band width and solvent dependence (34). In the ions $M_2Cl_9^{3-}$ (M is Cr, Mo, M) (2), the

2

chromium case is localized (Cr–Cr 3.1 A) with a weak magnetic inter-action, the tungsten case is strongly metal–metal bonded (W–W 2.4 A), and the molybdenum case is intermediate (Mo–Mo 2.7 A) (50). For the ion $Mo_2Br_9^{3-}$ (Mo–Mo 2.8 A), there is a temperature-dependent para-magnetism that suggests a weak metal–metal interaction (51). Inter-mediate cases can be expected to have unusual properties, and an under-standing of them will probably require new insight. The possible thermal accessibility of more than one state for such systems is also intriguing.

Acknowledgments

I should like to acknowledge my colleagues whose work it is that I have described here; they include John Ferguson, Ajao Adeyemi, Gilbert Brown, Tom Weaver, Randy Bock, Steve Wilson, Eugene Johnson, Bob Callahan, Dennis Salmon, Joey Hughey, and Michael Powers.

Literature Cited

1. Meyer, T. J., *Prog. Inorg. Chem.* (1975) 19, 1.
2. Levenson, R. A., Gray, H. B., Ceasar, G. P., *J. Am. Chem. Soc.* (1970) 92, 3653.
3. Wrighton, M. S., *Chem. Rev.* (1974) 74, 401.
4. Hughey IV, J. L., unpublished data.
5. Hughley IV, J. L., Bock, C. R., Meyer, T. J., *J. Am. Chem. Soc.* (1975) 97, 4440.
6. Wrighton, M. S., Ginley, D. S., *J. Am. Chem. Soc.* (1975) 97, 4246.
7. Ferguson, J. A., Meyer, T. J., *Inorg. Chem.* (1971) 10, 1025.
8. Johnson, E. C., Winterton, N., Meyer, T. J., *Inorg. Chem.* (1971) 10, 1673.
9. Dessy, R. E., Weissman, P. M., Pohl, R. L., *J. Am. Chem. Soc.* (1966) 88, 5117.
10. Ferguson, J. A., Meyer, T. J., *Chem. Commun.* (1971) 623.
11. Ferguson, J. A., Meyer, T. J., *J. Am. Chem. Soc.* (1972) 94, 3409.
12. Trinh-Toan, Fehlhammer, W. P., Dahl, L. F., *J. Am. Chem. Soc.* (1972) 94, 3389.
13. Trinh-Toan, Fehlhammer, W. P., Dahl, L. F., "Abstracts of Papers," 161st National Meeting, ACS, 1971, INORG 130.
14. Frankel, R. B., Reiff, W. M., Meyer, T. J., Cramer, J. L., *Inorg. Chem.* (1974) 13, 2515.

15. Braddock, J. N., Meyer, T. J., *Inorg. Chem.* (1973) **12**, 723.
16. Dessy, R. E., King, R. B., Waldrop, M., *J. Am. Chem. Soc.* (1966) **88**, 5112.
17. Dessy, R. E., Kornmann, R., Smith, C., Hayter, R., *J. Am. Chem. Soc.* (1968) **90**, 2001.
18. Dessy, R. E., Rheingold, A. L., Howard, C. D., *J. Am. Chem. Soc.* (1972) **94**, 746.
19. Dessy, R. E., Wieczorek, L., *J. Am. Chem. Soc.* (1969) **91**, 4963.
20. Haines, R. J., duPreez, A. L., *Inorg. Chem.* (1972) **11**, 330.
21. Ferguson, J. A., Meyer, T. J., *Chem. Commun.* (1971) 1544.
22. Haines, R. J., de Preez, A. L., *Inorg. Chem.* (1972) **11**, 330.
23. Salmon, D., Scaringe, R., unpublished data.
24. Connelly, N. G., Dahl, L. F., *J. Am. Chem. Soc.* (1970) **92**, 7472.
25. Brown, G. M., Meyer, T. J., Cowan, D. O., LeVanda, C., Kaufmann, F., Roling, P. V., Rausch, M. D., *Inorg. Chem.* (1975) **14**, 506.
26. Adeyemi, S. A., Braddock, J. N., Brown, G. M., Meyer, T. J., Miller, F. J., *J. Am. Chem. Soc.* (1972) **94**, 300.
27. Adeyemi, S. A., Johnson, E. C., Meyer, T. J., Miller, F. J., *Inorg. Chem.* (1973) **12**, 2371.
28. Johnson, E. C., Ph.D. Dissertation, University of North Carolina, 1975.
29. Callahan, R. W., Brown, G. M., Meyer, T. J., *J. Am. Chem. Soc.* (1974) **96**, 7829.
30. Callahan, R. W., Brown, G. M., Meyer, T. J., *Inorg. Chem.* (1975) **14**, 1443.
31. Tom, G. M., Creutz, C., Taube, H., *J. Am. Chem. Soc.* (1974) **76**, 7827.
32. Hush, N. S., *Prog. Inorg. Chem.* (1967) **9**, 391; *Electrochim acta* (1968) **13**, 1005.
33. Cowan, D. O., LeVanda, C., Park, J., Kaufman, F., *Acc. Chem. Res.* (1973) **6**, 1.
34. Creutz, C., Taube, H., *J. Am. Chem. Soc.* (1973) **95**, 1086.
35. Robin, M. B., Day, P., *Adv. Inorg. Chem. Radiochem.* (1967) **10**, 247.
36. Citrin, P., *J. Am. Chem. Soc.* (1973) **95**, 6472.
37. Callahan, R. W., Meyer, T. J., unpublished data.
38. Weaver, T. R., Meyer, T. J., Adeyemi, S. A., Brown, G. M., Eckberg, R. P., Hatfield, W. E., Johnson, E. C., Murray, R. W., Untereker, D., *J. Am. Chem. Soc.* (1975) **97**, 3039.
39. Phelps, D. W., Kahn, E. M., Hodgson, D. J., unpublished data.
40. Brown, G. M., Ph.D. Dissertation, University of North Carolina, 1974.
41. Griffith, W. P., *Coord. Chem. Rev.* (1970) **5**, 459.
42. Schugar, H. J., Rossman, G. R., Barraclough, C. G., Gray, H. B., *J. Am. Chem. Soc.* (1972) **94**, 2683.
43. Carty, P., Walker, A., Matthew, M., Palenik, G. J., *Chem. Commun.* (1969) 1374.
44. Eckberg, R., Johnson, E. C., Callahan, R. W., Hatfield, W. E., Meyer, T. J., unpublished data.
45. Magnuson, R. H., Taube, H., *J. Am. Chem. Soc.* (1972) **94**, 7213.
46. Spencer, A., Wilkinson, G., *J. Chem. Soc. Dalton Trans.* (1972) 1570.
47. *Ibid.* (1974) 786.
48. Cotton, F. A., Norman, Jr., J. G., *Inorg. Chim. Acta* (1972) **6**, 411.
49. Wilson, S. T., Salmon, D. J., Meyer, T. J., *J. Am. Chem. Soc.* (1975) **97**, 2285.
50. Saillant R., Wentworth, R. A. D., *Inorg. Chem.* (1968) **7**, 1606.
51. *Ibid.* (1969) **8**, 1226.

RECEIVED January 24, 1975. Work supported by the Army Research Office, Durham; the National Science Foundation; the Petroleum Research Foundation; and DARPA through the Materials Research Center of the University of North Carolina.

8

Electron Exchange between Pairs of Vanadium Atoms in Novel Geometric Isomers of Heteropoly Tungstates

MICHAEL T. POPE, STEPHEN E. O'DONNELL, and RONALD A. PRADOS

Georgetown University, Washington, D.C. 20057

The heteropoly anions $PV_2Mo_{10}O_{40}^{5-}$ and $PV_2W_{10}O_{40}^{5-}$ are each shown to exist as mixtures of the five possible stereo-isomers by ^{31}P NMR spectroscopy. Controlled potential electrolytic reduction of $H_5PV_2W_{10}O_{40}$ yields $PV^{IV}V^VW_{10}-O_{40}^{6-}$ (I) and $PV_2^{IV}W_{10}O_{40}^{7-}$ (II) which were isolated as potassium salts. The ESR spectrum of anion I consists of superimposed 8- and 15-line components with $\langle g \rangle = 1.952$ and $\langle a \rangle = 104.5$ and 53 G. The relative intensities of the 8- and 15-line spectra are in quantitative agreement with the assumption that electron exchange between neighboring vanadium atoms (V–O–V) is rapid, but that in isomers with remote vanadium atoms (V–O–W–O–V, etc.) the electron is effectively trapped on a single vanadium. The ESR spectrum of anion II has a normal 15-line pattern arising from the triplet state. The intervalence optical transition in anion I occurs at 8.8 kK.

Heteropoly transition metal oxocomplexes generally have structures based on edge- and corner-shared MO_6 octahedra (1, 2). These structures consequently resemble discrete fragments of close-packed metal oxide lattices, and heteropoly complexes frequently exhibit properties associated with electron delocalization and magnetic exchange that are found in metal oxides. The so-called heteropoly blues for example, are heteropoly anions in which some of the d^0 metal atoms (tungsten, molybdenum, or vanadium) have been reduced to the d^1 oxidation state (3). In such compounds, the extra electrons appear to be delocalized over all structurally and chemically equivalent metal atoms by a hopping process (4) (*cf.* semiconduction in mixed valence metal oxides).

The present work was undertaken in order to define more exactly the processes of electron delocalization in mixed valence heteropoly anions. The work has also proved the existence of geometric isomers that result from the various ways of arranging two kinds of metal atoms within a given structure (5, 6).

The complexes studied have the general formula $PV_xZ_{12-x}O^{(3+x)-}$ where Z is Mo or W, and $x = 1$ and 2. The structure of these anions (Figure 1) is an arrangement of 12 edge- and corner-shared ZO_6 octahedra surrounding a central PO_4 tetrahedron. The site symmetry of each metal atom is approximately C_{4v}. We demonstrated elsewhere (7, 8, 9), that the $PVW_{11}O_{40}^{4-}$ and $PVMo_{11}O_{40}^{4-}$ complexes undergo one-electron reductions to give complexes that are intensely colored but that nevertheless have trapped valences (i.e., they are $PV^{IV}W_{11}O_{40}^{5-}$ and $PV^{IV}Mo_{11}-O_{40}^{5-}$) according to ESR spectroscopy. When more than one tungsten or molybdenum atom has been replaced by vanadium, the possibility of isomerism arises. Since the 12 metal atoms in the structure depicted in Figure 1 (the Keggin structure) are equivalent, there are $12!/x!(12 - x)!$

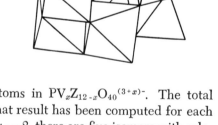

Figure 1. Keggin structure for $MZ_{12}O_{40}^{n-}$ heteropoly anions showing arrangement of ZO_6 octahedra around the central MO_4 tetrahedron

ways of arranging the vanadium atoms in $PV_xZ_{12-x}O_{40}^{(3+x)-}$. The total number of distinguishable isomers that result has been computed for each possible value of x (5). For the case $x = 2$, there are five isomers with relative statistical abundances of 6, 12, 12, 12, and 24 (see Figures 2 and 3).

Experimental

Preparation of Compounds. Decamolybdodivanadophosphoric acid was prepared by the method of Tsigdinos and Hallada (10). Anal. for $H_5PV_2Mo_{10}O_{40} \cdot 17H_2O$: calcd: V 4.99, Mo 46.95; found: V 4.91, Mo 47.43, Mo/V 5.13.

Decatungstodivanadophosphoric acid was prepared by Kokorin's method as described by Smith and Pope (7). Anal. for $H_5PV_2W_{10}O_{40} \cdot 12H_2O$: calcd: V 3.60, W 64.90; found: V 3.73, W 65.10, W/V 4.83.

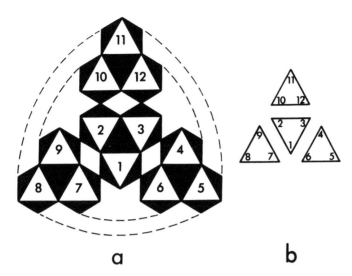

Figure 2. (a) Unfolded Keggin structure showing numbering of Z-atoms; (b) simplified version of (a) in which only edge-shared octahedra are depicted as linked

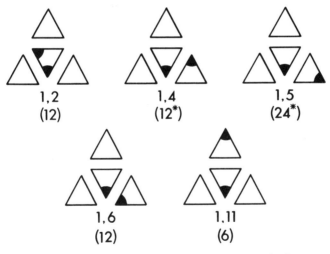

Figure 3. The five isomers of $PV_2Z_{10}O_{40}^{5-}$ with their respective degeneracies (in parentheses)

Species labelled with an asterisk are dissymmetric, i.e. they have non-superimposable mirror images. Both enantiomorphs are included in the degeneracy count of such species.

Voltammograms of solutions of this acid agreed with those reported previously (7, 8).

Potassium 10-tungstodivanado(IV,V)phosphate was prepared by controlled potential reduction of a solution of the oxidized complex in an acetate buffer, pH 5. A graphite cloth cathode (Union Carbide Corp.)

at $+0.29$ V vs. SCE was used. A slight excess of potassium chloride was added to the reduced solution. Dark green crystals of the product separated after the solution had been stored overnight in the refrigerator. Anal. for $K_6PV_2W_{10}O_{40} \cdot 12H_2O$: calcd: V 3.33, W 60.05, equiv. wt 3062.0; found: V 3.33, W 59.92, equiv. wt (coulometrically) 3066, W/V 4.98.

Potassium 10-tungstodivanado(IV)phosphate was prepared by controlled potential reduction at -0.10 V of the vanadium(IV,V) complex at pH 9 or 10. The dark brown potassium salt was isolated from the reduced solution as described above. The product gave satisfactory IR spectra and voltammograms.

Physical Measurements. For the electrolyses, a Wenking potentiostat model 70TS1 and a Koslow Scientific coulometer model 541 were used. Voltammetry with wax-impregnated graphite and rotating platinum electrodes was performed as described elsewhere (7, 8). IR and electronic spectra were measured on Perkin-Elmer 225 and Cary 14 instruments. X-band ESR spectra were recorded at room temperature on a JEOL MES-3X spectrometer. Phosphorus-31 NMR spectra were recorded in the pulse mode on a Varian XL-100 instrument at 40.5 MHz using a deuterium lock, or on a Bruker HFX-90 instrument at 36.43 MHz using a fluorine lock.

Results and Discussion

NMR Spectroscopy. The ^{31}P NMR spectra of solutions of H_5PV_2-$Mo_{10}O_{40}$ and $H_5PV_2W_{10}O_{40}$ are presented in Figures 4 and 5. The peaks at 3.09 and 3.79 ppm in Figure 4 have integrated intensities in the ratio 3:8, and they are consistent with the presence of the five isomers with relative abundances of $(6 + 12):(24 + 12 + 12)$. The spectrum of $H_5PV_2W_{10}O_{40}$ (Figure 5) has a similar pattern of resonances at higher field, 13.90, 14.20, and 14.27 ppm. These peaks have integrated intensities in the ratio 1:3:1, and they are consistent with the presence of four of the expected isomers with relative abundances of $12:(24 + 12):12$. The fifth (1,11) isomer, with a relative abundance of 6, may be indicated by the resonance at 12.37 ppm. A more complete discussion of the NMR spectra of mixed heteropoly anions is given elsewhere (11, 12).

Reduction of $PV_2W_{10}O_{40}{}^{5-}$. According to previously reported voltammetric measurements (7, 8), the $PV_2W_{10}O_{40}{}^{5-}$ anion is stable in solution up to pH 5. At that pH, a voltammogram indicates two well-defined, reversible one-electron reductions at ca. $+0.3$ and 0.0 V. In more acidic solutions, the second reduction shifts to more positive potentials and overlaps with the first (7, 8). Controlled potential electrolysis at $+0.29$ V of a solution of the anion at pH 5 consumed one faraday per mole of heteropoly complex. The reduced solution was brown-green, and a rotating platinum electrode voltammogram of this solution demonstrated that the first wave was totally anodic. The ESR spectrum of the reduced solution (Figure 6). consists of overlapping eight- and 15-line com-

ponents. The relative intensities of all the features of the ESR spectrum
did not change during the electrolysis although the overall spectrum

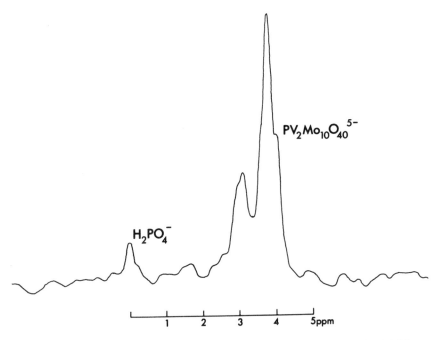

Figure 4. ^{31}P *NMR spectrum of 0.1M* $H_5PV_2Mo_{10}O_{40}$ *in 0.25M* H_2SO_4

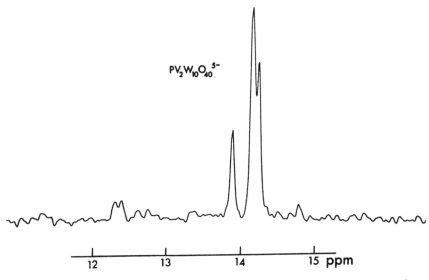

Figure 5. ^{31}P *NMR spectrum of 0.1M* $H_5PV_2W_{10}O_{40}$ *in 0.25M* H_2SO_4

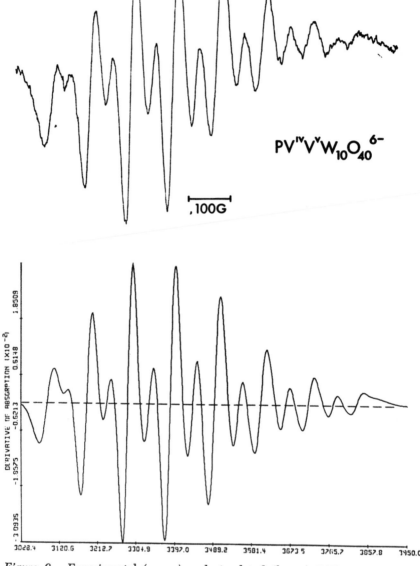

$PV^{IV}V^{V}W_{10}O_{40}{}^{6-}$

$\vdash$ 100G $\dashv$

Figure 6. Experimental (upper) and simulated (lower) ESR spectra of an aqueous solution of $PV^{IV}V^{V}W_{10}O_{40}{}^{6-}$ anion at pH 5

gained in intensity as the electrolysis proceeded. Solutions with similar ESR spectra could also be obtained by chemical reduction of $PV_2W_{10}O_{40}{}^{5-}$ using the mild reducing agents hydroquinone, catechol, and tiron. If the controlled potential electrolysis was carried out at pH 3, the reduced solution contained the decomposition products VO^{2+} and $PVW_{11}O_{40}^{5-}$ as identified by ESR (7, 8) and voltammetry (7, 8). Solid samples of the potassium salt of the reduced anion $PV^{IV}V^{V}W_{10}O_{40}^{6-}$ were isolated and characterized as described above. A solution of this salt in an acetate buffer at pH 5 gave an ESR spectrum identical to that in Figure 6. The IR and electronic spectra of the reduced anion are discussed below.

According to ESR and voltammetry, the $PV^{IV}V^{V}W_{10}O_{40}^{5-}$ anion is stable between pH 4 and 11. A controlled potential reduction of a solution of this anion at pH 9–10 was complete after the transfer of one faraday per mole of heteropoly anion. The resulting solution was dark brown. During electrolysis the ESR spectrum (originally as in Figure 6) underwent an overall reduction in intensity accompanied by a gradual disappearance of the eight-line component spectrum. The spectrum of the fully reduced

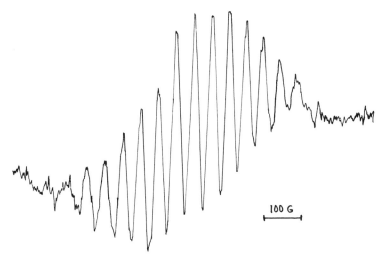

Figure 7. ESR spectrum of an aqueous solution of $PV_{2}^{IV}W_{10}O_{40}^{7-}$ anion at pH 10

anion is presented in Figure 7. The potassium salt of the fully reduced anion ($PV_{2}^{IV}W_{10}O_{40}^{7-}$) was isolated as described above. A polycrystalline sample of the salt gave an ESR spectrum with an intense broad line (*ca.* 400 G) at $g \sim 2$ and a weaker line at $g \sim 4$. Solutions of the reduced anion gave voltammograms like those for $PV_{2}W_{10}O_{40}^{5-}$ and $PV^{IV}V^{V}W_{10}O_{40}^{6-}$. The fully reduced anion was unstable in solution below pH 8.

Spectra of Reduced Anions. IR spectra of the $PV_{2}W_{10}O_{40}^{n-}$ anions are very similar to those reported for other heteropoly tungstates with the Keggin structure. Of particular interest however is the T_2 mode of the central PO_4 group which occurs as an intense sharp absorption at 1080

cm^{-1} in $H_3PW_{12}O_{40}$ (*13*). The same absorption in $PV_2W_{10}O_{40}^{5-}$ appears at 1064 cm^{-1} but is split into four components (1100, 1082, 1065, and 1048 cm^{-1}) in the spectrum of $PV^{IV}V^{V}W_{10}O_{40}^{6-}$ and into three components (1083, 1062, and 1042 cm^{-1}) in that of $PV_2^{IV}W_{10}O_{40}^{7-}$ (*see* Figure 8).

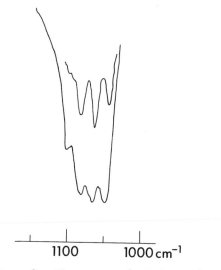

$\vdash$ $\vdash$ $\vdash$

1100   **1000 cm⁻¹**

Figure 8. IR spectra in the P–O stretch-
ing region of $K_6PV^{IV}V^{V}W_{10}O_{40}$ (lower)
and $K_7PV_2^{IV}W_{10}O_{40}$ (upper)

Although it is conceivable that the threefold degeneracy of the T_2 mode is totally removed in $PV_2^{IV}W_{10}O_{40}^{7-}$, the presence of four bands in the spectrum of $PV^{IV}V^{V}W_{10}O_{40}^{6-}$ can only indicate the presence of two or more isomers.

The ESR spectrum of the mixed valence anion $PV^{IV}V^{V}W_{10}O_{40}^{6-}$ could be simulated (Figure 6) by combining an eight-line and a 15-line spectrum in the intensity ratio 7:4 respectively. The appropriate parameters, correct to second order, are $\langle g \rangle = 1.952$ (both spectra) and $\langle a \rangle = 104.5$ G (eight-line) and 53 G (15-line). The relative intensities of the eight- and 15-line spectra were chosen on the assumption that all five stereoisomers were present in their expected statistical amounts. It was further assumed that the unpaired electron in the isomers with vicinal vanadium atoms (1,2 and 1,6 in Figure 3) interacted with both vanadium nuclei ($I = 7/2$) through rapid intramolecular hopping ($> 3 \times 10^8$/sec) to give a 15-line spectrum. The remaining isomers (1,4, 1,5, and 1,11) contain vanadium atoms that are separated by one or more tungsten atoms (*i.e.* they have V–O–W–O–V sequences), and the electron is effectively trapped (life-time $>$ *ca.* 10^{-8} sec) on a single vanadium atom thus giving an eight-line spectrum (the explanation is analogous to that proposed for the paracyclophane anions in Ref. *14*). The relative intensities of the 15- and eight-line spectra should therefore be $(12 + 12):(24 + 12 + 6)$ or 4:7. However, there is no convincing evidence in the NMR spectrum that the 1,11 isomer is indeed formed in the case of the tungstate anions. It could be argued that the peak at 12.37 ppm in Figure 5 is caused by traces of impurities. If this were so, the relative intensities of the 15- and 8-line spectra would be $(12 + 12):(24 + 12)$ or 4:6 (instead of 4:7). Although it is not possible to distinguish between these two ratios by simulation, the ESR findings provide strong evidence for the presence of several isomers as well as for the electron hopping process of delocalization in mixed valence polyanions of this type.

The ESR spectrum of the fully reduced anion, $PV_2^{IV}W_{10}O_{40}^{7-}$ (Figure 7), has the 15-line pattern ($\langle g \rangle = 1.95$, $\langle a \rangle = 53$ G) expected for the interaction of the unpaired electrons with two vanadium nuclei. Observation of the half-field ($\Delta m_s = 2$) transition in the spectrum of the solid potassium salt confirms the triplet state of the species responsible for the spectrum in Figure 7. It seems very likely that those isomers (1,2 and 1,6 with V–O–V groups) which are responsible for the 15-line spectrum of the mixed valence complex would have antiferromagnetically coupled spins in the fully reduced anion. The remote isomers with vanadium atoms separated by O–W–O (1,4 and 1,5) or O–W–O–W–O (1,11) sequences are therefore probably responsible for the triplet state ESR spectrum. Further ESR and magnetic susceptibility measurements are planned on individual isomers.

Electronic spectral data for the two reduced anions are compared in Table I with those reported for $PV^{IV}W_{11}O_{40}^{5-}$ (*7, 8, 15*). The assign-

Table I. Electronic Transitions of Reduced Tungstovanadophosphate Anions

PV^{IVa}	$PV^{IV}V^V$	PV_2^{IV}	*Tentative Assignment*
	8.8 (290)		$V^{IV} \rightarrow V^V$
12 sh,b $(130)^b$	12 sh,b (260)	11 sh,b (150)	*d-d*
14.5 sh,b (300)	15.5 (360)	15.5 sh (430)	*d-d*
20 (750)	19 sh (510)	19.5 sh (950)	$V^{IV} \rightarrow W^{VI}$
25 sh (600)	c	26.5 sh (1430)	$V^{IV} \rightarrow W^{VI}$

^{a} $PV^{IV} = PVW_{11}O_{40}^{5-}$ at pH 2 (*15*); $PV^{IV}V^V = PV^{IV}V^VW_{10}O_{40}^{6-}$ at pH 5; $PV_2^{IV} = PV_2^{IV}W_{10}O_{40}^{7-}$ at pH 10.
^{b} Energy in kK (molar absorbance, $M^{-1}cm^{-1}$), sh: shoulder, b: broad.
^{c} Obscured by intense $O \rightarrow V^V$ charge transfer absorption.

ments listed in Table I are based on considerations discussed previously (*7, 8, 15, 16*). Most of the spectral features are broad, and the energies given are subject to uncertainties of ± 0.5 kK. Further discussion of the electronic structures of these heteropoly anions must await measurements on single isomers.

Acknowledgments

We thank E. Sokoloski (National Institutes of Health) and A. English (E. I. duPont de Nemours & Co., Inc.) for assistance in obtaining the NMR spectra, and Susanne Raynor for assistance in programming the spectra simulation.

Literature Cited

1. Evans, Jr., H. T., *Perspect. Struct. Chem.* (1971) 4, 1.
2. Weakley, T. J. R., *Struct. Bonding Berlin* (1974) 18, 131.
3. Pope, M. T., *Inorg. Chem.* (1972) 11, 1973.
4. Prados, R. A., Meiklejohn, P. T., Pope, M. T., *J. Amer. Chem. Soc.* (1974) 96, 1261.
5. Pope, M. T., Scully, T. F., *Inorg. Chem.* (1975) 14, 953.
6. Pope, M. T., O'Donnell, S. E., Prados, R. A., *J. Chem. Soc. Chem. Commun.* (1975) 22.
7. Smith, D. P., Pope, M. T., *Inorg. Chem.* (1973) 12, 331.
8. Smith, D. P., So, H., Bender, J., Pope, M. T., *Inorg. Chem.* (1973) 12, 685.
9. Altenau, J. J., Pope, M. T., Prados, R. A., So, H., *Inorg. Chem.* (1975) 14, 417.
10. Tsigdinos, G. A., Hallada, C. J., *Inorg. Chem.* (1968) 7, 437.
11. O'Donnell, S. E., Ph.D. Dissertation, Georgetown University, 1975.
12. O'Donnell, S. E., Pope, M. T., manuscript in preparation.
13. Lange, G., Hahn, H., Dehnicke, K., *Z. Naturforsch. Teil B* (1969) 24, 1498.
14. Weissman, S. I., *J. Amer. Chem. Soc.* (1958) 80, 6462.
15. Flynn, C. M., Jr., Pope, M. T., *Inorg. Chem.* (1973) 12, 1626.
16. So, H., Pope, M. T., *Inorg. Chem.* (1972) 11, 1441.

RECEIVED January 24, 1975. Work supported by NSF grants GP-10538 and -40991X. The ESR spectrometer was purchased with the aid of NSF equipment grant GP-29184.

9

Complex Compounds with a Large Charge Separation

JOSEPH CHATT

School of Molecular Sciences, University of Sussex, Brighton BN1 9QJ, U.K.

Asymmetrical coordination compounds can have very high electric dipole moments. Covalent compounds with moments of up to 14.8 Debye units are known. The charge separation in the coordinate bonds to the metal from the ligand atoms of so-called good electron-donor neutral ligands is at least as important in many complex compounds as that of the bonds from the metal to the anionic ligands. It is shown how existing knowledge could be used to tailor covalent molecules with very high dipole moments, perhaps as much as 20 Debye units.

The purpose of this review is to draw attention to the large charge separation which occurs in some coordination compounds and to its source. The review is the result of the author's experience in the use of dipole moment measurements to determine the configuration of complex compounds before unambiguous spectroscopic means were available. His research group has measured a great number of dipole moments during the past 38 years, yet few coordination chemists appear to be aware of the large dipole moments which can be developed in assymetric covalent compounds as a result of the presence of coordinate bonds.

In the pre-Pauling era, one was led to believe that the main separation of charge in a molecule of a complex compound was associated with the coordinate bonds between the metal and the ligand atoms of the so-called uncharged ligands, rather than in the bonds from the metal to the formally anionic ligands. For many years this assumption served well in assigning configurations to tertiary phosphine complexes of Groups VII and VIII metal halides on the basis of their dipole moments. Many years and a considerable number of measured dipole moments later, the author is still convinced that the charge separation in the bond between a metal and a ligating atom of a good donor ligand is at least as great as that between the metal and an anionic ligand. Some recent XPS measurements (1) support this contention and indicate, in conjunction with dipole moments, that in tertiary phosphine complexes of the type $[MCl_n(PR_3)_n]$ the metal is about neutral. The chlorine atom would carry a charge of about $-0.3\ e$ and the phosphorus about $+0.3\ e$.

This knowledge can be used to tailor highly dipolar molecules that are soluble in organic solvents or fusible without change. This is illustrated by a representative selection of compounds whose dipole moments were measured in benzene solution. The practically important factors are cited.

In 1936 K. A. Jensen (2, 3) measured the moments of a number of platinum(II) dihalido complexes with tertiary phosphines and other such ligands. He showed that the cis complexes had moments of up to about 11.0 Debye units (D) and that atom polarization in complex compounds is much higher than in organic compounds. He also showed that all the complexes of this type had had their cis or trans configurations incorrectly assigned on the basis of color, in analogy with the diammine dichloro platinum(II) complexes.

Factors Affecting the Size of Dipole Moments in Mononuclear Complexes

Empirically there are now sufficient data to show which factors affect the charge separation in complex compounds. These are discussed below. The references to the Tables list many more values of dipole moments, and they also cite references for others.

Dependence of Dipole Moment on the Metal. Generally the group moments, e.g. of the group P–M–Cl (M = transition metal), in cis-$[MCl_n(PR_3)_n]$-type complexes do not depend strongly on the metal or on its oxidation state, provided that only good donor ligands, whether neutral or anionic, are involved. Such ligands are tertiary phosphines and chloride ions in the examples in Table I. The dipole moments given are those of some axially symmetrical complex compounds that contain a variety of metals. On the assumption that the ligand atom bonds are at right angles, which is not quite justified, the moments have been resolved along the P–M–Cl directions, and the inferred group moments are listed. The group moment is generally 6–7 D (with a little scatter outside this range), and the aliphatic phosphines tend to produce slightly higher moments than the more electronegative aromatic phosphines. The exact metal involved is obviously not of great significance in determining the group moment.

Dependence on So-called Electrically Neutral Ligands. Good donors —e.g. tertiary phosphines, arsines, and stibines—give analogous complexes with almost equal moments when they are associated with halide

Table I.   Some Molecular Dipole Moments and the P–M–Cl
Group Moments Inferred from Them

Compound[a]	Molecular Dipole Moment, D	P–M–Cl Group Moment, D
cis-[RuCl₂(depe)₂][a]	9.8	7.0
cis-[RuCl₂(dppe)₂][a]	9.5	6.8
cis-[OsCl₂(depe)₂][a]	9.3	6.6
cis-[OsCl₂(dppe)₂][a]	8.3	5.9
mer-[ReCl₃(PEt₂Ph)₃][b]	6.3	6.3
mer-[RhCl₃(PEt₃)₃][c]	7.0	7.0
mer-[IrCl₃(PEt₃)₃][d]	6.9	6.9
cis-[PtCl₂(PEt₃)₂][e]	10.9	7.6

[a] Legend: depe = Et₂PCH₂CH₂PEt₂; dppe = Ph₂PCH₂CH₂PPh₂.
[b] Ref. *16*.
[c] Ref. *9*.
[d] Ref. *17*.
[e] Ref. *18*.
[f] Ref. *19*.

Table II.   Dipole Moments of Some *cis*-Platinum(II) Chloro Complexes

Complex	Dipole Moment, D
[PtCl₂(PEt₃)₂][a]	10.9
[PtCl₂(AsEt₃)₂][a]	10.9
[PtCl₂(SbEt₃)₂][a]	10.45
[PtCl₂(PF₃)₂][b]	4.4
[PtCl₂(CO)₂][b]	4.65

[a] Ref. *19*.
[b] Ref. *20*.

in a complex (Table II). Pyridine and other amines give similar charge separation. Poor donors—e.g., good back-bonding ligands such as carbon monoxide and phosphorus trifluoride—give complexes with considerably lower moments. The examples in Table II show that the neutral ligand is certainly as important as the anionic in determining total charge separation. Also, since the carbonyl and phosphorus trifluoride ligands are much less positively charged than are tertiary phosphine ligands, the metal itself must carry more positive charge in these complexes than it does in the analogous tertiary phosphine and similar complexes. Thus, despite the constant formal oxidation state of the metal in the examples in Table II, its absolute charge must vary widely. Dinitrogen and so-called NO⁺ are even more negative in their complexes than are carbon monoxide and phosphorus trifluoride (*1*). It is interesting too that the separation of charge in the very labile tertiary stibine platinum(II) complex is almost as great as that in the stable tertiary phosphine complex (Table II). If there is any weakening of electron donation from the stibine, as compared with that from the phosphine, it is almost compensated for by the greater Pt–Sb bond distance.

Since the ligand atoms of neutral ligands can carry very different charges in their complexes that range from strongly positive to neutral and occasionally to quite negative (*e.g.* so-called NO^+), it is possible to obtain very dipolar complexes that contain only so-called neutral ligands. The monodentate phosphine and arsine in the examples in Table III give complexes with moments of about 3.5 D, but the chelate diphosphines, which close the natural tetrahedral angle on the nickel atom, give compounds with considerably greater moments (5.0 D). This illustrates that a large part of the charge separation occurs in the Ni–P bonds where the arsenic and phosphorus are positive relative to the $Ni(CO)_x$ residue.

Table III.   Dipole Moments of Some Complexes that
Contain No Anionic Ligands

Complex	Dipole Moment, D
$[Ni(CO)_3(PPh_3)]$[a]	3.8
$[Ni(CO)_3(AsPh_3)]$[a]	3.6
$[Ni(CO)_2(PPh_3)_2]$[a]	3.8
$[Ni(CO)_2(Ph_2PCH_2CH_2PPh_2)]$[a]	4.8
$[Ni(CO)_2\{o\text{-}C_6H_4(PEt_2)_2\}]$[a]	5.4
$[Ni(CO)\{PhP(o\text{-}C_6H_4PEt_2)_2\}]$[b]	5.2

[a] Ref. 21.
[b] Ref. 22.

Dependence on Anionic Ligand. Axially symmetrical complexes with different anionic ligands at the two ends of the axis often have considerable dipole moments. The moments of such compounds are almost a direct measure of the differences between the tendencies of those anions (A⁻) to give up electronic charge to the metal ion, or, as an alternative view, for those atomic ligands (A˙) to receive electronic charge from the metal atom. Thus, in complexes of the *trans*-$[Pt(Me)(A)(PEt_3)_2]$ type, the moment (D) increases in the order: A = Me (0.0) < Cl (3.4) < Br (3.7 < I (4.1) < NO₃ (6.0) < NCS (6.5) (4). Presumably, the thiocyanate is N-bonded. It seems from this series that the moment increases as the anion becomes harder and perhaps also as it becomes larger. In the halide series, the increase in size appears to be outweighed by the effect of increasing polarizability.

Similarly, in the series of *trans*-$[PtCl(A)(PEt_3)_2]$ complexes, the moment (D) increases in the order: A = Cl (0.0) < Ph (2.6) (5) < Me (3.4) (4) < H (4.2) (6). These values illustrate that large dipole moments can be developed along uniterminal axes of symmetry between different mono-anionic ligands, with the ligands increasing in negative contribution in the order: H < Me < Ph < Cl < Br < I < NO₃ < NCS. It follows that, of these anions, H as opposed to N-bonded NCS should produce the greatest interanion moments. If the moments were additive,

Table IV.   Comparison of Directly Measured and
Inferred Values of the Cl–M–CO Group Moment

	Dipole Moment, D	
	Molecules	*Cl–M–CO Group*
trans-[IrCl(CO)(PEt$_2$Ph)$_2$][a]	2.1	2.1
trans-[RhCl(CO)(PEt$_2$Ph)$_2$][b]	2.4	2.4
trans-[ReCl(CO)(PMe$_2$Ph)$_4$][c]	2.0	2.0
mer-trans-[IrCl$_3$(CO)(PEt$_3$)$_2$][d]	2.8	2.8
mer-trans-[RhCl$_3$(CO)(PEt$_2$Ph)$_2$][b]	3.6	3.6
fac-cis-[IrCl$_3$(CO)(PBu$_3$)$_2$][d]	12.4	8.1 [e]
cis-[PtCl$_2$(CO)$_2$][f]	4.7	3.4 [e]
cis-[PtCl$_2$(CO)(PEt$_3$)][g]	10.0	6.5 [e]
cis-[PtCl$_2$(CO)(PPr$_3$)][g]	10.2	6.6 [e]

[a] Ref. *23*.
[b] Ref. *24*.
[c] Ref. *25*.
[d] Ref. *26*.
[e] Inferred on the basis that the P–M–Cl group moment is 6.6 D with phosphorus positive.
[f] Ref. *20*.
[g] Ref. *27*.

a compound such as trans-[PtH(NCS)(PEt$_3$)$_2$] would have a moment of $\sim$ 7.45 D; the observed moment is 7.4 D (*6*). Evidently these large moments are vectorially additive in axially symmetrical complexes, but this is not true of less symmetrical complexes. The effect of low symmetry on the vector addition is shown by comparing the directly observed and the inferred values of the Cl–M–CO group moments (Table IV). Whereas the Cl–M–CO group moment generally has a value of 2–3 D in axially symmetrical complexes when it is equal to the molecular moment, in less symmetrical complexes, when it must be inferred by resolution of the group moments, the value is generally 6–8 D. This anomaly may be explained by supposing that the absolute charge on the metal varies from complex to complex even when the formal oxidation state is constant. When organophosphine and chloride ligands are attached in equal numbers to a transition metal atom, the charge on the metal is around zero, whatever the oxidation state. This would explain the somewhat similar chemistry of the series of compounds [AuCl(PR$_3$)], [PtCl$_2$(PR$_3$)$_2$], and [IrCl$_3$(PR$_3$)$_3$] although the formal charge increases from I to III along the series. As soon as the phosphine and chloride ligands get out of balance, a charge develops on the metal, with the charge depending on the dissimilarity of the other ligands. The metal generally tends to become negative if the P-to-Cl ratio increases, and positive if it decreases or if the phosphines are replaced by poorer electron-donor ligands. This charge on the metal would affect the bond moments associated with other ligands attached to the metal atom so that the simple vector summation of the bond moments would cease to be valid unless the dissimilar ligands were arranged symmetrically around the axis of the moment. It follows that the concept of vector additivity of bond moments must be used with great care in attempts to determine the configurations of highly asymmetric complex compounds when ligands of widely different types are present together.

Dependence of the Magnitude of Formal Charge on Mono-Atomic Anionic Ligands (A′). It might be thought that if a chloride were replaced by the formally more highly charged oxide or nitride ligands, a greater dipole moment along the M–A′ bond would result. In fact the oxide and nitride ligands make no more contribution to the moment (μ) of the M–A′ bond than does the chloride.

This is shown for the oxo ligand by comparing the moments of the two complexes [ReCl$_3$X(PEt$_2$Ph)$_2$] (X = Cl or X = 0) of configuration

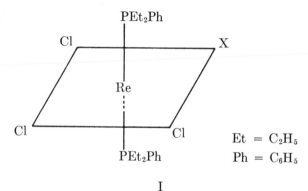

I

I, somewhat distorted in the oxo-complex (7); when X = Cl, $\mu = 0.87$ D (8), and when X = 0, $\mu = 1.7$ D (9). If the Cl–Re–Cl and Cl–Re–O group moments have the same sign, which is highly probable, they are equal to within 0.8 D. Such a difference is insignificant in this context. This is confirmed by the moment of fac-cis-[ReCl$_3$O(PEt$_2$Ph)$_2$] which is 10.8 D and is close to that of the cis-[PtCl$_2$(PR$_3$)$_2$]-type complexes; the oxo ligand obviously makes no exceptional contribution to the dipole moment. It has been suggested that it is even less negative than a chloro ligand and that the analogous phenylimido ligand (PhN⁻) may make a slight positive contribution (9).

Very few dipole moments of nitrido complexes have been measured, but complex II has a moment of 6.4 D (10). This is a characteristic P–Re–Cl group moment, and it indicates that the Cl–Re–N group moment is near zero. The slight distortions from regular octahedral arrangement (11) are insufficient to affect this conclusion. The trans-[ReCl$_2$N-(PPh$_3$)$_3$] complex has a distorted tetragonal pyramidal structure (12)

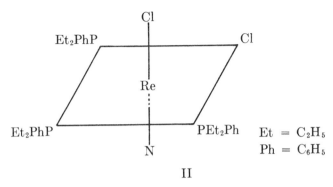

$$Et = C_2H_5$$
$$Ph = C_6H_5$$

II

with the nitride ligand at the apex. It has a moment of 1.6 D (*10*), and that of the corresponding bromide is 1.5 D which again indicates that there is no great separation of charge in the Re–N bond.

To obtain stable oxo, imido, and nitrido complexes, it is necessary to have metal atoms such as rhenium in rather high oxidation states. Then covalency is so strong that the M–N and M–O bonds are essentially triple in character and no more dipolar than the M–Cl bond.

It is evident that the greatest charge separation occurs across a metal center in complexes with good electron-donor ligands (*e.g.* tertiary phosphines or amines) on one side and hard, somewhat extended anionic ligands on the other. However, the presence of formally highly charged mono-atomic anionic ligands does nothing to enhance the moment.

Dipole Moments of Polynuclear Complexes

If one could arrange two *cis*-[PtCl$_2$(PR$_3$)$_2$] complexes in parallel in a dinuclear complex system, one should obtain a compound with a moment approaching twice that of the mononuclear complex. A partial approach to this is found in some dinuclear thiolato-bridged platinum(II) complexes formed as by Reaction 1 (*13*).

Compound III is the stable isomer. The trans isomer is unknown. This is remarkable because with all the electropositive ligand atoms along one side and chloride ligands along the other, complex III must resemble a small capacitor. Its electrical energy could be discharged to some extent by interchanging the terminal chloride and phosphine ligands on one platinum atom. The dithiolato complex (IV) is no more dipolar than the *cis*-[PtCl$_2$(PR)$_2$]-type complex, but its moment can be enhanced by putting a *p*-nitrothiophenolato ligand into the bridge trans to the phosphine molecules so that the moment of the NO$_2$ group adds to that of complex IV. In this way, one obtains complex V with $\mu = 14.8$ D, which is the most dipolar covalent molecule known to the author (*14*). Its moment is greater than that of tetrabutylammonium acetate (11.2 D)

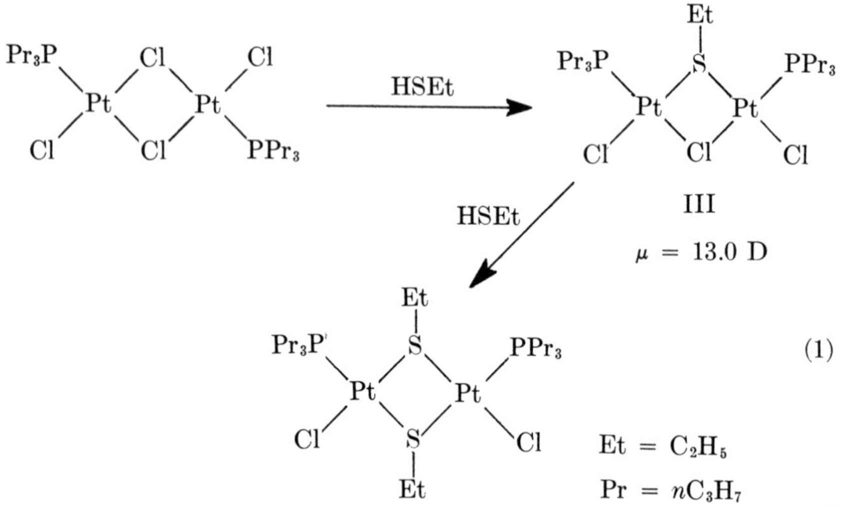

$$\mu = 13.0 \text{ D}$$

$$\mu = 10.3 \text{ D}$$

$$\text{Et} = C_2H_5$$
$$\text{Pr} = nC_3H_7$$

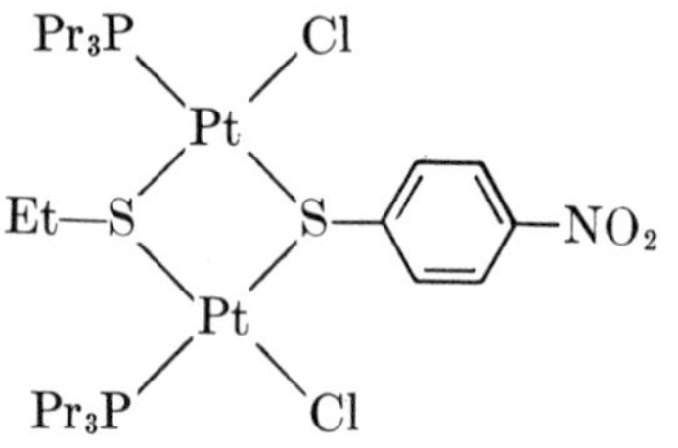

(1)

$$\text{Et} = C_2H_5$$
$$\text{Pr} = nC_3H_7$$

(15). Perhaps it could be enhanced further by replacing the two chloride ions with NCS⁻.

To obtain compounds with large dipole moments, it would seem best to string highly dipolar centers with their moments in parallel by means of constraining ligands. This leads to the suggestion that a com-

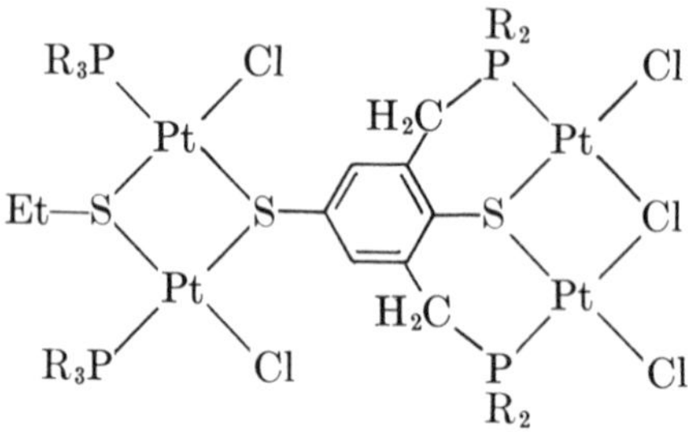

$$\text{Et} = C_2H_5$$
$$\text{R} = \text{alkyl}$$

VI

pound such as VI would have a dipole moment of over 20 D. This could be enhanced a little further by replacing the bridging chloride ligand to the right of the molecule with a *p*-nitrothiophenolato ligand. Such molecules may tend to be insoluble when the alkyl groups are small. However, by choosing sufficiently long alkyl chains (R) on the phosphines, the complexes could be rendered soluble in organic solvents or could be obtained as oils or waxes with the highly dipolar centers buried inside them.

No use appears to have been found for substances with very large dipole moments such as those described, but, if they were needed, they could undoubtedly be made and could possibly be tailored to fit that need.

Literature Cited

1. Chatt, J., Elson, C. M., Hooper, N. E., Leigh, G. J., *J. C. S. Dalton*, in press.
2. Jensen, K. A., *Z. Anorg. Chem.* (1936) **229**, 225.
3. *Ibid.* (1936) **229**, 250.
4. Chatt, J., Shaw, B. L., *J. Chem. Soc.* (1959) 705.
5. *Ibid.* (1959) 4020.
6. Chatt, J., Shaw, B. L., *J. Chem. Soc.* (1962) 5075.
7. Ehrlich, H. W. W., Owston, P. G., *J. Chem. Soc.* (1963) 4368.
8. Chatt, J., Garforth, J. D., Johnson, N. P., Rowe, G. A., *J. Chem. Soc.* (1964) 601.
9. Chatt, J., Rowe, G. A., *J. Chem. Soc.* (1962) 4019.
10. Chatt, J., Garforth, J. D., Johnson, N. P., Rowe, G. A., *J. Chem. Soc.* (1964) 1012.
11. Corfield, P. W. R., Doedens, R. J., Ibers, J. A., *Inorg. Chem.* (1967) **6**, 197.
12. Doedens, R. J., Ibers, J. A., *Inorg. Chem.* (1967) **6**, 204.
13. Chatt, J., Hart, F. A., *J. Chem. Soc.* (1953) 2363.
14. Chatt, J., Hart, F. A., *J. Chem. Soc.* (1960) 2807.
15. Geddes, J. A., Kraus, C. A., *Trans. Faraday Soc.* (1936) **32**, 585.
16. Chatt, J., Hayter, R. G., *J. Chem. Soc.* (1961) 896.
17. Chatt, J., Johnson, N. P., Shaw, B. L., *J. Chem. Soc.* (1964) 2508.
18. Chatt, J., Field, N. E., Shaw, B. L., *J. Chem. Soc.* (1963) 3371.
19. Chatt, J., Wilkins, R. G., *J. Chem. Soc.* (1952) 4300.
20. Chatt, J., Williams, A. A., *J. Chem. Soc.* (1951) 3061.
21. Chatt, J., Hart, F. A., *J. Chem. Soc.* (1960) 1378.
22. Chatt, J., Hart, F. A., *J. Chem. Soc.* (1965) 812.
23. Chatt, J., Johnson, N. P., Shaw, B. L., *J. Chem. Soc.* (1967) 604.
24. Chatt, J., Shaw, B. L., *J. Chem. Soc. A* (1966) 1437.
25. Chatt, J., Crabtree, R. H., Jeffery, E. A., Richards, R. L., *J. C. S. Dalton* (1973) 1167.
26. Chatt, J., Johnson, N. P., Shaw, B. L., *J. Chem. Soc.* (1964) 1625.
27. *Ibid.* (1964) 1662.

RECEIVED February 6, 1975.

10

Room Temperature Fused Salts: Liquid Chlorocuprates(I)

W. W. PORTERFIELD

Hampden-Sydney College, Hampden-Sydney, Va. 23943

J. T. YOKE

Oregon State University, Corvallis, Ore. 97331

A liquid with the composition of the presumed compound triethylammonium dichlorocuprate(I) is prepared by mixing stoichometric amounts of the solid precursors CuCl and Et₃NHCl; similar liquids can be prepared using other tertiary amine and phosphine hydrochlorides. Spectroscopic and electrochemical studies indicate that the liquid nature of the systems is attributable to the presence of multiple chlorocuprate(I) species through chloride donor–acceptor equilibria. Two applications that utilize the liquid property of the compounds are: (a) use as a solvent for transition-metal chlorides in examining the effect of drastically lowered temperatures on the stoichiometry and coordination geometry of the chlorometallate complexes previously observed in molten LiCl/KCl and AlCl₃, and (b) construction of a voltaic cell in which the liquid chlorocuprate(I) serves as solvent and as both electroactive species.

In studies of copper(I) coordination about 12 years ago (1), Yoke and co-workers observed that the crystalline solids copper(I) chloride and triethylammonium chloride react quickly when brought into contact to produce a light green oil. The oil is quite oxygen sensitive, and it was assumed that the compound triethylammonium dichlorocuprate(I) had been formed which was liquid because of the low lattice energy inherent to the bulky cation. Many previous preparations of polyalkylammonium chlorocuprates (2, 3, 4, 5), however, had yielded only crystalline solids. Consequently, the unusual liquid nature of the system inspired structural studies and the synthesis of comparable compounds.

Experimental

All chlorocuprate syntheses were carried out in Schlenk glassware, and 10-mm borosilicate glass cuvettes were adapted to Schlenk use in order to avoid oxidation during filling. The cell design (Figure 4) was also adapted for direct filling from the Schlenk apparatus. Raman spectra were obtained on a Cary 82 instrument at Oregon State University and on a Jarrell–Ash instrument at the University of North Carolina. Visible range spectra ($14,000$–$26,000$ cm^{-1}) were recorded on a Bausch and Lomb 505 spectrophotometer.

Synthesis and Structural Studies

Reaction 1 is the synthesis reaction and Reactions 2, 3, and 4 are the chloride-transfer equilibria that are involved in the formation of triethyl-ammonium dichlorocuprate(I). Because of the sensitivity of the liquid to oxidation, all operations are carried out in Schlenk ware. The synthesis is effected simply by mixing stoichiometric quantities of the two solids which react endothermically to form the liquid in about two minutes.

$$Et_3NHCl + CuCl \rightarrow Et_3NHCuCl_2 \tag{1}$$

$$CuCl_2^- + CuCl \rightleftharpoons Cu_2Cl_3^- \tag{2}$$

$$2CuCl_2^- \rightleftharpoons Cu_2Cl_3^- + Cl^- \tag{3}$$

$$CuCl_2^- + Cl^- \rightleftharpoons CuCl_3^{2-} \tag{4}$$

Results were entirely equivalent with the cations N-ethylpiperidinium, triethylphosphonium, and triethylchlorophosphonium.

The liquid nature of these systems, however, is not simply the result of lattice-energy effects. Solid tetraphenylborates and dibromocuprates(I) of the alkylammonium cations have been prepared even though such anions should yield even smaller lattice energies. Significant hydrogen bonding occurs between the ammonium hydrogen and the chlorocuprate species as was demonstrated by previously ^{1}H NMR studies and IR spectra (6) and by the conductance studies described below. However, the existence of triethylchlorophosphonium dichlorocuprate (I) as a liquid in which no hydrogen bonding can possibly occur rules out strong, specific hydrogen bonding as a possible cause of the low melting points of these systems.

The three equilibria (Reactions 2, 3, and 4) probably provide significant quantities of at least four anionic species simultaneously in the liquid mixture, thereby lowering the freezing point below room temperature. No well defined freezing point exists, however, since the system becomes more viscous and glasses on cooling; it remains recognizably liquid down to about 0°C. At elevated temperatures, the liquid is stable to about 110°C, above which temperature triethylamine distills out. The $CuCl_3^{2-}$ species is questionable for the ammonium systems but is inferred for the phosphonium systems from the greater solubility of the triethylphosphonium chloride (up to the 3:1 Cl:Cu mole ratio).

Evidence for the existence of the binuclear $Cu_2Cl_3^-$ is provided by the Raman spectra of the liquids (Figure 1). The peak at 302 cm^{-1} is in

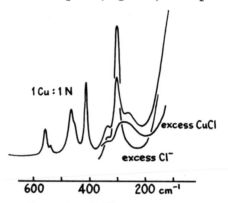

Figure 1. $CuCl_2^-$ Raman spectra

good agreement with earlier work (7) that assigned this frequency to the $CuCl_2^-$ symmetric stretch, but there is an additional broad peak or group of unresolved peaks at 270–310 cm^{-1} that is enhanced at the expense of the $CuCl_2^-$ ion by addition of excess CuCl but is suppressed by addition of excess Cl$^-$. For a bent Cl–Cu–Cl–Cu–Cl structure, there should be nine Raman-active modes, of which four correspond to bond stretches and might reasonably occur in this region of the spectrum (6).

On the basis of this formulation of the liquid's structure, these substances should be viewed as molten salts; their specific conductance is of interest in this context (see Table I). In a separate conductance study

Table I.   Specific Conductances

Salt	State	T, °C	Specific Conductance, ohm^{-1} cm^{-1}
Et$_3$NH$^+$ CuCl$_2^-$	molten	25°	3.84 × 10^{-3}
		85°	3.16 × 10^{-2}
KCl^a	0.01M aqueous	25°	1.413 × 10^{-3}
	molten	872°	2.407 × 10^0

^{a} From Ref. 8.

in acetonitrile solution, an overall dissociation constant of 9 × 10^{-2} was obtained for the dichlorocuprate solution; this is somewhat less than

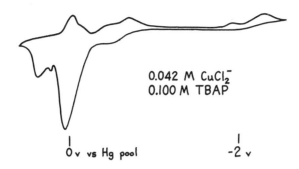

Figure 2. $CuCl_2^-$ CVA in acetonitrile

half the value for the corresponding tetraphenylborate. This reinforces the spectroscopic evidence that hydrogen bonding provides a significant degree of association within the liquid.

Cyclic voltammetry studies of the chlorocuprate liquid which were also carried out in acetonitrile solution indicate multiple electroactive species (*see* Figure 2). The number of cathodic peaks is affected by the Cl:Cu stoichiometry and also to some extent by the age of the chlorocuprate liquid before dissolving (though not reproducibly). This tends to confirm the presence of mobile equilibria and suggests the possibility of a slow formation of other polynuclear chlorocuprates in the liquid. Some of the reductions are irreversible whereas others are quasireversible (*9*), which agrees with previous studies (*10*). If this conclusion can be legitimately extended to the liquid chlorocuprate, it has an obvious bearing on the possibility of constructing a voltaic cell that delivers a high current density.

Applications

One interesting application of liquid chlorocuprates is based on their ability to serve as molten-chloride solvents for other transition-metal chlorides. Gruen and co-workers (*11, 12, 13*) studied extensively chlorometallate complexes in molten LiCl/KCl mixtures and molten $AlCl_3$. These have proved to be excellent monatomic-ligand coordinating solvents that permit spectroscopic studies over an extended frequency range because of their transparency at nearly all wave numbers below 50,000 cm^{-1}. Triethylammonium dichlorocuprate(I) has a UV absorption cutoff at about 26,000 cm^{-1}, but it is otherwise transparent except for the triethylammonium IR peaks. It offers, in addition, the convenience of operating at room temperature and the possibility of improved resolution of electronic spectra because of the narrower vibrational envelopes for each electronic transition at lower temperatures.

The metal chlorides whose spectra were obtained in liquid chloro-cuprate solvent at 14,000–26,000 cm⁻¹ were $TiCl_3$, VCl_3, $CrCl_3$, $MnCl_2$, $FeCl_2$, $FeCl_3$, $RuCl_3$, $CoCl_2$, $NiCl_2$, and $CuCl_2$. There are some solubility limitations, but most transition metal chlorides seem sufficiently soluble to permit spectroscopic studies. A more important limitation is the redox reactivity of the solvent; even relatively weak oxidizing agents such as $FeCl_3$ oxidize the solvent to copper(II) species. Strongly reducing chlorides would presumably reduce the solvent to copper metal, but $TiCl_3$, at least, seems to be stable toward electron-transfer reactions in the chlorocuprate medium. Given adequate solubility and a suitable $E°$, quite useful spectra can be obtained (e.g. Figure 3). This spectrum is

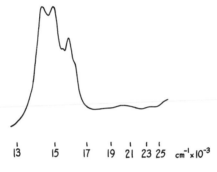

13	15	17	19	21	23	25	cm⁻¹×10⁻³

Figure 3. $CoCl_2^-$ in $CuCl_2^-$ solution

closely similar to Gruen's and may be interpreted as that of the tetrahedral $CoCl_4^{2-}$ ion. Some of the metals studied are known to be involved in temperature-sensitive equilibria between octahedral and tetrahedral species, and, in general, the markedly lower temperature used in this work has the expected effect of shifting these equilibria toward the higher coordination number.

The coordination geometries that were tentatively assigned, according to Gruen's theoretical treatment, to the species existing in the liquid chlorocuprate solvent are given in Table II. The geometry of the Ti^{3+} species is uncertain since the near-IR peaks that characterize the species Gruen studied in molten chloride solvent are not now accessible to us; in addition, the unsymmetrical peak at 22,000 cm⁻¹ is inexplicable for any coordination geometry given the position of Cl^- in the spectrochemical series. The near-IR range of the spectrum will be studied further as soon as equipment is available. The same discussion applies to the solutes $FeCl_2$ and $CuCl_2$. For the other solutes, agreement was reasonably good between our spectra in chlorocuprate solvent and Gruen's spectra in molten chloride solvent, and we have adopted his assignments for the observed transitions.

Table II. Visible Spectra of Chlorometallate Species

Absorption Peaks, cm^{-1}

Solute	Solvent: $CuCl_2^-$	Solvent: Molten Cl^{-a}	Assigned Coordination Geometry in $CuCl_2^-$
$TiCl_3$	22,000	10,000[b] 13,000[b]	octahedral $TiCl_6^{3-a}$
VCl_3	19,900	11,000[b] 18,500	octahedral VCl_6^{3-}
$CrCl_3$	18,800 23,500	12,500[b] 18,500	octahedral $CrCl_6^{3-a}$
$MnCl_2$	22,400 ($\varepsilon = 0.2$) 23,200	23,400 ($\varepsilon = 0.2$) 28,200	tetrahedral $MnCl_4^{2-}$
$FeCl_2$	24,100	5,100 ($FeCl_4^{2-}$)	octahedral $FeCl_6^{4-a}$
$CoCl_2$	14,300 14,900 15,800	5,600[b] 14,300 15,100 16,400	tetrahedral $CoCl_4^{2-}$
$NiCl_2$	14,300 15,400	8,000[b] 14,200 15,300	dist. tetrahedral $NiCl_4^{2-}$
$CuCl_2$	23,600	9,500[b] ($CuCl_4^{2-}$) strong above 20,000	unknown

[a] From Refs. *11, 12,* and *13.*
[b] In the IR range.

Another application of the molten-salt character of the liquid chloro-cuprates that offers some interest is the possibility of constructing a voltaic cell in which the chlorocuprate would serve as the high conductance charge-transfer medium and as the source of both electroactive species. Figure 4 depicts our cell design and notes the half-reactions which are written in the spontaneous sense. Such a cell should have a fairly high energy density because of the absence of an inert solvent and the resultant high weight fraction of electroactive components. The cell is most easily prepared in the totally discharged form with pure liquid chlorocuprate surrounding inert electrodes. At less than 1% of full charge, the cell develops 0.85 V, a value that is independent of any further degree of charge. However, the solubility of $CuCl_2$ in the chlorocuprate solvent causes rapid decay of the measured potential after charging stops. Since a full charge would produce an entirely solid system, this difficulty with solubility of the electrode material should be ameliorated when the degree of charge is high enough to provide a saturated solution. Although gelling of the liquid was observed, it has not been possible to achieve full charge

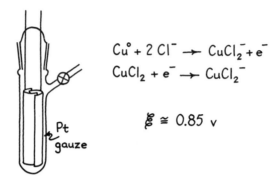

$$Cu^{\circ} + 2\, Cl^{-} \rightarrow CuCl_{2}^{-} + e^{-}$$
$$CuCl_{2} + e^{-} \rightarrow CuCl_{2}^{-}$$

$$\mathscr{E} \cong 0.85 \text{ v}$$

Figure 4. $CuCl_{2}^{-}$ galvanic cell

of the cell because of the growth of copper metal dendrites that cause internal shorting of the cell.

As the cyclic voltammetry studies suggest, the electrode reactions are not reversible, and the resultant low exchange current together with the viscosity of the liquid cause substantial polarization of the cell at low load impedances. The discharge curve of the cell, even under polarized conditions, demonstrates that the potential is very nearly constant until the solid $CuCl_{2}$ electrode material is exhausted.

Our preliminary examination of these two applications seems to confirm the molten-salt interpretation of the liquid chlorocuprate(I) structure. There may be circumstances in which this property may find other application, or in which a liquid copper(I) compound might be useful in organic syntheses designed to use a solid copper(I) halide as a reductant or hydride acceptor (*see* Refs. 14 and 15 and the references cited therein). Our aim in presenting this report is to encourage others to take advantage of this series of inorganic compounds with an unusual property.

Acknowledgments

Credit is due to several Hampden-Sydney College students who carried out most of this work during the past three years: Bennie Good, Paul Page, Glen Williams, Pierce Brown, and Mike Carroll.

Literature Cited

1. Yoke, J. T., *et al., Inorg. Chem.* (1963) 2, 1210.
2. "Gmelin's Handbuch der anorganischen Chemie," Lieferung, Gmelin Institut, System 60, Teil B, p. 317, Verlag Chemie, Weinheim, 1958.
3. *Ibid.* Lieferung 3, 1965, p. 1092.
4. *Ibid.* Lieferung 3, 1965, p. 1411.
5. Remy, H., Laves, G., *Ber.* (1933) 66, 571.

6. Axtell, D. D., *et al., J. Amer. Chem. Soc.* (1973) **95**, 4555.
7. Creighton, J. A., Lippincott, E. R., *J. Chem. Soc.* (1963) 5134.
8. Bockris, J. O'M., Reddy, A. K. N., "Modern Electrochemistry," Plenum, New York, 1970.
9. Brown, E. R., Large, R. F., in "Physical Methods of Chemistry," A. Weissberger and B. W. Rossiter, Eds., Vol. IIA, pp. 423–530, Wiley, New York, 1971.
10. Manahan, S. E., Iwamoto, R. T., *Inorg. Chem.* (1965) **4**, 1409.
11. Gruen, D. M., McBeth, R. L., *Pure Appl. Chem.* (1963) **6**, 23.
12. Øye, H. A., Gruen, D. M., *Inorg. Chem.* (1964) **3**, 836.
13. *Ibid.* (1965) **4**, 1173.
14. Cohen, T., *et al., J. Amer. Chem. Soc.* (1974) **96**, 7753.
15. Massamune, S., *et al., J. Amer. Chem. Soc.* (1974) **96**, 6452.

RECEIVED January 24, 1975. Work supported by National Science Foundation URP and RPCT grants.

11

Non-Stoichiometric Liquid Enclosure Compounds ("Liquid Clathrates")

JERRY L. ATWOOD

Department of Chemistry, University of Alabama, University, Ala. 35486

JIM D. ATWOOD

School of Chemical Sciences, University of Illinois, Urbana, Ill. 61801

Compounds of the general formulation $M[Al_2(CH_3)_6X]$— where M = alkali metal, tetraalkylammonium ion and X = N_3^-, SCN^-, $SeCN^-$, Cl^-, Br^-, I^-—react in a visually dramatic fashion with small aromatic molecules to form non-stoichiometric liquid complexes. In the most favorable situation, as many as 16 benzene molecules are trapped per anionic unit. Because of similarity to the well known solid-state clathrates, the designation "liquid clathrate" is used for the new substances. The term is not meant to imply order as in the solid state, but to convey the method of interaction of $M[Al_2(CH_3)_6X]$ with aromatic molecules. The liquid clathrate may consist of either roughly spherical or layerlike domains; the latter is supported by the x-ray crystallographically determined structure of $K[CH_3Se\{Al(CH_3)_3\}_3] \cdot 2C_6H_6$.

The interactions of organometallic molecules with small hydrocarbon substrates have been studied extensively both in solution and in the solid state. In the latter, the interactions are necessarily quite strong, and they may be divided into two extreme categories. There are molecules which involve a strong directional bond between the metal ion and the aromatic moiety [*e.g.*, in $Cr(C_6H_6)_2$ (*1*, *2*) or in the $AgClO_4 \cdot$ aromatic systems (*3*)], and there are arrays for which there is no specific interaction between the metal atom itself and the aromatic center [*e.g.*, the case presented by simple molecules of solvation (*4, 5*)]. On the other hand, the diversity of systems reported for the liquid state is so great as to defy summation, but some of the most important have involved subtle interpretations of detailed nuclear magnetic resonance (NMR) studies (*6, 7, 8*).

The class of substances we refer to as "liquid clathrates" exhibits a new type of strong organometallic–aromatic involvement (*9, 10, 11*). However, before this behavior is described in detail, it is necessary to review the origin and characterization of the parent molecules.

Preparation of M[Al$_2$R$_6$X]

Ziegler and co-workers (*12*) first reported in 1960 that aluminum alkyls react with alkali metal halides to form either 1:1 or 2:1 complexes

$$AlR_3 + MX \rightarrow M[AlR_3X] \tag{1}$$

$$2AlR_3 + MX \rightarrow M[Al_2R_6X] \tag{2}$$

(Reactions 1 and 2). These substances were initially of interest because they provided a neat inorganic study of the relative importance of the lattice energies of MX and M[AlR$_3$X] or M[Al$_2$R$_6$X]. Thus, Lehmkuhl (*13*) concluded that complexation occurs most readily when the ionic radius of the alkali metal is large, the alkyl chain is short, and the complex ion is small.

Although the logical extension of the preparative method to include alkali metal pseudohalides was mentioned in the original paper (*12*), no well defined complexes of this sort were noted until the 1971 reports by our group (*14*) and subsequently by Dehnicke *et al.* (*15*).

Anionic Structure of M[Al$_2$R$_6$X]

In 1963, Allegra and Perego (*16*) first reported structure of a 2:1 halide complex, that of K[Al$_2$(C$_2$H$_5$)$_6$F]. They found that the anion exhibited a completely linear Al–F–Al linkage, and they postulated both sp hybridization of the fluorine atom and d orbital participation by the aluminum atom in order to explain the observed bond lengths and angles.

Our group has found precisely the same geometry (Structure **1**) about the fluorine atom in K[Al$_2$(CH$_3$)$_6$F] · C$_6$H$_6$ (*5*), but a different situation must be envisioned for the Cl⁻, Br⁻, and I⁻ bridged complexes. Although the 2:1 complexes of these ions were not studied by x-ray crystallography, simple hybridization arguments suggest that the chlorine, bromine, and iodine atoms would use essentially p orbitals in a bridging situation (Structure **2**). Structures such as that of the recently determined bromonium ylide (Structure **3**) (*17*) provide structural verification of this idea.

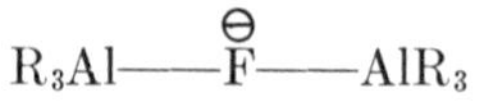

$$R_3Al\!-\!\!-\!\!-\!\overset{\ominus}{F}\!-\!\!-\!\!-\!AlR_3$$

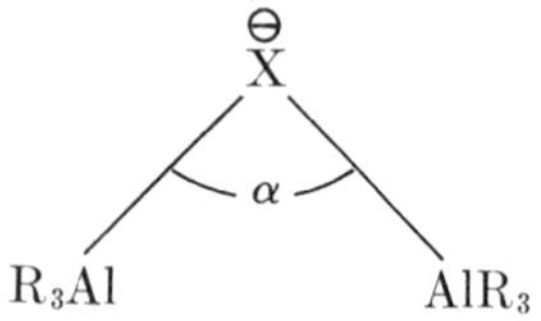

$$\overset{\ominus}{X}$$

$$R_3Al \qquad\qquad AlR_3$$

$$X = Cl^-,\ Br^-,\ I^-;\ \alpha \sim 90°$$

1   **2**

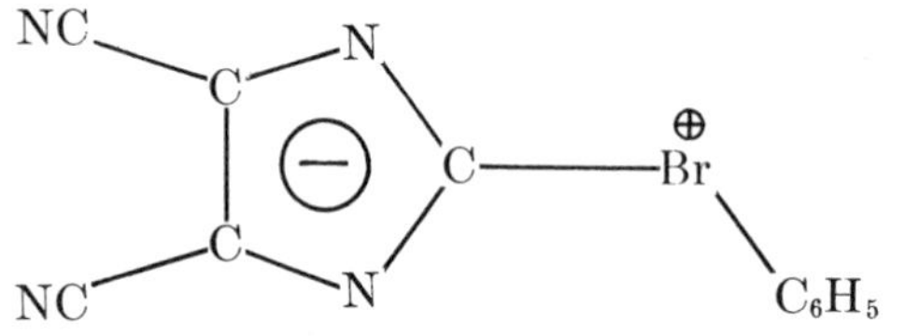

3

For the 2:1 pseudohalide complexes, two cases may be differentiated with reference to $[Al_2R_6SCN]^-$. Structure **4** appears to be favored in the solid state for $R = -CH_3$ from detailed IR and Raman spectral studies (*18*) and by analogy to the structure of $K[Al_2(CH_3)_6N_3]$ (*9, 10*).

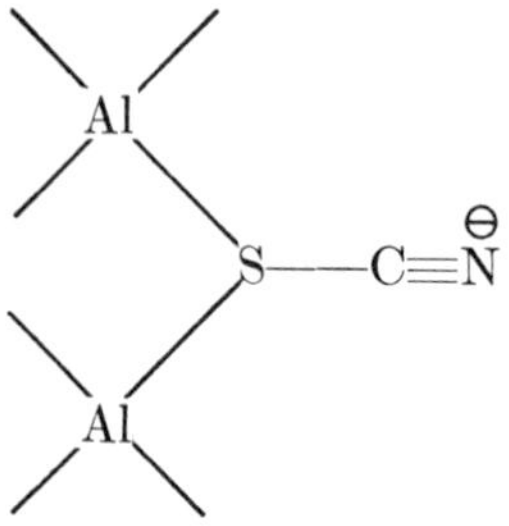

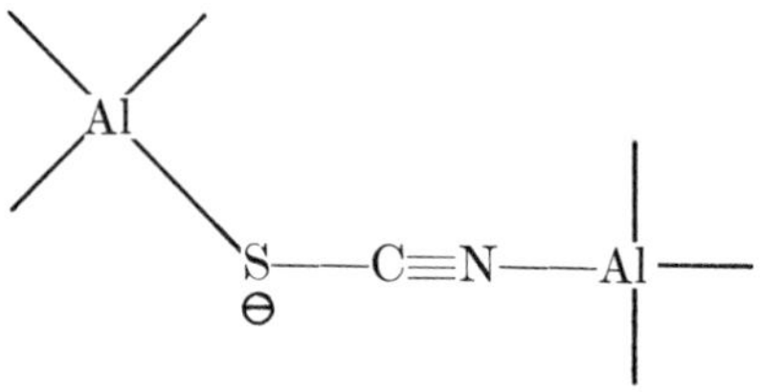

4   **5**

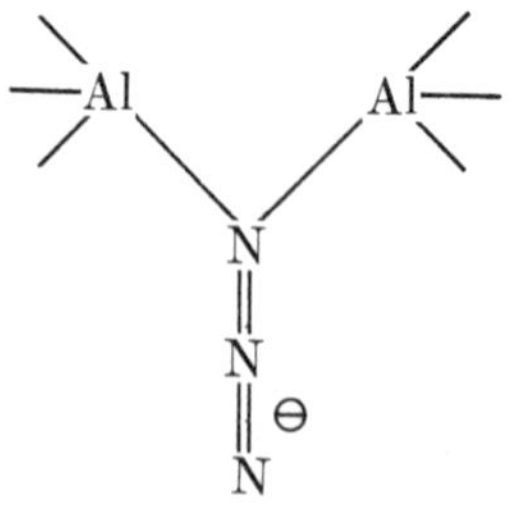

6

However, synthetic work now indicates that Structure 5 is dominant at high temperatures (*19*) and for R = —C_2H_5 or higher alkyl chain length (*20*).

Direct structural evidence is provided by the x-ray diffraction study of $K[Al_2(CH_3)_6N_3]$ (*9, 10*) in which (as was indicated previously) the azide ion has a single nitrogen atom bridge (Structure 6).

An interesting sidelight is the information revealed about the methyl group conformation; there are two anionic possibilities which have C_{2v} and C_s point symmetries. Viewed down the nitrogen atom chain they appear as:

$$C_{2v} \qquad\qquad C_s$$

Weller and Dehnicke (*18, 21*) determined from spectroscopic studies that the anions in $[N(CH_3)_4][Al_2(CH_3)_6N_3]$ and in $[N(CH_3)_4]$-$[Al_2(CH_3)_6SCN]$ are of C_{2v} symmetry. However, in the crystallographic asymmetric unit of $K[Al_2(CH_3)_6N_3]$, there are two nonequivalent $[Al_2(CH_3)_6N_3]^-$ ions: one of C_{2v} symmetry and one of C_s symmetry. The bond distances and angles in both forms are quite similar, and the existence of both types is indicative of the closeness in overall energy of the two configurations in solution.

Definition of Liquid Clathrate Behavior

The term "liquid clathrate" is used to designate that group of non-stoichiometric compounds which form upon the interaction of aromatic molecules with certain $M[Al_2(CH_3)_6X]$ moities. With M = alkali metal ion and X = N_3^-, SCN^-, or $SeCN^-$, $M[Al_2(CH_3)_6X]$ is an air-sensitive solid, whereas with M = tetraalkylammonium ion and X = Cl^-, Br^-, or I^-, $M[Al_2(CH_3)_6X]$ is an air-sensitive liquid. In both cases, interaction with the appropriate aromatic substance produces a liquid which is immiscible with excess aromatic. The aromatic molecules in the liquid clathrate appear to be trapped much as they would be in a solid-state clathrate; they can be freed by lowering the temperature and thus be reclaimed unchanged. These observations may be summarized by the equilibrium

$$M[Al_2(CH_3)_6X] + n \text{ aromatic} \underset{\substack{\text{low} \\ \text{temp.}}}{\overset{\substack{\text{high} \\ \text{temp.}}}{\rightleftharpoons}} M[Al_2(CH_3)_6X] \cdot n \text{ aromatic} \qquad (3)$$

$$\text{liquid clathrate}$$

Proposed Shape Theory of Liquid Clathrate Formation

Information about the nature of liquid clathrates may be derived by tabulating those halides and pseudohalides that promote the effect (Table I). All the anions which may be used to form a liquid clathrate are believed to have an angular geometry (Structure 7), whereas a linear

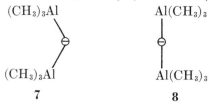

geometry (Structure 8) is characteristic of those anions which do not form liquid clathrates. The angular structure has two noteworthy features: the anion has an appreciable dipole moment, and there is a definite separation of organic and inorganic regions. Thus, the negative charge is largely localized on the very accessible inorganic end of the anion, while the rather massive organic region is quite far away. We reported previously (11) that anions with this geometry will react with molecules such as benzene or toluene to form liquid complexes which contain 1.5–13.0 aromatic molecules per anionic unit.

Table I.   Classification of Halide and Pseudohalide Ions in
$[Al_2(CH_3)_6X]^-$ as to the Formation of LiquidClathrates
with Benzene

Does Form Liquid Clathrate	*Does Not Form Liquid Clathrate*
Cl⁻	F⁻
Br⁻	CN⁻
I⁻	
SCN⁻	
SeCN⁻	
N₃⁻	
NO₃⁻	

The nature of the interaction has been related to the shape of the anion in $M[Al_2(CH_3)_6X]$; it was proposed (10, 11) that a cage or layer-like structure of oriented anions is set up with counter ions and aromatic molecules trapped inside.

Constitution of Liquid Clathrates

The NMR spectra of all liquid clathrates have one feature in common: the entire spectrum is shifted 0.2–0.5 ppm downfield relative to the pure aromatic substance. However, this is thought to be a bulk diamagnetic effect, since in every case the liquid clathrate is more dense than the aromatic moiety.

For each individual $M[Al_2(CH_3)_6X]$ · aromatic, there is a maximum aromatic/anion ratio. The tabulation of liquid clathrate data in Table II illustrates the following (*11*): (a) the larger the cation, the greater the number of aromatic molecules in the clathrate; (b) the more electro-negative the halide ion, the greater the number of aromatic molecules in

Table II.   Compositions of Various Liquid Clathrates

Compound	Aromatic	Maximum Ar/An Ratio[a]	Al–CH₃ Proton Chemical Shift[b]
$K[Al_2(CH_3)_6N_3]$	benzene	5.8	7.78
$Rb[Al_2(CH_3)_6N_3]$		6.1	7.78
$Cs[Al_2(CH_3)_6N_3]$		7.4	7.71
$K[Al_2(CH_3)_6NO_3]$		7.0	7.64
$Cs[Al_2(CH_3)_6NO_3]$		12.0	7.63
$[N(C_2H_5)_4][Al_2(CH_3)_6NO_3]$		9.8	7.69
$[N(CH_3)_4][Al_2(CH_3)_6Cl]$		8.1	7.67
$[N(CH_3)_4][Al_2(CH_3)_6I]$		6.5	7.52
$[N(C_2H_5)_4][Al_2(CH_3)_6I]$		7.3	7.45
$[N(C_3H_7)_4][Al_2(CH_3)_6I]$		9.0	7.43
$[N(C_4H_9)_4][Al_2(CH_3)_6I]$		9.9	7.42
$[N(C_5H_{11})_4][Al_2(CH_3)_6I]$		13.0	7.41
$K[Al_2(CH_3)_6SCN]$	toluene	2.5	7.86
$K[Al_2(CH_3)_6N_3]$		3.8	7.77
$Rb[Al_2(CH_3)_6N_3]$		5.7	7.77
$Cs[Al_2(CH_3)_6N_3]$		6.3	7.67
$[N(C_2H_5)_4][Al_2(CH_3)_6NO_3]$		6.2	7.64
$[N(CH_3)_4][Al_2(CH_3)_6Cl]$		5.6	7.62
$[N(CH_3)_4][Al_2(CH_3)_6Br]$		5.5	7.53
$[N(CH_3)_4][Al_2(CH_3)_6I]$		5.0	7.42
$[N(C_2H_5)_4][Al_2(CH_3)_6I]$		6.0	7.42
$[N(C_3H_7)_4][Al_2(CH_3)_6I]$		6.4	7.39
$[N(C_4H_9)_4][Al_2(CH_3)_6Br]$		9.3	7.49
$[N(C_4H_9)_4][Al_2(CH_3)_6I]$		7.0	7.40
$[N(C_5H_{11})_4][Al_2(CH_3)_6I]$		11.0	7.36
$[N(C_6H_5)(CH_3)_3][Al_2(CH_3)_6I]$		8.4	7.30
$[N(C_2H_5)_4][Al_2(CH_3)_6I]$	ethylbenzene	4.6	7.42
$[N(C_3H_7)_4][Al_2(CH_3)_6I]$		5.0	7.49
$[N(C_4H_9)_4][Al_2(CH_3)_6I]$		5.9	7.48
$[N(C_5H_{11})_4][Al_2(CH_3)_6I]$		11.0	7.47
$Cs[Al_2(CH_3)_6N_3]$	p-xylene	4.3	7.57
$[N(C_5H_{11})_4][Al_2(CH_3)_6I]$		7.0	7.32
$[N(C_5H_{11})_4][Al_2(CH_3)_6I]$	m-xylene	6.0	~7.26
$[N(C_5H_{11})_4][Al_2(CH_3)_6I]$	mesitylene	3.7	7.10

[a] The Ar/An ratio is defined as the number of aromatic molecules per anionic unit in the liquid clathrate.
[b] The ppm relative to the aromatic resonance in the liquid clathrate.

the clathrate; and (c) the larger the aromatic molecules, the smaller the number of aromatic molecules in the liquid clathrate. Points a and c are most vividly demonstrated in the data for the tetraalkylammonium iodide series depicted in Figure 1.

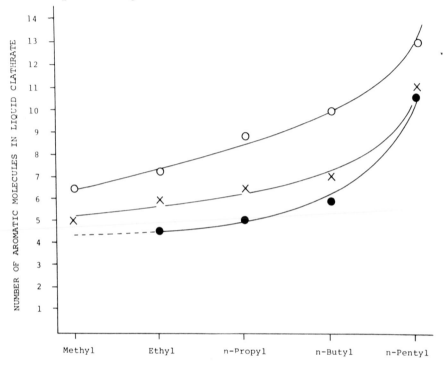

Figure 1. Maximum aromatic/anion ratio liquid clathrates of the form $[N(alkyl)_4][Al_2(CH_3)_6I]$ · aromatic. $\bigcirc$ = benzene, $\times$ = toluene, and $\bullet$ = ethylbenzene.

Figure 1 also illustrates one interesting anomaly. The large number of ethylbenzene molecules associated with $[N(C_5H_{11})_4][Al_2(CH_3)_6I]$ may signal the approach of the end of liquid clathrate behavior for larger tetraalkylammonium ions. One could expect that for the tetrahexyl-ammonium iodide–ethylbenzene or the tetrapentylammonium iodide–propylbenzene clathrate, the aromatic/anion ratio might be so great as to give solutions with no liquid layering (i.e., they are miscible with pure solvent).

Liquid Clathrates Based on the Nitrate Ion

The group of liquid clathrates in Table II which are not based on hailde- or pseudohalide-containing anions differs from the others in two ways. First, the addition of the slightest excess of trimethylaluminum (beyond the normal 2:1 stoichiometry) changes the color of the nitrate-containing liquid clathrate from colorless to yellow. Presumably, this is caused by shifting the energy of an electronic transition by coordination of the additional trimethylaluminum molecule.

Second, the tetraethylammonium nitrate/tri(n-propyl)aluminum liquid clathrate is the first clearly defined liquid cathrate of an aluminum alkyl other than trimethylaluminum. The inability of longer chain AlR$_3$ to form liquid clathrates was previously believed to be attributable to the size requirement of the alkyl groups. Thus, for R $\neq$ –CH$_3$, the angle α is in all probability much greater than 90° (*see* Structure 2) for Cl$^-$, Br$^-$, or I$^-$ complexes, while the essential angular geometry could well be destroyed for the pseudohalide complexes (Structure 9). For the nitrate ion, however, the steric requirements are much less severe (*see* Structure 10), and the necessary shape of the anion is preserved.

$$R_3Al-N=N=N-AlR_3 \quad (\ominus)$$

9

$$R_3Al-O-N(\ominus)(O)-O-AlR_3$$

10

Reactions of Liquid Clathrates

The isolated liquid clathrate undergoes two possible reactions (Reactions 4 and 5). M[Al$_2$(CH$_3$)$_6$X] with M = tetraalkylammonium ion is

$$M[Al_2(CH_3)_6X] + n \text{ aromatic} \quad (4)$$

$$M[Al_2(CH_3)_6X] \cdot n \text{ aromatic} \rightleftarrows$$
$$M[Al(CH_3)_3X] + Al(CH_3)_3 + n \text{ aromatic} \quad (5)$$

a liquid for all compounds reported so far. Reaction 4 is unimportant for these substances. All M[Al(CH$_3$)$_3$X] compounds are solids. All K[Al$_2$(CH$_3$)$_6$N$_3$] · n aromatic systems are unstable with respect to conversion by Reaction 4 at room temperature, as are the K[Al$_2$(CH$_3$)$_6$NO$_3$] · n aromatic and the Cs[Al$_2$(CH$_3$)$_6$NO$_3$] · n aromatic systems. On the other hand, Cs[Al$_2$(CH$_3$)$_6$N$_3$] · n aromatic undergoes Reaction 5 spontaneously (22). The nature of the aromatic group is also important; for example, [N(CH$_3$)$_4$][Al$_2$(CH$_3$)$_6$I] cannot be produced in the presence of ethylbenzene at room temperature because of Reaction 5. Therefore, just as the formation of M[Al$_2$R$_6$X] is governed predominantly by a series of lattice energy considerations (13), so also is the stability of $M[Al_2-(CH_3)_6X] \cdot n$ aromatic.

There are a few reactions of liquid clathrates which are based on the decomposition of $M[Al_2(CH_3)_6X]$. The most thoroughly investigated system (23) involves the thermolysis or photolysis of $M[Al_2(CH_3)_6N_3]$ (Reaction 6).

$$M[Al_2(CH_3)_6N_3] \xrightarrow[\text{3000 A}]{\text{180°C}} M[Al(CH_3)_4] + N_3Al(CH_3)_2 \qquad (6)$$

A particularly important series of reactions currently under investigation is the substitution of foreign molecules into the liquid clathrate. For example, the maximum aromatic/anion benzene clathrate of $[N(CH_3)_4][Al_2(CH_3)_6I] - [(CH_3)_4][Al_2(CH_3)_6I] \cdot 6.4C_6H_6$—substitutes toluene slowly (when placed in contact with toluene) at room temperature. The rate is such that, after two hours, the composition of the clathrate is $[N(CH_3)_4][Al_2(CH_3)_6I] \cdot 5.4C_6H_6 \cdot 0.7C_6H_5CH_3$. If, however, the benzene liquid clathrate is heated with toluene in a sealed tube at 80°C for 24 hours, the composition of the clathrate appears to be directly proportional to the benzene/toluene ratio in the sealed tube. (For the solution referred to above, the composition was $[N(CH_3)_4]-[Al_2(CH_3)_6I] \cdot 2.2C_6H_6 \cdot 3.7C_6H_5CH_3$.)

It is also possible to substitute non-aromatic molecules into the liquid clathrates. So far we have studied trimethylaluminum, cyclohexane, and ferrocene. Although the details of these experiments are not yet available, the general trend is a limited substitution of the foreign molecules together with the expulsion of a related volume of aromatic molecules. It should be noted that in all cases the liquid clathrate is immiscible with the substituent, but exchange can be effected either thermally or by mechanical agitation.

IR Spectroscopic Studies of Liquid Clathrates

Benzene-containing liquid clathrates were chosen for these studies because of their high symmetry and the simplicity of the IR spectrum. The benzene portions of the spectra of all liquid clathrates were quite similar, but there were two significant differences between the guest benzene and the neat benzene (24): the combination bands at 1818 and 1962 cm^{-1} were broadened, and a very weak peak appeared at 990 cm^{-1}. The data in Table III reveal that the broadening of the IR bands at 1818 and 1962 cm^{-1} is a measure of the extent of interaction of the cation with the benzene molecules. The smaller Rb$^+$ and Cs$^+$ cations, in fact, have a splitting of the degenerate modes whereas the larger ammonium

Table III.   IR Data for Selected Liquid Clathrates[a]

Clathrate	Band Width at Half Height, cm^{-1}		Absorbance		A/B Ratio
	At 1818 cm^{-1}	At 1962 cm^{-1}	A: at 990 cm^{-1}	B: at 1035 cm^{-1}	
$Rb[Al_2(CH_3)_6N_3]\cdot 6.1C_6H_6$	$\begin{Bmatrix}1816\\1836\end{Bmatrix}$[b]	$\begin{Bmatrix}1960\\1974\end{Bmatrix}$[b]	0.029	0.448	0.065
$Cs[Al_2(CH_3)_6N_3]\cdot 7.4C_6H_6$	$\begin{Bmatrix}1815\\1833\end{Bmatrix}$[b]	$\begin{Bmatrix}1960\\1972\end{Bmatrix}$[b]	0.036	0.650	0.055
$[N(CH_3)_4][Al_2(CH_3)_6I]\cdot 6.5C_6H_6$	28	24	0.011	0.245	0.045
$[N(C_2H_5)_4][Al_2(CH_3)_6I]\cdot 7.3C_6H_6$	28	25	c	c	c
$[N(C_4H_9)_4][Al_2(CH_3)_6I]\cdot 9.7C_6H_6$	27	25	0.014	0.271	0.052
$[N(C_5H_{11})_4][Al_2(CH_3)_6I]\cdot 11.0C_6H_6$	25	22	0.020	0.407	0.049
$[N(C_5H_{11})_4][Al_2(CH_3)_6I]\cdot 13.0C_6H_6$	24	18	0.014	0.234	0.060

[a] From Ref. *11*.
[b] Split into distinct peaks with maxima at wave numbers given in brackets.
c The absorbance at 990 cm^{-1} is obscured by a strong absorbance at 995 cm^{-1} that is caused by the $N(C_2H_5)_4^+$ groups.

ions only broaden the peaks. Within the tetraalkylammonium series, the trend is also mirrored, and the expected reduction in broadening with an increased aromatic/anion ratio can be observed for the tetra-*n*-pentyl-ammonium clathrates.

The IR spectrum of benzene has been thoroughly studied (*24*). The absorption at 1818 cm^{-1} is assigned as a combination of $E_u^+ + E_g^-$ fundamentals which has E_u^- symmetry. The band at 1962 cm^{-1} is assigned as a combination of the fundamentals of symmetry $E_u^+ + B_{2g} = E_u^-$. Both bands thus have the E_u^+ fundamental in common. Furthermore, they are both degenerate and can be broadened by a partial lifting of the degeneracy. The E_u^+ fundamental is symmetry forbidden, and it is not observed where it was predicted, at 970–985 cm^{-1}, in the neat benzene spectrum. However, a very weak band was observed in all spectra of the clathrates at 990 cm^{-1}. Since it appears in such a diversity of samples, it must be assigned either to the guest benzene or to the trimethylaluminum units. The correctness in assigning this peak to the benzene is reflected by the absorbance ratios with the peak at 1035 cm^{-1} (which is assigned to the E_u^- fundamental of benzene) (*24*). These ratios (*see* Table III) are almost identical for widely varying mole ratios of benzene and trimethyl-aluminum.

The E_u^+ mode is illustrated as:

The fact that this is the only mode affected by the formation of the liquid clathrate indicates that the interaction is localized over the carbon–carbon bonds rather than being symmetrically centered in the π system.

Solid State Clues to Liquid Clathrate Behavior

Temperature is all important to the existence of liquid clathrates. $K[Al_2(CH_3)_6N_3]$ · benzene is reasonably stable (Reaction 4) at 25°C and completely stable at 40°C. $Cs[Al_2(CH_3)_6NO_3]$ · benzene is exceedingly unstable (Reaction 4) at 25°C but completely stable at 80°C. $K[Al_2(CH_3)_6N_3]$ · p-xylene does not exist at 25°C, but it is stable at 130°C. It does not seem unreasonable that the proper combination of cation, anion, and guest might form a liquid clathrate at slightly elevated temperatures, but at room temperature it would form a solid with a structure similar to that of the liquid clathrate. We now have what may be an example of this in the crystal structure of $K[CH_3Se\{Al(CH_3)_3\}_3]$ · $2C_6H_6$.

The thermal decomposition of the benzene liquid clathrate of $K[Al_2(CH_3)_6SeCN]$ produces a number of products (Reaction 7). We

$$K[Al_2(CH_3)_6SeCN] \cdot nC_6H_6 \rightarrow$$

$$K[CH_3Se\{Al(CH_3)_3\}_3] \cdot 2C_6H_6 + K[Al_2(CH_3)_6CN] + \ldots \quad (7)$$

examined the structure of the methylselenide; for an excellent set of x-ray data, the conventional agreement index, R, is 0.092 for the $K[CH_3Se\{Al(CH_3)_3\}_3]$ portion of the molecule. All bond lengths and angles are quite normal (see Table IV).

Table IV.   Important Bond Lengths and Angles for
$K[CH_3Se\{Al(CH_3)_3\}_3]$ · $2C_6H_6$

Bond	Bond Length, A	Bond	Bond Length, A	Bonds	Bond Angle, °
Se–Al1	2.596(4)	Al2–C21	2.00(2)	C(m)–Se–Al1	104.0(5)
Se–Al2	2.570(4)	Al2–C22	1.96(2)	C(m)–Se–Al2	103.9(4)
Se–Al3	2.566(4)	Al2–C23	2.01(2)	C(m)–Se–Al3	102.6(5)
Se–C(m)	1.93(4)	Al3–C31	1.96(2)	Al1–Se–Al2	115.5(4)
Al1–C11	1.96(2)	Al3–C32	1.99(2)	Al1–Se–Al3	114.5(5)
Al1–C12	2.03(2)	Al3–C33	1.99(2)	Al2–Se–Al3	114.2(4)
Al1–C13	1.98(2)				

The amazing feature of the structure is that, although the R factor indicates that the problem is essentially solved, the benzene molecules (which comprise some 35% of the total electron density) have not been included. Moreover, it is not possible to locate accurately and to refine

the positional parameters of the benzene carbon atoms. In all attempts
to do so, the temperature factors have blown up ($B \sim 30\,\text{Å}^2$). All these
observations are correlated when one realizes that the benzene molecules
are behaving essentially as a liquid (thereby contributing nothing to the
bulk of the x-ray data) while the $K[CH_3Se\{Al(CH_3)_3\}_3]$ units behave
normally and dominate the x-ray diffraction pattern. (We believe that
high thermal motion better accounts for the crystallographic data than
does disorder. On a difference Fourier map, very weak peaks can be
seen in good positions for an ordered six-membered ring.)

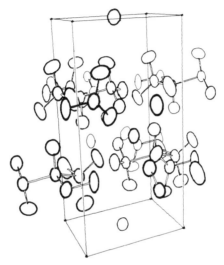

*Figure 2.   Unit cell packing diagram
for $K[CH_3Se\{Al(CH_3)_3\}_3] \cdot 2C_6H_6$*

Careful examination of the unit cell packing illustration (Figure 2),
together with crude placement of the benzene molecules, affords a micro-
scopic picture of the crystalline solid in which there are layers of solid
material alternating with layers of liquid (Structure **11**). (There are two

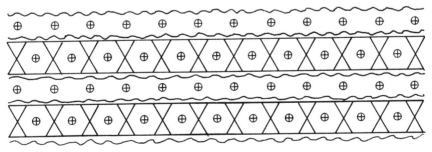

11

crystallographically independent potassium ions in the asymmetric unit; one is associated with the solid layer, the other with the liquid layer.) Care must be taken that the term "liquid" is not misinterpreted. The layer is liquid in that the thermal motion of the benzene molecules is very high, but it is certainly solid in that the benzene molecules have no true translation freedom because of the presence of the potassium ions (*see* Structure 12 for a top view of six unit cells of the liquid layer).

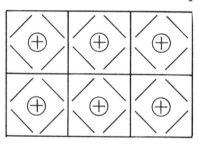

12

Careful inspection of Figure 2 reveals that the structure is that of a reverse clathrate: the aromatic molecules are associated with the organic regions of the anions. $K[CH_3Se\{Al(CH_3)_3\}_3] \cdot 2C_6H_6$ may well present liquid clathrate formation at an elevated temperature, for the anion does have the proper shape (Structure 13). However, this has not been veri-

$$
\begin{array}{c}
Al(CH_3)_3 \\
(CH_3)_3Al \diagdown \quad \diagup Al(CH_3)_3 \\
Se \\
| \quad \ominus \\
CH_3
\end{array}
$$

13

fied because of the difficulty in preparing a pure sample of sufficient size. At present it is not possible to assess the relevance of solid-state structures to liquid clathrate behavior, but we feel that this is an important area for future investigation.

Conclusions

Compounds of the type $M[Al_2(CH_3)_6X]$ react with small aromatic molecules to form nonstoichiometric liquid complexes for those cases in which the anion structure has both a separation of organic and inorganic areas and an appreciable dipole moment. For example, the structure of the anion of a substance which exhibits this type of behavior, $K[Al_2(CH_3)_6N_3]$ (*10*), was found to be:

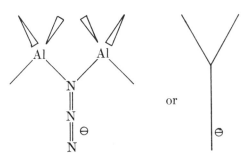

The liquid complexes have many of the properties ascribed to enclosure compounds, and the term liquid clathrate seems to be an appropriate designation.

Since the IR spectrum of the guest benzene is only slightly perturbed from that of neat benzene, the aromatic molecules must interact only weakly with the cations in the liquid clathrate.

The structure of the liquid clathrate may reasonably consist of either roughly spherical or layer-like domains. These alternatives may be represented in two dimensions as Structures **14** and **15** respectively. At present,

14   **15**

from a consideration of the properties of a range of liquid clathrates, it is believed that the layer-like structure (Structure **15**) is the more likely. A reverse layered solid-state structure is shown for $K[CH_3Se[Al(CH_3)_3]_3]$ · $2C_6H_6$.

Experimental

Preparations. The two different ways by which the liquid clathrates may be prepared are best illustrated by reference to the $K[Al_2(CH_3)_6N_3]$ complex. By the method described previously (*10*), 0.010 mole $Al(CH_3)_3$ was added to 0.005 mole KN_3 in N_2 atmosphere dry box. The mixture was then sealed in a bomb tube, heated to 80°C, returned to the dry box, and opened; another 0.005 mole $Al(CH_3)_3$ was added to the powdered contents. After three cycles of grinding, adding $Al(CH_3)_3$, and heating, the white crystalline product was dried under vacuum. A typical analysis gave: calcd for $K[Al_2(CH_3)_6N_3]$: Al 23.97, C 31.99, H 8.05; found: Al 23.16, C 29.21, H 7.20. Addition of benzene ($\sim$0.10 mole) followed by heating 1 hr at 60°C afforded the liquid clathrate $K[Al_2(CH_3)_6N_3]$ · $5.8C_6H_6$.

A decidedly improved method for producing the compounds involves simply adding 0.005 mole KN_3 and 0.010 mole $Al(CH_3)_3$ to $\sim$0.10 mole C_6H_6 in the dry box. A liquid clathrate identical in composition to the one prepared by the previous method was obtained in 1 hour. All liquid clathrates reported here were synthesized this way.

It should be noted that the formation of the liquid clathrate is a visually dramatic event. As the reaction proceeds, a separation of two liquid layers (liquid clathrate and excess aromatic) becomes obvious; upon shaking, the layers resemble oil and water.

Analysis. The liquid clathrates were analyzed by the integration of NMR spectra recorded on a Perkin-Elmer R20-B instrument. The aromatic stoichiometries in Table II are in most cases the average of three preparations and integrations; a realistic standard deviation would be ±0.2 molecules. The chemical shifts are accurate to better than 0.02 ppm.

The liquid clathrates, although water and oxygen sensitive, are much less reactive than the pure parent organometallic compounds. None of the azides was found to present an explosion hazard.

IR Data. The IR spectra (in the 600–3000 cm^{-1} region) were taken on neat samples in 0.025-mm NaCl cells in the double beam mode with air as reference. The spectra were run on a Beckman IR-7 instrument with a scan rate of 80 cm^{-1}/min and 25 cm^{-1}/in. All half-width data were taken with a scale expansion of 10 cm^{-1}/in. at a scan rate of 20 cm^{-1}/min. All liquid clathrate spectra were analyzed by direct comparison with the spectrum of benzene taken under identical conditions.

X-ray Diffraction Data. Lattice parameters for $K[CH_3Se\{Al-(CH_3)_3\}_3]$ · $2C_6H_6$ were determined from a least-squares refinement of the angular settings of 12 reflections accurately centered on an Enraf-Nonius CAD-4 diffractometer. They are $a = 10.144(4)$, $b = 10.156(4)$, and $c = 17.165(6)$ A, and $\alpha = 75.90(4)$, $\beta = 80.58(4)$, and $\gamma = 60.72(4)°$. The crystal system is triclinic, and the space group is $P\bar{1}$. Data were collected on the diffractometer by a collection and reduction process (25).

The structure was solved by the Patterson method, and refined by anisotropic least-squares techniques. For 2114 observed reflections, the final R values were:

$$R_1 = \Sigma|F_o| - |F_c| / \Sigma|F_o| = 0.086$$

$$R_2 = \{\Sigma w(|F_o| - |F_c|)^2 / \Sigma w|F_o|^2\}^{1/2} = 0.092$$

Tables of positional and thermal parameters and of structure factor amplitudes are available upon request.

Acknowledgment

We thank past students S. K. Seale and W. R. Newberry for their consultation during the early phases of this research and present students

R. T. Carlin, K. D. Crissinger, and K. E. Stone for their on-going efforts in this area.

Literature Cited

1. Bailey, M. F., Dahl, L. F., *Inorg. Chem.* (1965) **4**, 1299.
2. *Ibid.* (1965) **4**, 1314.
3. Rodesiler, P. F., Griffith, E. A. H., Amma, E. L., *J. Am. Chem. Soc.* (1972) **94**, 761.
4. Kobayashi, Y., Iitaka, Y., Yamazaki, H., *Acta Crystallogr. Sect. B* (1972) **28**, 899.
5. Atwood, J. L., Newberry, W. R., *J. Organomet. Chem.* (1974) **66**, 15.
6. Williams, K. C., Brown, T. L., *J. Am. Chem. Soc.* (1966) **88**, 5460.
7. Schaschel, E., Day, M. C., *J. Am. Chem. Soc.* (1968) **90**, 503.
8. Jeffery, E. A., Mole, T., *Aust. J. Chem.* (1969) **22**, 1129.
9. Atwood, J. L., Newberry, W. R., *J. Organomet. Chem.* (1972) **42**, C77.
10. *Ibid.* (1974) **65**, 145.
11. Atwood, J. L., Atwood, J. D., unpublished data.
12. Ziegler, K., Koster, R., Lehmkuhl, H., Reinert, K., *Ann.* (1960) **629**, 33.
13. Lehmkuhl, H., *Angew. Chem. Int. Ed. Engl.* (1964) **3**, 107.
14. Atwood, J. L., Milton, P. A., Seale, S. K., *J. Organomet. Chem.* (1971) **28**, C29.
15. Weller, F., Wilson, I. L., Dehnicke, K., *J. Organomet. Chem.* (1971) **30**, C1.
16. Allegra, G., Perego, G., *Acta Crystallogr.* (1963) **16**, 185.
17. Atwood, J. L., Sheppard, W. A. *Acta Crystallogr. Sect. B*, in press, 1975.
18. Weller, F., Dehnicke, K., *J. Organomet. Chem.* (1972) **36**, 23.
19. Seale, S. K., Atwood, J. L., *J. Organomet. Chem.* (1974) **73**, 27.
20. Atwood, J. L., Bowles, L. K., unpublished data.
21. Weller, F., Dehnicke, K., *J. Organomet. Chem.* (1972) **35**, 237.
22. Atwood, J. L., Newberry, W. R., *J. Organomet. Chem.* (1975) **87**, 1.
23. Atwood, J. L., Hrncir, D. C., *J. Organomet. Chem.* (1973) **61**, 43.
24. Bailey, C. R., Ingold, C. K., Poole, H. G., Wilson, C. L., *J. Chem. Soc.* (1946) 222.
25. Atwood, J. L., Smith, K. D., *J. Am. Chem. Soc.* (1973) **95**, 1488.

RECEIVED January 24, 1975. Work partially supported by National Science Foundation grant NSF-GP-24852.

12

The Thermally Equilibrated Excited (Thexi) State Chemistry of Some Co(III) Ammines

ARTHUR W. ADAMSON

University of Southern California, Los Angeles, Calif. 90007

The photochemistry of coordination compounds is discussed in terms of thermally equilibrated excited (thexi) states as the chemically reacting species. Such states are in thermodynamic equilibrium with their surroundings and are essentially high energy isomers of the ground state. By contrast, the species obtained by light absorption of wavelength around a ligand field band maximum are Franck–Condon states that have a nonthermodynamic distribution of vibrational excitations; such states are treated as pseudo pure electronic states in ligand field theory. New theory is needed to treat thexi states. Data on the ligand field substitutional photochemistry of trans- and cis-[Co(en)₂(NH₃)Cl]²⁺ are reported. Data are discussed mechanistically as part of the thexi state chemistry of Co(III) ammines.

The emphasis in this paper is on an aspect of the photochemistry of coordination compounds that is often rather understated in the literature. Other aspects are important, but I believe that the present emphasis is essential to an understanding of the field. We will be dealing with excited states reached by essentially d–d transitions from the ground state. The typical transition is that which occurs upon absorption of light in the wavelength of the first (L_1) or second (L_2) ligand field band of the complex, particularly when such bands are of usual intensity and are apparently not complicated by admixture with charge transfer (CT) character. A common method of displaying qualitative ligand field state energies is by the well known Tanabe–Sugano diagrams (1, 2, 3, 4, 5).

As was noted (6), there is a tension between ligand field concepts and the implications of photochemical and photophysical studies. These implications concern the nature of emitting and generally chemically reactive excited states; they present some major theoretical problems.

Ligand field theory, like its parent, crystal field theory, makes the approximation in treating ground and excited states that the symmetry and the intensity of the ligand field are invariant. These states correspond to variously energetic arrangements of metal d electrons (relative to the free ion) in the partially degenerate (usually) set of d orbitals, modified by suitable accountancy for interelectronic repulsions. As was mentioned, the conceptual model is that of a metal ion imbedded in a crystal lattice so rigid that the electrostatic environment of the metal ion is fixed. The corresponding molecular orbital treatment allows for bonding considerations but still retains the above basic approximation.

At this conventional level of approximation, the energy difference between ligand field states is given experimentally by the wavelength of the appropriate absorption band maximum. It is this energy which is used in obtaining the 10 Dq of ligand field theory. Thus for a d^3 octahedral complex, 10 Dq is just the energy corresponding to the L_1 band maximum; for a d^6 complex, there is a correction of 35 F_4, F_4 being one of the Condon–Shortely interelectronic repulsion parameters ($1, 2, 3, 4, 5$). It is thus the energy of band maxima which gives the well known spectrochemical series; the same is true for the nephelauxetic series (3). The 10 Dq values obtained from band maxima are used to calculate crystal field stabilizations and to estimate activation energies for ligand substitution reactions (7).

The rigid crystal lattice approximation is similarly used for octahedral type complexes that have more than one kind of ligand. Band splittings on going from O_h to D_{4h} symmetry, for example, are treated in terms of fixed pseudo-octahedral axes of non-equal ligand field strengths ($1, 2, 3, 4, 5$) and, again, positions of band maxima are used to obtain the energy separations of the so-deduced excited-state term system. The approximation is used for other geometries, such as square planar and tetrahedral. In brief, conventional ligand field theory assumes that energies at band maxima represent specific pure electronic excited states in a system of fixed geometry.

It might be expected, in terms of ligand field theory, that d–d transitions would be very sharp. The actual situation is illustrated in Figure 1. Such absorption bands may sharpen somewhat and develop indications of vibrational structure at very low temperature, but the effect is typically not very dramatic [there are exceptions, as with $Cr(CN)_6^{3-}$ ($8, 9$)]. Ligand field theory does not specifically treat the matter of band width. It has been ascribed in a general way to perturbations of an electronic nature (such as Jahn–Teller effects) (2).

It is my opinion that the above general picture was irretrievably compromised by the observations ($10, 11$) that fluorescence (that is, spin-allowed) emission from $Cr(III)$ complexes is dramatically red-

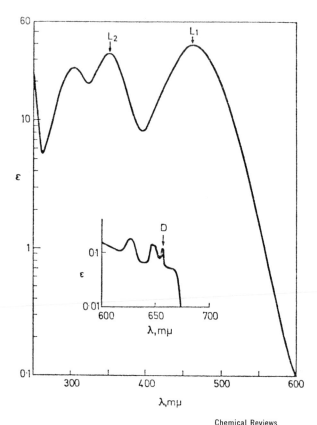

Figure 1. Absorption spectrum for aqueous
$[Cr(NH_3)_6]^{3+}$ (15)

shifted relative to the absorption band. The classic case of $Cr(urea)_6^{3+}$ is plotted in Figure 2. The inescapable conclusion is that a so-called ligand field excited state (of energy given by band maxima) is in reality a vibrationally excited state of a species whose chemical (that is, electronic only) energy is lower by 10–20 kcal/mole than that given by the band maximum. This chemical-only energy can be estimated from the crossing region of the absorption and emission bands and by other ways in the absence of emission (6).

To be blunt, it is fictional to treat band maxima energies as pure electronic energies as does ligand field theory. The broadness of typical ligand field bands is now recognized as reflecting primarily the Franck–Condon overlap factor (8, 9); the most probable transition is to that vibrationally excited level of the electronically excited state for which the probable nuclear positions are close to those of the most populated ground state vibrational level (a more detailed way of stating this is suggested below). The (o,o) (ground state $v = o$ to excited state $v = o$) transition is improbable because of the small overlap of the nuclear wave functions, and transitions involving energies much greater than that of the band maximum are likewise improbable.

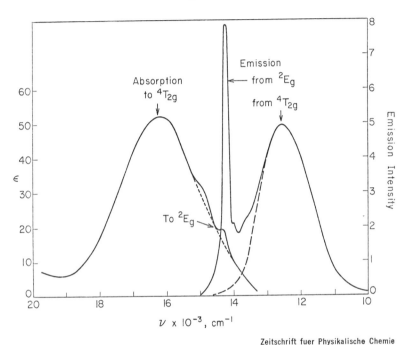

Zeitschrift fuer Physikalische Chemie

Figure 2.   Absorption and (low temperature) emission for $[Cr(urea)_6]^{3+}$
(10)

The further and centrally important implication is that the probable nuclear positions for the pure electronic excited state are significantly different from those in the ground state. In other words, the two states are distorted relative to each other. The situation is illustrated schematically in Figure 3, which also depicts the origin of the red shift on fluorescence. (The commonly seen parabolic potential energy diagrams are misleading in that they do not allow the showing of more than one vibrational ladder, corresponding to alternative modes of distortion.) Absorption of a light quantum produces a vibrationally excited species which thermally equilibrates to ambient temperature vibrational–rotation energy. Because of the consequent shifts in nuclear positions, the most probable fluorescent transition is the one that terminates at a vibrationally excited ground state.

None of the above inferences are new by now; they are in fact generally accepted (8, 9). Yet certain important consequences are rarely

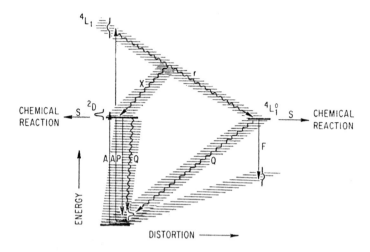

"Inorganic Reaction Mechanisms"

Figure 3. Energy level diagram for a Cr(III) complex (6)
The distortion coordinate is schematic; various ladders represent radiationless processes which may be either vibrational only or vibronic in nature. A, absorption; P, phosphorescent emission; F, fluorescent emission; Q, radiationless deactivation; and X, intersystem crossing.

stressed; they are uncomfortable to conventional ligand field theory. First, the electronic-only excited state energies of coordination compounds may bear little relationship to the usual ligand field energy schemes. This is true with respect to both absolute energies and energy ratios; even the ordering of states obtained by application of ligand field theory can be wrong. A clear illustration is that provided by the Cr(III) family of complexes. There are several complexes for which the thermally equilibrated (or electronic-only) excited first quartet state is below the first doublet state in energy (*see* Ref. *11*). Ligand field theory, ordering these states according to the appropriate absorption band maxima, places the first quartet well above the doublet state. In general, any pair of states for which the metal-to-ligand bonding characteristics are apt to be different are also apt to have energies that are quite different, both absolutely and relatively, from those obtained from ligand field theory and band maxima positions.

A further consequence is that the concepts of the spectrochemical and nephelauxetic series become confused and are not generally usable. Since the thermally equilibrated excited state may be of different geometry from the ground state and have different bond lengths, the rigid lattice approximation is not correct, and no unique ligand field exists. The term ligand field has thus lost its usual experimental and theoretical meaning. Moreover, the term designations of conventional ligand field theory may not be correct for the thermally equilibrated excited state since the point group symmetry may not be the same as for the ground state. Thus a hexacoordinated O_h, D_{4h}, or C_{4v} ground state complex might be C_{5v} in a thermally equilibrated excited state; a square planar complex might be tetrahedral. Yet not only does emission appear to come from thermally equilibrated excited states, but, as is developed further below, so does the substitutional photochemistry of coordination compounds.

Because of the several qualitative differences between the concepts of ligand field and thermally equilibrated excited states, it is useful to develop a distinguishing vocabulary. Conventional ligand field excited states will be called Franck–Condon states since the energies are those of band maxima and the transitions between them are essentially those vertical ones with maximum Franck–Condon overlap. The abbreviation thexi state has been proposed (*12*) for a thermally equilibrated excited state. The gist of the foregoing is that conventional ligand field theory treats hypothetical electronic excited states which are in reality Franck–Condon states, whereas it is thexi states that are important in photophysical and photochemical processes. New theory or new extensions of present theory are clearly needed to treat the latter type of state.

Full appreciation of the differences between a Franck–Condon state and a thexi state has likely been hindered by the fact that the word "state" has quite different meanings in the two expressions. The distinction is that between a spectroscopic state and a thermodynamic state. The former has to do with the detailed quantum mechanical description of an individual molecule (as, for example, by giving electronic, vibrational, and rotational quantum numbers); the latter has to do with the phenomenological specification of an ensemble of molecules in thermal and mechanical equilibrium with their surroundings. Absorption of light of wavelength around a band maximum produces a collection of vibrationally and electronically excited species each with a spectroscopic specification; the collection is not, however, a thermodynamic ensemble. Only after thermal equilibration does the collection of molecules constitute a thermodynamic state. It will now have a conventional molar free energy, enthalpy, and entropy; it will have a standard redox potential. None of these quantities have meaning for a Franck–Condon state. Alternatively, a thexi state corresponds conceptually to a chemical isomer of the ground state. It possesses equilibrium structure and distinctive, activated reaction kinetics. The thexi state is a chemical species.

Some of the evidence for the importance of thexi states in the photochemistry of coordination compounds is given in the next section, followed by a discussion of the reaction chemistry of the thexi state important for Co(III) ammines. The concluding section deals with some further aspects of Franck–Condon and thexi states.

Some Evidences for the Realtiy and Importance of Thexi States

It is possible to surmise some of the attributes of the typical energetic species which undergoes ligand substitution reaction following irradiation in the wavelength region of an L_1 or L_2 band. The species is not generally a vibrationally hot ground state molecule, first of all, because of the prevalence of photoreactions different from thermal ones. We call such behavior antithermal. The energetic species is therefore almost certainly an electronically excited one. With the Cr(III) family of complexes, there are some very direct indications that the photosubstitution occurs from a low lying quarted excited state (*13, 14, 15*).

Next, there are several indications that the photoreactive species is not in a Franck–Condon state, but rather it is in a thermally equilibrated excited state. One indication is that quantum yields as well as the nature of the photoreaction do not vary appreciably as the irradiating wavelength traverses the width of a ligand field band (although variations may occur on going from one band to another) (*13, 14, 15*). It appears that a common reactive state is reached, regardless of the degree of vibrational excitation of the initially produced Franck–Condon state. The simplest explanation is that this common state is a thexi state.

As another line of approach, there is at least indirect evidence that the photoreactive species is much longer lived than would be expected for a Franck–Condon state. The lifetime of this last, that is, the relaxation time for thermal equilibration to ambient temperature, should not be much greater than a few hundred vibrational periods and it is probably much less. Typical complexes are studied in polar, hydrogen-bonding solvents, and the ligands must be in excellent vibrational communication with the solvent medium. From the fact that quantum yields typically are well below unity, it may also be inferred that excess vibrational energy is dissipated quickly to the surrounding medium. This means that most of the absorbed energy is dissipated nonradiatively, and it almost certainly passes through highly vibrationally excited ground states (*see* Figure 3 and Ref. 9). The energy being dissipated can easily be much greater than the activation energy for a known thermal substitution reaction of the complex. The latter is typically 20–30 kcal/mole, and typical energies of absorbed light quanta and hence energies being dissipated are 50–70 kcal/mole. Clearly, vibrationally excited ground states must lose excess energy to the medium in less than the time required for that energy to find its way into those particular nuclear motions that lead to thermal chemical reaction. The same situation is likely to be true for Franck–Condon excited states. (A caveat is discussed below.)

By contrast, lifetimes of a nanosecond or longer can be inferred for the actual photoreactive state. First, since quantum yields are rarely greater than a few tenths, the dominant dissipative process is not one of chemical reaction; it must usually be that of radiationless deactivation (emission normally being of trivial importance under typical photochemical conditions). Neglecting some possible complications, the quantum yield for chemical reaction, ϕ, is then given approximately by k_{cr}/k_{nr}, the ratio of the rate constants for chemical reaction and for nonradiative return to the ground state, respectively. The temperature dependence of ϕ then gives an apparent activation energy, E_ϕ, that is equal to $(E_{cr} - E_{nr})$. Radiationless deactivation also completes with emission, and $\phi_e = 1/\tau_o k_{nr}$ where τ_o is the natural (temperature-independent) emission lifetime. Thus E_{nr} may be inferred from the temperature dependence of the emission yield; typical values are 3–5 kcal/mole (*9, 14, 15*) and they may be larger. Approximately, however, $E_{cr} = E_\phi + 3$ in kcal/mole.

The temperature dependence of ϕ was reported for a number of systems. With Cr(III) complexes, values of E_ϕ range from near zero up to 10 or more kcal/mole (*14, 15*). E_{cr} values of 12 or more kcal/mole thus seem not uncommon. One may then make the following analysis. The rate constant for a first-order reaction should be about 10^{13} exp-$(-E^*/RT)$, neglecting activation entropy. For $E_{cr} = E^* \simeq 12$ kcal/mole, the rate constant k_{cr} would be about 10^4 sec^{-1} at room temperature. Even for an E_{cr} of 4 kcal/mole, k_{cr} is about 10^{10} sec^{-1} which still corresponds to an excited-state lifetime of thousands of vibrational periods. In summary, the prevalence of activated photochemistry strongly suggests that the photoreactive state is much longer lasting than would be expected were it a Franck–Condon state. The situation is that expected for a thexi state.

Another indication that the typical reacting exciting state is relatively long-lived is in the selectivity, especially the stereoselectivity of photochemical substitution reactions. As one example, *trans*-[Cr(en)$_2$Cl$_2$]$^+$ photoaquates chloride, but it gives *cis*-[Cr(en)$_2$(H$_2$O)Cl]$^{2+}$ rather than the trans thermal reaction product. Further, if the ethylenediamines are connected by the belt ligand cyclam so that trans-to-cis isomerization becomes impossible, the photoaquation yield drops a thousand fold (*15*). Some illustrations involving Co(III) ammines are presented below. Such specificity presents no problem for a thexi state. By contrast, a Franck–Condon excited state should be too short-lived to engage in more than simple bond fission processes that have little stereoselectivity.

Final evidence that photoreactive species are too long-lived to be Franck–Condon states comes from the occasional systems for which emission has been reported under photochemical conditions. There are several Cr(III) complexes which show spin-forbidden emission at room temperature (*16*), and lifetimes of microseconds have been measured (*17*). (Emission lifetimes of 1.3, 1.7, and 46 μsec were found (*18*) for

$[Cr(en)_3]^{3+}$, $[Cr(NH_3)_6]^{3+}$, and $[Cr(bipyr)_3]^{3+}$ respectively in room temperature, aqueous solution.) To the extent that the emitting doublet state is (a) itself chemically reactive or (b) in thermal equilibrium with a reactive quartet excited state, one is led to conclude that the photochemistry involves thexi states. In case (a) the reacting and emitting state are the same and it is therefore known that the lifetime of the former is too long for it to be a Franck–Condon state. In case (b) the very postulation of ordinary kinetic interconversion between two electronic states implies that both are thexi states.

Emission lifetimes for spin-allowed d–d transitions of coordination compounds apparently have not yet been measured under photochemical conditions. The radiative lifetime, τ_0, can be estimated (with serious potential error) from the area under the absorption band (9); the values are in the order of microseconds. The low-temperature fluorescence lifetime of $Cr(urea)_6^{3+}$ was reported as 50 μsec (9, 19). If one assumes that E_{nr} averages about 3 kcal/mole, an order of magnitude calculation suggests a room temperature lifetime of about 0.02 nsec, which is still long compared with vibrational periods. This may be delayed fluores-

Table I.   Photochemistry

Compound	Photochemistry[b]
	Product
$[Co(NH_3)_6]^{3+}$	$[Co(NH_3)_5(H_2O)]^{3+}$
$[Co(NH_3)_5(H_2O)]^{3+}$	$[Co(NH_3)_4(H_2O)_2]^{3+}$
$[Co(NH_3)_5F]^{2+}$	$[Co(NH_3)_4(H_2O)F]^{2+}$
	$[Co(NH_3)_5(H_2O)]^{3+}$
$[Co(NH_3)_5Cl]^{2+}$	$[Co(NH_3)_4(H_2O)Cl]^{2+}$
	$[Co(NH_3)_5(H_2O)]^{2+}$
$trans$-$[Co(en)_2Cl_2]^+$	$trans$-$[Co(en)_2(H_2O)Cl]^{2+}$
	cis-$[Co(en)_2(H_2O)Cl]^{2+}$
$trans$-$[Co(cyclam)Cl_2]^{2+}$	$trans$-$[Co(cyclam)(H_2O)Cl]^{2+}$
cis-$[Co(en)_2Cl_2]^+$	$trans$-$[Co(en)_2(H_2O)Cl]^{2+}$
	cis-$[Co(en)_2(H_2O)Cl]^{2+}$
cis-$[Co(en)_2(H_2O)Cl]^{2+}$	$trans$-$[Co(en)_2(H_2O)Cl]^{2+}$
cis-α-$[Co(trien)Cl_2]^+$	cis-α-$[Co(trien)(H_2O)Cl]^{2+}$
cis-α-$[Co(trien)(H_2O)Cl]^{2+}$	cis-α-$[Co(trien)(H_2O)_2]^{3+}$
cis-β-$[Co(trien)Cl_2]^+$	cis-β-$[Co(trien)(H_2O)Cl]^{2+}$
	$trans$-$[Co(trien)(H_2O)Cl]^{2+}$
cis-β-$[Co(trien)(H_2O)Cl]^{2+}$	cis-β-$[Co(trien)(H_2O)_2]^{3+}$
	$trans$-$[Co(trien)(H_2O)Cl]^{2+}$
$trans$-$[Co(trien)(H_2O)Cl]^{2+}$	cis-$[Co(trien)(H_2O)Cl]^{2+}$
$[Co(tren)Cl_2]^{2+}$	$[Co(tren)(H_2O)Cl]^{2+}$
β-$[Co(tren)(H_2O)Cl]^{2+}$	$[Co(tren)(H_2O)_2]^{3+}$

[a] From Refs. 20, 21, and 22; irradiations around L_1 band and at either 0°C or 25°C.

cence, however. Prompt fluorescence from *trans*-$Cr(NH_3)_2(NCS)_4^-$ is reported to be < 3 psec (M. Windsor, G. Porter and A. D. Kirk, private communication).

Ligand Field Photochemistry of Some Co(III) Ammines

Until recently only scattered data were reported for the irradiation of Co(III) ammines in the wavelength region of the L_1 band (*14, 15*). With acidoammine complexes, the acido group may be reported to photo-aquate, but no ammonia aquation was looked for and the photochemistry appeared to be merely a catalysis of the thermal reaction. Reported quantum yields were low (except where CT character was present, in which case redox decomposition was also observed), and the ligand field photochemistry of this class of complexes seemed rather uninteresting.

Recent work in this laboratory has demonstrated that such photochemistry is in fact very rich and varied (*20, 21, 22*). The ligand aquated is not always the same as that in the thermal reaction, and, moreover, the photoreaction is stereospecific, often differently from the ground state reaction (*see* Table I). The findings can be accounted for in detail if one

of Co(III) Ammines[a]

Photochemistry[b] ϕ (× 10^4)	*Thermal Chemistry,* k (× 10^6) sec^{-1} at 25°C
2.1	very small
1.3	very small
19.6	very small
5.5	0.086
50.7	very small
17.1	1.7
7.9	32
3.1	very small
4.0	1.1
17.5	very small
6.5	240
42	
<0.1[c]	160
<0.1[c]	0.54
<0.1[c]	1500
80	very small
<0.1[c]	23
45	very small
<0.1[c]	1700
150	3000
<0.1[c]	270

[b] Negligible Co(II) formation in all cases.
[c] For any reaction.

assumes that the position that is labilized is determined by the following rules: (a) if the octahedral complex is represented by three essentially mutually perpendicular axes, the axis labilized will be that of the lowest average ligand field strength, and (b) if the ligands occupying the labilized axis differ, the ligand of greater ligand field strength will be the one preferentially labilized (14, 15).

With complexes having bi- and polydentate ligands such as ethylene-diamine, trien, and tren, it was necessary to assume that even if one of the coordinated nitrogen atoms was labilized, recoordination rather than complete loss of the ligand would occur. Secondary or tertiary nitrogens can recoordinate only to the original position, thereby annulling the photochemical reaction. Primary nitrogens, being at the end of a short hydrocarbon chain and hence having some mobility, can recoordinate either to the original position or to an adjacent octahedral position in an intramolecular displacement reaction. These mechanistic stipulations allow prediction not only of the photostereochemistry but also of other-wise unexpected cases of photoinertness. For example, cis-α-[Co(trien)-Cl$_2$]$^+$ is photoinert because the only possible labilization, by the rules, is

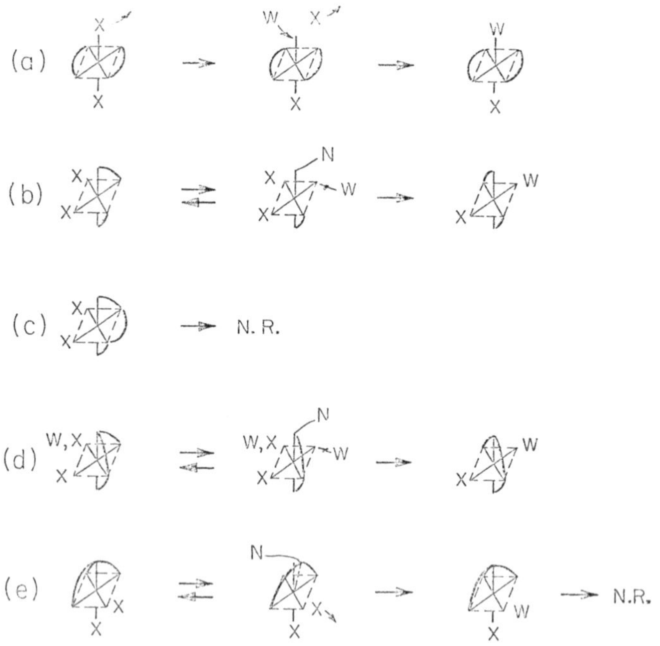

Inorganic Chemistry

Figure 4. Photolysis rules as applied to (a) trans-[Co(en)$_2$-Cl$_2$]$^+$, (b) cis-[Co(en)$_2$Cl$_2$]$^+$, (c) cis-α-[Co(trien)Cl$_2$]$^+$, (d) cis-β-[Co(trien)Cl$_2$]$^+$ or cis-β'-[Co(trien)(H$_2$O)Cl]$^{2+}$, and (e) [Co(tren)Cl$_2$]$^+$ (22)

that of a secondary nitrogen, whereas *cis-β*-[Co(trien)Cl₂]⁺ photoaquates to *trans*-[Co(trien)(H₂O)Cl]²⁺. With the latter, a primary nitrogen is expected to be labilized, and it can then undergo an intramolecular displacement of a chloride. Several of these mechanistic analyses are depicted in Figure 4.

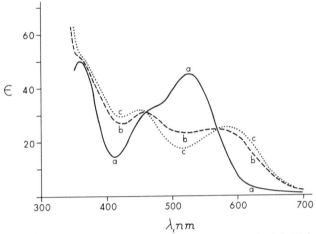

Figure 5.   Photolysis at 488 nm of trans-[Co(en)₂(NH₃)-Cl]²⁺

(a) *Initial spectrum;* (b) *after 75% photolysis;* (c) *calculated final spectrum (which agrees with that expected for 80%* trans-[Co(en)₂(H₂O)Cl]²⁺, *20%* trans-[Co(en)₂(NH₃)(H₂O)]²⁺)

Comparable (although different) stereospecificity is exhibited by Cr(III) ammine complexes. It was important to determine whether the specific ligand that is indicated by the rules as the one to be labilized is indeed the particular one which undergoes substitution. With Cr(NH₃)₅-Cl²⁺, for which the rules predict ammonia aquation, it was shown that it is indeed the ammonia trans to the chloride that is photoreactive (23). A similar check was desirable for the Co(III) series. However, the ¹⁵N-labelling procedure used with the Cr(III) was not convenient for use with Co(NH₃)₅Cl²⁺, and the alternative of studying the photochemistry of *cis*- and *trans*-[Co(en)₂(NH₃)Cl]²⁺ was adopted. Spectroscopically, these complexes are quite similar to Co(NH₃)₅Cl²⁺; ligand field theory relies very much on the general observation that to a first approximation only the nature of the atoms directly coordinated to the central metal ion need be considered. The spectra of *cis*- and *trans*-Co(en)₂(NH₃)Cl²⁺ are presented in Figures 5 and 6. For comparison, the L₁ and L₂ bands for Co(NH₃)₅Cl occur at 512 nm and 375 nm, respectively, and the respective extinction coefficients are 36 and 39 M^{-1} cm⁻¹. All three spectra are thus rather similar except for relatively small intensity differences.

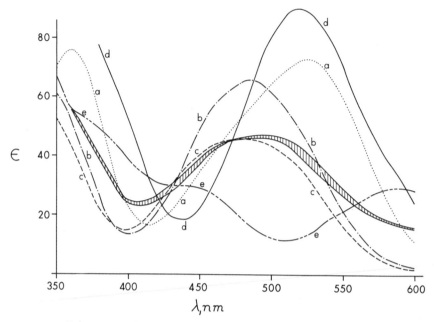

Figure 6.   Photolysis at 488 nm of cis-[Co(en)₂(NH₃)Cl]²⁺
(a) Initial spectrum; (b) cis-[Co(en)₂(NH₃)(H₂O)]³⁺;   (c) trans-[Co(en)₂(NH₃)(H₂O)]³⁺;
(d) cis-[Co(en)₂(H₂O)Cl]²⁺; (e) trans-[Co(en)₂(H₂O)Cl]²⁺; (f) hashed spectrum is that
extrapolated from runs of 15–20% reaction to 100% reaction with hashing repre-
senting the experimental error range of the extrapolation

Table II.   Ligand Field Photolysis of *cis* and
***trans*-[Co(en)₂(NH₃)Cl]²⁺ [a]**

Quantum Yield (× 10⁴)

Isomer	ϕ_{NH_3}	ϕ_{Cl}	Product
trans	11.5 ± 0.5		trans-[Co(en)₂(H₂O)Cl]²⁺
		2.9 ± 0.2	trans-[Co(en)₂(NH₃)(H₂O)]³⁺
cis	2.0 ± 0.1		70 ± 15% trans-[Co(en)₂(H₂O)Cl]²⁺
		3.1 ± 0.1	60 ± 20% cis-[Co(en)₂(NH₃)(H₂O)]³⁺

[a] Irradiation was at 488 nm and 25°C. Incident light intensities were determined by Reineckate actinometry (*see* Refs. *14* and *15*). Ammonia and chloride aquation yields were determined by direct analysis for the released ligands; the isomeric composition of the products was determined from the changes in absorption spectra after irradiation.

The data (*24*) are summarized in Table II. The analysis is as follows. Considering first the case of *trans*-[Co(en)₂(NH₃)Cl]²⁺, we assume that the perturbation of replacing four equatorial ammonias by two ethylenediamines is minor with respect to ligand labilizations although absolute quantum yield values might be affected by changes in k_{cr} or k_{nr}. We do observe about a five-fold decline in quantum yields for *trans*-[Co(en)$_2$(NH$_3$)Cl]$^{2+}$ relative to [Co(NH$_3$)$_5$Cl]$^{2+}$, but the important result is that the ϕ_{NH_3}/ϕ_{Cl} ratio is comparable in the two cases, as would be expected if the rules did in fact specify that it is predominantly the nitrogen ligand trans to the chloride that is labilized. Were the ammonia aquation in [Co(NH$_3$)$_5$Cl]$^{2+}$ random, the value of 3 for $\phi_{NH_3}/$ ϕ_{Cl} would mean that the quantum yield for photoaquation of ammonia from any one position would be 0.6 relative to ϕ_{Cl}. Were this true, ϕ_{NH_3}/ϕ_{Cl} should have a value of just 0.6 for *trans*-[Co(en)$_2$(NH$_3$)Cl]$^{2+}$. The much higher ratio found strongly supports the rules-based mechanism.

The case of *cis*-[Co(en)$_2$(NH$_3$)Cl]$^{2+}$ is more complicated. If one assumes that one end of an ethylenediamine is aquated, recoordination can occur by intramolecular displacement at an adjacent site; the possible reactions are given in Figure 7. By the rules, only Reactions 1 and 2 should occur to yield *cis*-[Co(en)$_2$(NH$_3$)(H$_2$O)]$^{3+}$ and *trans*-[Co(en)$_2$-

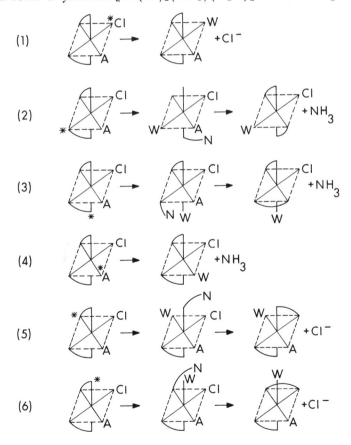

Figure 7.   Rules-based mechanism for photolysis of Co(III) ammines

$(H_2O)Cl]^{2+}$. Reaction 2 should be relatively inefficient, however, since, according to the assumed mechanism, much of the time recoordination would occur to the original site, thereby annulling the reaction. The observed ϕ_{NH_3}/ϕ_{Cl} ratio of 0.6 is thus quite reasonable, as is the finding that the ammonia yield decreases on going from the trans to the cis form of $Co(en)_2(NH_3)Cl^{2+}$ whereas the chloride yield remains about the same.

The spectral changes accompanying an irradiation of cis-$[Co(en)_2$-$(NH_3)Cl]^{2+}$ (Figure 6) were calculated as the hypothetical product spectrum. That is, the actual spectral change was found experimentally for 15–20% reaction (as determined by ammonia analysis), and the change was extrapolated to 100% reaction. Included in the figure are the spectra of the four possible aquation products. Analysis of these spectra reveals that the isomeric compositions are about 60 ± 20% cis- relative to $trans$-$Co(en)_2(NH_3)(H_2O)^{3+}$ and about 70 ± 15% $trans$- relative to cis-$Co(en)_2(H_2O)Cl^{2+}$. The isomeric composition conforms approximately to that predicted by the rules-based mechanism since cis aquoamine and trans aquochloro predominate. More accurate chromatographic product analyses are required in order to establish the isomer ratios more exactly.

Were the ethylenediamine labilization random rather than rules-based, Reactions 2, 3, 4, 5, and 6 (Figure 7) would occur with about equal probability, and one would predict 2/3 cis- relative to $trans$-$[Co(en)_2(H_2O)Cl]^{2+}$ which agrees very poorly with observation. Furthermore, in order to obtain the observed value of 0.6 for ϕ_{NH_3}/ϕ_{Cl} by random nitrogen labilization, Reaction 1 would have to have a probability of about 0.6 relative to that for nitrogen labilization, or a probability of about 3 relative to Reactions 2, 3, 4, 5, or 6. One then expects about 4/5 cis-relative to $trans$-$[Co(en)_2(NH_3)(H_2O)]^{2+}$, which is about that observed; no distinction between the random and the labilization rules-based mechanism is thus possible in this case.

In summary, the photochemistry of $trans$-$[Co(en)_2(NH_3)Cl]^{2+}$ definitely supports the rules-based mechanism. That of cis-$[Co(en)_2(NH_3)$-$Cl]^{2+}$ is approximately consistent with it. There is little support for random labilization.

Thexi vs. Franck–Condon States as Direct Precursors to Photosubstitution Reactions of Co(III) Ammines

Two characteristics of thexi states which distinguish them from Franck–Condon states are that (a) their properties should be independent of the manner of their formation and (b) their lifetimes should be long compared with vibrational times. Where emission is observable, characteristic (a) can be confirmed by determining that the emission spectrum is independent of irradiating wavelength. In the case of $[Ru(bipyr)_3]^{2+}$, there is a strongly emitting state which can be generated either spectroscopically or chemically (by reduction of $[Ru(bipyr)_3]^{3+}$ (*see* Ref. 28). Characteristic (b) can also be established if emission is observed by determining the emission lifetime.

These direct approaches are not feasible as yet with the Co(III) ammines. No emission has been reported for this family of complexes. There is some less direct evidence however. First, because of its small lifetime, a Franck–Condon state is most likely capable only of simple metal–ligand bond cleavage—the limiting S_N1 reaction of the coordination chemistry (7). The result of such cleavage would be a pentacoordinated intermediate, perhaps square pyramidal in geometry initially, but with immediate equilibration toward a trigonal bipyramidal shape. By this mechanism, solvent water then coordinates with expansion of the coordination sphere back to six-coordination to give the final product. The problem with the limiting S_N1 mechanism is that it is difficult to account for substitutional stereospecificity; the pentacoordinated intermediate should be rather stereolabile. This problem arises in thermal substitution reactions, and there it has seemed necessary to assume both solvent assistance to bond breaking and some directing effect of the departing ligand (7). In effect, an S_N2 mechanism is called for (which is often described as associative or nonlimiting S_N1).

The findings summarized in Tables I and II demonstrate that the photochemistry of Co(III) ammines are quite stereospecific, and I take this as an indication that a concerted process is involved. The rules-based mechanism is of this type if it is stipulated that the labilized ligand leaves in a concerted manner with incoming solvent so that the octahedral framework is retained. Such a concerted process involves nuclear motions not corresponding to simple vibrations (the solvent water which coordinates is hydrogen-bonded to other waters as is the leaving group so that many nuclear motions would accompany the aquation process). The process is not likely to occur unless the reacting state has a lifetime that is long compared with a vibrational period. Thus the specificity of the photochemistry of Co(III) ammines suggests that a reactive, but essentially normal, chemical species is involved—a thexi rather than a Franck–Condon state.

Another indication that the reactive state is moderately long lived is that the photoaquation of *trans*-$[Co(en)_2Cl_2]^+$ shows an apparent activation energy of 5.2 kcal/mole (20). Since, as was noted earlier, the quantum efficiency depends inversely on k_{nr}, which should also be activated, the implication is that E_{cr} is significantly greater than 5.2 kcal/mole. A species with such an activation energy should, by reaction rate theory, have a lifetime that is relatively long compared with vibrational times.

There are thus two indirect indications that the photochemistry of Co(III) ammines derives from thexi rather than from Franck–Condon states. The spectroscopic nature of this state has not been determined. It could be related (after thermal equilibration and possible descent in symmetry) to the first excited ligand field state in O_h, $^1T_{1g}$. It could be triplet or quintet in spin multiplicity, however.

An important seeming paradox remains to be discussed. On the one hand, the photochemistry of both Cr(III) and Co(III) can be largely rationalized in terms of the rules. These rules invoke pseudo-octahedral axes and make use of the spectrochemical series. Several recent papers have applied conventional ligand field theory to explain the rules in terms of changes in sigma- and pi-bonding on ligand field excitation (24, 25, 26). Exceptions to the rules have been rationalized by ligand field theory as well (25, 26, 27). On the other hand, great emphasis has been placed in this paper on the fact that photochemistry proceeds through thexi states and that such states are not treated by conventional ligand field theory. Distortions change the symmetry from that of ground state; the concept of ligand field strength becomes difficult. How is it, then, that the rules seem to be useful in describing thexi state chemistry?

In the case of the Co(III) ammines, however, the rules-based mechanism essentially retains the pseudo-octahedral axes; no stereochemical change occurs if the labilized ligand is monodentate, as with trans-[Co(en)$_2$Cl$_2$]$^+$, and there is little effect if the two ethylenediamines are replaced by the belt ligand cyclam. The more complex stereochemical behaviors that occur when the labilized position is occupied by one arm of a multi-dentate ligand are explained in terms of subsequent intramolecular displacement reactions. It would appear that the difference in geometry between ground and thexi state is one of tetragonal distortion. Thermal equilibration is then a fulfillment of the bonding changes predicted for the hypothetical ligand field excited state—and the rules should indeed have direct applicability.

The case of the Cr(III) ammine family is not so easy since stereochemical change occurs even when it is a monodentate ligand that is labilized and, moreover, stereomobility seems to be a prerequisite to excited-state reaction. Also, the rules lose crisp validity when three or more types of ligands are present (29). Because of the presence of only three metal d electrons, there is orbital availability for increased coordination, and I regard it as very likely that thermal equilibration of a Cr(III) ammine leads to a geometry quite different than that of the ground state. A possibility is that of distortion towards a pentagonal pyramidal geometry with concerted coordination of solvent to make the thexi state a hepta-coordinated species. It is entirely possible to write a mechanistic recipe involving such an intermediate—one which retains uniqueness for the rules-indicated labilized group and which also gives the observed stereochemistry. If this explanation is correct, the rules are approximately valid for Cr(III) photochemistry for quite different reasons than for Co(III) photochemistry. For example, simple ligand field treatments of excited-state bonding changes may be relevant in the latter case, but would be specious in the former case. (Note, however, that Zink (*30*) believes that the stereochemistry of Cr(III) photoreactions probably can be accounted for in terms of ligand field analyses of excited *vs.* ground state bonding.) Currently planned high energy, pulse laser photolysis experiments may help to unravel the situation; we expect to determine directly the lifetime of the thexi state precursor to chemical reaction and perhaps also its spectral characteristics.

Some Further Aspects of Franck–Condon and Thexi States

There are some interesting aspects to thermal equilibration if the approach is from a general physical chemical point of view rather than from a purely spectroscopic one. The typical situation involves a strongly solvated complex, yet the role of the solvent cage is rarely considered. The distortion that occurs on thermal equilibration must require some rearrangement of the immediate solvation shell. If the net effect is equivalent to a shift of just one solvent molecule from one solvent position to another, or a displacement of, say, 2 A, phenomenological diffusion theory predicts that about $(2 \times 10^{-8})^2/(2)(1 \times 10^{-5})$ or about 0.02 nsec would be required (with 1×10^{-5} cm^2/sec as the solvent diffusion coefficient). A much longer time would be needed in a low temperature, rigid medium. Two types of thermal equilibration can be postulated. The first occurs very rapidly, on a time scale such that the solvent environment is approximately that of a rigid lattice. The second occurs more slowly, at a rate determined largely by that at which the solvent cage can adjust to the final thexi state geometry.

These two stages may be rather distinct in low temperature, rigid lattices. As suggested in Figure 3, intersystem crossing may be most probable during the first stage. Note, for example, that the phosphorescence emission from a Cr(III) ammine shows a break in its temperature dependence at the glass point of the medium (*31*). From concommittant photochemical studies it was in fact inferred that thermal equilibration to the reacting quartet state competed with rather than led to intersystem crossing. Forster and co-workers (*see* Ref. 8) reported lifetime and spectral complexities for the phosphorescence emission from Cr(III) complexes that seem best explained in terms of different solvation species retaining their differences through the intersystem crossing process.

The concept of a two-stage thermal equilibration process may be expanded. We assume that solvent molecules can shift only in diffusion-like jumps in accommodation to the geometry changes between the Franck–Condon and the final thexi state. The period between each jump could be dozens of vibrational times at room temperature, and hundreds or thousands at low temperatures. During the period between jumps, the complex is in a quasi-rigid lattice and has more or less defined vibrational states. These states change with each solvent jump, however, so the picture is that indicated schematically in Figure 8 where series of parabolic potential wells marks the progression towards thermal equilibration. There is an analogy in rheology, that of visco-elastic behavior. Such a system is elastic on a short time scale but viscous on a longer one. The vibrations of a Franck–Condon state may be of this visco-elastic type, with a time-dependent restoring force. The effect would be to erase vibrational detail to give the smooth absorption band envelope that is seen. If the equilibrium geometry of the excited state is nearly that

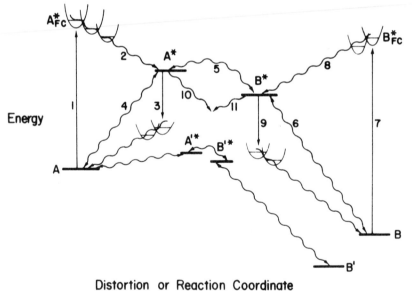

Distortion or Reaction Coordinate

"Concepts of Inorganic Photochemistry"

Figure 8.   Generalized scheme for the photochemical and thermal reaction A ⇌ B (31)

(1) Absorption to Franck–Condon state A*$_{FC}$; (2) thermal equilibration to the thexi state A*; (3) emission to Franck–Condon state of A; (4) radiationless deactivation of A* to A, or Boltzmann activation of A to give A*; (5) activated reaction of A* to give thexi state B*; (6) radiationless deactivation of B* to B, or Boltzmann activation of B to give B*; (7) absorption to Franck–Condon state B*$_{FC}$; (8) thermal equilibration to thexi state B*; (9) emission from B* to Franck–Condon state of B; (10) radiationless deactivation or reaction of A* directly to B; (11) radiationless deactivation or reaction of B* directly to A. The path through A'* and B'* is a possible alternative thermal reaction path.

of the ground state, then the viscous component would be small and vibrational states should become better defined. Thus the ^2E_g state in an octahedral d^3 complex involves only rearrangement of non-bonding electrons, and its equilibrium geometry should be close to that of the ground state. In confirmation, the phosphorescence emission from Cr(III) complexes is only slightly red-shifted from the absorption band (*note* Figure 2). Moreover, both emission and absorption spectra typically have fairly sharp vibrational detail (*note* also Figure 1).

Figure 8 is drawn to emphasize that a thexi state is essentially a high energy isomer of the ground state. It is also very close in concept to the transition state of reaction kinetics; both are postulated to be in thermodynamic equilibrium with their surroundings. The special assumption that a transition state has an open coordinate along which translation on the reaction path occurs is a convenience for calculation rather than a necessity. The transition state can, in fact, more gracefully be considered a high energy, thermodynamic species for which a small additional increment of energy allows reaction to occur (33) in which case the conceptual distinction between a thexi and a transition state vanishes.

The figure also illustrates that a thexi state can be obtained in several ways: (a) spectroscopically by direct light absorption, (b) by thermal activation or the reverse of radiationless deactivation, (c) by chemical reaction (as must be true in chemiluminescent processes), and (d) by photochemical reaction (as in chemical laser systems). Viewed in this light, both photochemical and photophysical processes (other than emission and light absorption) leave their usual spectroscopic milieu to join the broad field of reaction kinetics.

Literature Cited

1. Ballhausen, C. J., "Introduction to Ligand Field Theory," McGraw-Hill, New York, 1962.
2. Figgis, B. N., "Introduction to Ligand Fields," Interscience, New York, 1966.
3. Jørgensen, C. K., "Absorption Spectra and Chemical Bonding in Complexes," Pergamon, London, 1962.
4. Schläfer, H. L., Gliemann, G., "Einführung in die Ligandenfeldtheorie," Akademische Verlagsgesellschaft, Frankfurt-am-Main, 1967.
5. Cotton, F. A., "Chemical Applications of Group Theory," Interscience, New York, 1963.
6. Fleischauer, P. D., Adamson, A. W., Sartori, G. D., in "Inorganic Reaction Mechanisms," J. O. Edwards, Ed., Part II, Wiley, New York, 1972.
7. Basolo, F., Pearson, R. G., "Mechanisms of Inorganic Reactions," 2nd ed., Wiley, New York, 1967.
8. Forster, L. S., in "Concepts of Inorganic Photochemistry," A. W. Adamson and P. D. Fleischauer, Eds., chap. 1, Wiley, New York, 1975.
9. Porter, G. B., in "Concepts of Inorganic Photochemistry," A. W. Adamson and P. D. Fleischauer, Eds., chap. 2, Wiley, New York, 1975.
10. Porter, G. B., Schläfer, H. L., *Z. Phys. Chem.* (1963) **37**, 109.

11. Fleischauer, P. D., Fleischauer, P., Chem. Rev. (1970) 70, 199.
12. Adamson, A. W., Proc. Int. Conf. Coord. Chem. (1972) 14, 240.
13. Zinato, E., in "Concepts of Inorganic Photochemistry," A. W. Adamson and P. D. Fleischauer, Eds., chap. 4, Wiley, New York, 1975.
14. Balzani, V., Carassiti, V., "Photochemistry of Coordination Compounds," Academic, New York, 1970.
15. Adamson, A. W., Waltz, W. L., Zinato, E., Watts, D. W., Fleischauer, P. D., Lindholm, R. D., Chem. Rev. (1968) 68, 541.
16. Kane-Maguire, N. A. P., Langford, C. H., Chem. Commun. (1971) 895.
17. Wasgestian, H. F., Ballardini, R., Varani, G., Moggi, L., Balzani, V., J. Phys. Chem. (1973) 77, 2614.
18. Adamson, A. W., Pribush, R., Wright, R., Geosling, C., unpublished data.
19. Dingle, R., J. Chem. Phys. (1969) 50, 1952.
20. Pribush, R. A., Poon, C. K., Bruce, C. M., Adamson, A. W., J. Am. Chem. Soc. (1974) 96, 3027.
21. Sheridan, P. S., Adamson, A. W., J. Am. Chem. Soc. (1974) 96, 3032.
22. Sheridan, P. S., Adamson, A. W., Inorg. Chem. (1974) 13, 2482.
23. Zinato, E., Riccieri, P., Adamson, A. W., J. Am. Chem. Soc. (1974) 96, 375.
24. Adamson, A. W., Wright, W., Pribush, R. A., unpublished work.
25. Zink, J. I., Inorg. Chem. (1973) 12, 1018.
26. Zink, J. I., Mol. Photochem. (1973) 5, 151.
27. Wrighton, M., Gray, H. B., Hammond, G. S., Mol. Photochem. (1973) 5, 165.
28. Ford, P. C., Hintze, R. E., Petersen, J. D., in "Concepts in Organic Photochemistry," A. W. Adamson and P. D. Fleischauer, Eds., chap. 5, Wiley, New York, 1975.
29. Bifano, C., Linck, R. G., Inorg. Chem. (1974) 13, 609.
30. Zink, J. I., J. Am. Chem. Soc. (1974) 96, 4464.
31. Adamson, A. W., J. Phys. Chem. (1967) 71, 798.
32. Adamson, A. W., in "Concepts of Inorganic Photochemistry." A. W. Adamson and P. D. Fleischauer, Eds., chap. 10, Wiley, New York, 1975.
33. Adamson, A. W., "A Textbook of Physical Chemistry," chap. 15, Academic, New York, 1973.

RECEIVED January 24, 1975.

Luminescence as a Probe of Excited State Properties

G. A. CROSBY

Washington State University, Pullman, Wash. 99163

Quantitative investigations of the photoluminescence of inorganic compounds have led to experimental criteria for assigning orbital and spin labels to their low lying electronic excited states. For d^6 compounds, chemical modification of the sequencing of ligand-field, charge-transfer, and ligand-localized excited states has been demonstrated. The capability of prescribing the lowest excited states has produced a series of materials with unusual optical properties. Details of the charge-transfer-to-ligand excited configurations that have been obtained for ruthenium(II) and osmium(II) complexes provide a new perspective on the role of spin-orbit coupling in defining the properties of the associated states. Systematic study of excited state properties indicates possibilities for dictating the pathways of photochemical reactions, for relating spectroscopy to electrochemistry, and hopefully, for correlating excited state properties with thermal reactivities.

Emission spectroscopy has a long and venerable history of providing valuable information on the nature of the low lying excited states of organic molecules (*1, 2*), but, until recently, systematic use of photoluminescence as a probe of excited state properties of transition metal complexes was not widespread. For complexes that contain central metal ions of certain configurations, especially d^3 and d^6, the energy level schemes are propitious for occurrence and detection of luminescence, and the structural and environmental factors that control the properties of the low lying excited states are being defined through emission spectroscopy. Criteria have been developed for assigning orbital and spin labels to excited states of complexes (*3*), and some primitive attempts to engineer molecules with stipulated electronic properties were successful (4, 5). In this paper, attention is focussed on those features of the excited states of transition metal complexes that differentiate them from the well studied organic ones, on the magnificent versatility inherent in transition metal chemistry for designing molecules with prescribed electronic properties, and on the kinds of detailed information that can be obtained about the excited states of transition metal complexes by emission techniques. The usefulness of this kind of information for other fields of chemical research is considered briefly.

Chemical Tuning

The low lying excited states of d^6 strong field complexes that contain π-conjugated ligands can be conveniently classified into four orbital promotional types: (a) $\pi\pi^*$ states in which the excitation energy is localized essentially on the ligands and the characteristics of the states strongly reflect their ligand parentage, (b) $d\pi^*$ states in which the final states are derived from a configuration in which an electron has been transferred from the metal core to an antibonding orbital delocalized over the ligand π system, (c) πd states for which a transfer of charge from the ligand system to the metal ion can be visualized as the primary excitation process, and (d) dd states that are effectively metal-localized electronic excitations in which the ligands are involved only as contributors of the nonspherical static potential that determines the orbital promotional energy.

With the exception of πd excited states, detailed analyses of the emission characteristics of d^6 complexes of the second and third transition series have lead to criteria for classifying these states by means of emission spectroscopy (3, 6). Moreover, further investigation revealed that complexes can be engineered to possess a predetermined sequence of excited states with prescribed orbital types (4, 5). This capability, designated chemical tuning, has led to series of molecules whose low lying excited states were chosen specifically and whose spectroscopic properties were ordained as well.

Once the gross orbital types of the lowest excited states of complexes are determined, subtle alterations in properties can be effected by varying ligand substituents, by modifying the environment of the active species, or by subjecting the materials to external perturbations. This fine tuning of spectroscopic characteristics has produced a number of unusual electronic properties with potential use for exploitation in both a fundamental and a practical way. We direct attention particularly to charge-transfer excited states.

Unusual Properties of Charge-Transfer-To-Ligand (CTTL) Excited States

Systematic investigations of series of ruthenium(II) (*7, 8, 9, 10*), osmium(II) (*11*), and iridium(III) (*12*) complexes led to the experimental and theoretical characterization of CTTL excited states. Their properties, derived from analyses of spectra, decay times, and interactions with external fields, differ fundamentally from the excited states of organic materials and even from states of other orbital parentages within the same molecule.

Experimental Features of CTTL Excited States

The intense photoluminescence exhibited by complexes that display emission originating from $d\pi^*$ configurations was not well understood until the observation range was extended to temperatures below 77°K. For a single emitting level (or cluster of degenerate levels), the measured decay time and the quantum yield should be related by the equation $\tau = \phi\tau_0$ where τ_0, the limiting decay time near 0°K, is expected to be temperature independent. From measurements of τ and ϕ at 77°K on ruthenium(II) (*8*) and osmium(II) (*13*) complexes, values were predicted for series of complexes. When the low temperature experiments were performed, however, it was found that the measured decay times far exceeded the theoretical limits, a behavior that is strongly indicative of a manifold of emitting levels, each one with its own set of radiative

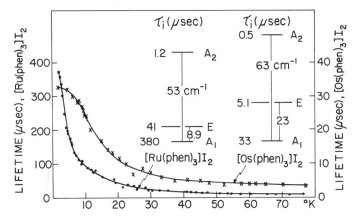

Figure 1. Temperature dependence of the calculated (———) and observed (· · · , × × ×) lifetimes of tris(1,10-phenanthroline)ruthenium(II) iodide and tris(1,10-phenanthroline)-osmium(II) iodide in poly(methyl methacrylate). Energy level splittings and individual mean decay times were determined from a computer fit of the experimental lata.

and radiationless decay constants whose separations are on the order of kT in the range of 5°–50°K. The situation is illustrated in Figure 1. Below 77°K, the decay time of each of the luminescent molecules rises monotonically with decreasing temperature, and it either fails to reach a limit at the lowest temperatures attained or it approaches a constant value that exceeds the predicted limit by many factors. These facts are inconsistent with the presence of a single luminescent level or a degenerate set of such levels.

A second pertinent feature of the decay kinetics of complexes that display charge-transfer luminescence is the exponentiality of the observed transients at all temperatures reached (~ 1.5°K). This behavior is in stark contrast to that exhibited by organic systems at low temperatures. For the latter, single exponential decays are maintained to ~ 10°K, but they are replaced by complicated kinetics at lower temperatures (2). Nonexponential decays have also been observed at low temperature for tris complexes of rhodium(III) that exhibit $^3\pi\pi^*$ states lowest (14).

Energy Level Schemes

Although the emission spectra that originate from $d\pi^*$ excited states of complexes have structure, the vibrational bandwidths exceed the splittings of the electronic levels. There is little sharpening at low temperature, and incorporation into a lattice does not appear to improve the resolution (15). Nonetheless, it is possible to obtain the level splittings from the decay data. If one assumes a manifold of excited states in thermal equilibrium at all temperatures, each decaying with temperature-independent decay constants, one arrives at an analytical expression for the temperature dependence of the mean decay time of the ensemble (8, 9, 10):

$$\tau(T) = \frac{\Sigma_i k_i e^{-\epsilon_i/kT}}{\Sigma_i e^{-\epsilon_i/kT}}$$

A computer fit of this expression yields the k_i's for the levels. Analyses of this type were made for series of $d\pi^*$ emitters and the energy level schemes derived from the decay curves are included in Figure 1.

The level schemes for [Ru(phen)$_3$]$^{2+}$ and [Os(phen)$_3$]$^{+2}$ (phen = 1,10-phenanthroline) are representative of the clusters of low lying electronic states that arise from $d\pi^*$ configurations of many ruthenium(II), osmium(II), and iridium(III) complexes. They are highly unusual since they have decay parameters that lie between the ranges expected for conventional singlet and triplet states and because of the magnitudes of the splittings themselves. These parameters control the nature of $d\pi^*$

photoluminescence and are responsible for its history of assignment and reassignment (*16*).

Coupling Model

In order to rationalize the behavior of the luminescence observed from nd^6 complexes that display $d\pi^*$ emission and to account for the exceptional splittings and decay parameters obtained for them, a model for the excited states has emerged that emphasizes the role of spin-orbit coupling in controlling their properties (*17, 18*). An excited $d\pi^*$ configuration is viewed as an abortive oxidation of the d^6 complex that produces a system with a d^5 core and a promoted (optical) electron distributed in an antibonding orbital encompassing the π-conjugated ligands. The core is viewed as a Kramers ion with a set of electronic states whose positions are defined by electrostatic, spin-orbit, and ligand-field interactions. These core states define a vector space. A second vector space is defined by the set of molecular spin orbitals open to the excited electron residing on the ligands. The final eigenspace for the total excited d^6 system is determined by diagonalizing the direct product space of the two subsystems under the full Hamiltonian of the six-electron problem.

For the lowest $d\pi^*$ configuration of a D_3 molecule, the model predicts a cluster of 8 levels (12 states) lying within 3 kK and located 15 kK above the ground state (*17, 18*). The cluster is separated into three groups, with the lowest one responsible for the observed luminescence. For trigonal systems, exemplified by tris(2,2'-bipyridine)ruthenium(II), the predicted level scheme for the emitting levels is given in Figure 2.

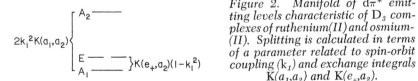

Figure 2. Manifold of $d\pi^*$ emitting levels characteristic of D_3 complexes of ruthenium(II) and osmium-(II). Splitting is calculated in terms of a parameter related to spin-orbit coupling (k_1) and exchange integrals $K(a_1,a_2)$ and $K(e_+,a_2)$.

Two features of the model calculation are of signal importance. The coupling scheme provides symmetry labels for the states that correlate with the observed selection rules (*8*), and the model reveals the key role played by spin-orbit forces in determining the properties of $d\pi^*$ excited states. Whereas the gross splittings of electronic excited states of organic molecules are controlled by electrostatic forces and spin-orbit coupling produces the fine structure (*1, 2*), the situation is reversed for $d\pi^*$ excited configurations of heavy metal complexes. In such systems, typified by the Ru(II) and Os(II) complexes of Figure 1, spin-orbit and electrostatic forces on the central metal atom define the separations of groups of levels, and the fine structure (< 100 cm^{-1}) is determined primarily by electrostatic (exchange) interactions. Exchange integrals have been determined for several ruthenium(II) molecules (19); they are in the range of 15 to 65 cm^{-1}. Since the spin-orbit coupling constant for ruthenium(II) is ⏜ 900 cm^{-1}, the importance of relativistic forces in controlling the dispositions of the $d\pi^*$ levels becomes manifest. The low values of the exchange integrals between the promoted electron and the core electrons reveal the large degree of charge delocalization away from the central ion that occurs upon $d\pi^*$ optical excitation. For an opposing point of view that is based on polarization measurements, the reader is referred to the article by Fujita and Kobayashi (20).

The Problem of Labels

Russell–Saunders coupling provides the appropriate labelling scheme for the terms of all the light atoms (21), but the nomenclature becomes progressively less accurate as the nuclear charge increases. For the rare earth elements, for example, the assignment of a unique spin and orbital designation provides a first description of the terms of a configuration, but the parentage of a state may be quite mixed (22). In organic molecules, the classification of states by orbital and spin labels can rarely be faulted, and the practice is deservedly widespread. For complexes, especially those that contain heavy metal atoms, both spin and orbital labelling may or may not be appropriate, and the proper nomenclature for an observed state may even depend on the host matrix of the system.

Spin-Based Nomenclature. For organic molecules, spin designations of $\pi\pi^*$ excited states are accurate, and, for most $\pi\pi^*$ states in complexes, the assignment of a unique spin label is usually quite acceptable. For d^6 and d^{10} complexes, many $^3\pi\pi^*$ states have been classified by emission spectroscopy (6, 24), and spectral criteria for recognizing them have been formulated (3). When unpaired spins are present in the ground state of a metal complex, however, the triplet designation fails and vector coupling schemes have been used to spin label the observed states. The trip-multiplets of metalloporphyrins are good examples (25, 26, 27). Spin still appears to be a good quantum number, however.

Attaching a unique spin label to $d\pi^*$ states in complexes is, according to the coupling model described above, entirely inappropriate (23). No unique spin label can be assigned. In fact, application of S^2 to any of the state eigenkets denoted in Figure 2, scatters the ket not only into others of the set, but also into levels that are several kilokaisers higher in energy. As the spin labelling fails, so do such commonly used terms as internal conversion and intersystem crossing (1, 2). For πd states, a similar inference can be drawn, but definitive experimental verification by emission spectroscopy has not yet been made. The description of *dd* states in complexes by a spin label is intermediate in usefulness. Just as for atoms, the label is progressively less meaningful as the atomic number of the metal ion increases.

Orbital Nomenclature. In the preceding discussion we implicitly assumed unique orbital descriptions for excited states. It is clear, however, that even orbital promotional designations may fail whenever a near degeneracy of uncoupled levels occurs. Configuration interaction becomes important, and a zeroth-order coincidence of levels of disparate orbital designations leads to some unusual excited state properties that are observable by emission techniques.

Environmental modulation of excited state qualities occurs for several complexes of iridium(III) (*28*). Although the energies of luminescing states are only slightly affected by solvent perturbations, the selection rules, and therefore the wave functions, are substantially modi-

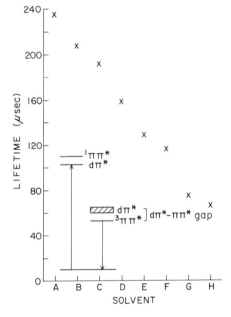

Figure 3. Solvent dependence of the measured decay time of dichlorobis(5,6-dimethyl-1,10-phenanthroline)iridium-(III) chloride at 77°K. Solvent: A, 1:1 glycerol–water; B, 4:1 glycerol–water; C, 4:1 methanol–water; D, glycerol; E, 2:1 glycerol–ethanol; F, 1:1 glycerol–ethanol; G, ethanol; and H, 4:1 ethanol–methanol. Insert: proposed solvent-dependent configuration interaction mechanism.

fied. The experimental situation is depicted in Figure 3. As the polarity of the solvent glass is progressively decreased, the measured decay time of the luminescence steadily decreases also. Our interpretation of this unusual sensitivity of the measured decay time to solvent interaction is presented in the figure insert. We postulate the existence of a cluster of $d\pi^*$ excited states located within ~ 1.0 kK of the lower $^3\pi\pi^*$ emitting level. The energy of the former is highly solvent dependent, whereas that of the latter is not. Closing the gap between the cluster of $d\pi^*$ levels and the near-degenerate set of $^3\pi\pi^*$ emitting states lends CT character to the latter, thus decreasing the measured decay time of the ensemble. Quantitative analysis of the described behavior (28, 29) provided an estimate of the configuration interaction parameter. If the configuration interaction model is indeed a faithful description of the physical situation, then the emitting levels must be described as $\psi_e = a\psi(d\pi^*) + b\psi(^3\pi\pi^*)$ where the a/b ratio is defined by the solvent. For a/b $\simeq$ 1, the emitting states can be assigned no unique orbital label and only a mixed terminology is appropriate. Furthermore, when the $d\pi^*$ configuration mixes strongly with the $^3\pi\pi^*$ states, the spin quantum number of the latter emitting multiplet progressively loses meaning. For strongly mixed $d\pi^*$–$^3\pi\pi^*$ levels, one can assign neither a unique orbital nor a unique spin label to the resultant states, and a state can be designated only by the irreducible representation label of the point group of the molecule.

External Magnetic Interactions

An external magnetic field directed along the principal axis of a D_3 complex should cause a first-order splitting of the degenerate E level (see Figure 4). This prediction has been verified (15) experimentally by a change in the luminescence spectrum with impressed magnetic field. At 1.65°K the emission spectrum in zero field becomes a new band when the sample is subjected to a field of 75 kG. This switching of the spectrum is accompanied by a substantial decrease in measured decay time of the luminescence. Theoretical analysis of this latter phenomenon yielded measurements of both the g factor of the E state and the off-diagonal matrix elements connecting the zero field levels (30).

Circularly Polarized Luminescence

The interaction of an external magnetic field with an ensemble of excited molecules provides a novel method for determining excited state parameters by inducing circularly polarized luminescence. The signal obtained from a sample of tris(2,2'-bipyridine)ruthenium(II) chloride

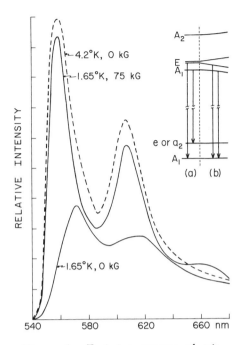

Figure 4. *Emission spectra of tris-(2,2'-bipyridine)ruthenium(II) cation in* [Zn(bipy)₃]SO₄ · 7H₂O *crystal lattice in zero magnetic field at 1.65° and 4.2°K and in 75-kG field at 1.65°K. (a) Energy level diagram of* [Ru-(bipy)₃]²⁺ *and (b) magnetic field dependence of the emitting levels with magnetic field parallel to principal axis; ↑, allowed transitions (15).*

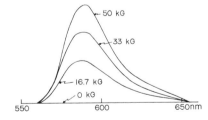

Figure 5. *Circularly polarized luminescence from tris(2,2'-bipyridine)ruthenium(II) cation in a KBr pellet at ∼20°K under constant UV excitation (31)*

contained in a KBr pellet under constant UV excitation at ∼ 20°K is plotted in Figure 5. The magnetic field-dependent signal is directly proportional to the difference between the intensities of the right and left circularly polarized emitted radiation. This preliminary finding is interpretable on the basis of the coupling model described above; it demonstrates the efficacy of the emission experiment as a sensitive probe of excited state properties (*31*).

Chemical Implications

Emission spectroscopy is providing new information on the excited states of transition metal complexes. In addition to defining the energies of the low lying levels, analyses of emitted light can yield detailed information on the wave functions through the behavior of the luminescence under intense magnetic fields, thermal stresses, and environmental perturbations. Recent developments in photochemistry (*32*) and electrochemistry (*33*) confirm that the lowest excited states are inextricably related to these phenomena and in some cases may be controlling the chemistry as well (*34*). From the accumulated data from emission spectroscopy, photochemistry, and electrochemistry, the principles for engineering molecules and solids with unusual and potentially useful electrooptical properties are becoming defined.

Literature Cited

1. Kasha, M., *Radiat. Res. Suppl.* (1960) **2**, 243.
2. El Sayed, M. A., *Acc. Chem. Res.* (1968) **1**, 8.
3. Crosby, G. A., Watts, R. J., Carstens, D. H. W., *Science* (1970) **170**, 1195.
4. Watts, R. J., Crosby, G. A., *J. Am. Chem. Soc.* (1971) **93**, 3184.
5. *Ibid.* (1972) **94**, 2606.
6. Carstens, D. H. W., Crosby, G. A., *J. Mol. Spectrosc.* (1970) **34**, 113.
7. Harrigan, R. W., Hager, G. D., Crosby, G. A., *Chem. Phys. Lett.* (1973) **21**, 487.
8. Harrigan, R. W., Crosby, G. A., *J. Chem. Phys.* (1973) **59**, 3468.
9. Hager, G. D., Crosby, G. A., *J. Am. Chem. Soc.* (1975) **97**.
10. Hager, G. D., Crosby, G. A., "Abstracts of Papers," 166th National Meeting, ACS, Aug. 1973, PHYS 083.
11. Pankuch, B. J., Crosby, G. A., unpublished data.
12. Watts, R. J., Crosby, G. A., unpublished data.
13. Lacky, D. E., Crosby, G. A., "Abstracts of Papers," 29th Annual Northwest Regional Meeting, ACS, June 1974.
14. Halper, W., DeArmond, M. K., *Chem. Phys. Lett.* (1974) **24**, 114.
15. Baker, D. C., Crosby, G. A., *Chem. Phys.* (1974) **4**, 428.
16. Demas, J. N., Crosby, G. A., *J. Am. Chem. Soc.* (1970) **92**, 7262.
17. Hipps, K. W., Crosby, G. A., *J. Am. Chem. Soc.* (1975) **97**.
18. Hipps, K. W., Hager, G. D., Crosby, G. A., "Abstracts of Papers," 166th National Meeting, ACS, Aug. 1973, PHYS 81.
19. Hager, G. D., Watts, R. J., Crosby, G. A., *J. Am. Chem. Soc.* (1975) **97**.
20. Fujita, I., Kobayashi, H., *Inorg. Chem.* (1973) **12**, 2758.
21. Condon, E. U., Shortley, G. H., "The Theory of Atomic Spectra," Cambridge University, Cambridge, England, 1957.
22. Dieke, G. H., "Spectra and Energy Levels of Rare Earths Ions in Crystals," Interscience, New York, 1968.
23. Crosby, G. A., Hipps, K. W., Elfring, Jr., W. H., *J. Am. Chem. Soc.* (1974) **96**, 629.
24. Hillis, J. E., DeArmond, M. K., *J. Lumin.* (1971) **4**, 273.
25. Ake, R. L., Gouterman, M., *Theor. Chim. Acta* (1969) **15**, 20.
26. Smith, B. E., Gouterman, M., *Chem. Phys. Lett.* (1968) **2**, 517.
27. Gouterman, M., Mathies, R. A., Smith, B. E., *J. Chem. Phys.* (1970) **52**, 3795.

28. Watts, R. J., Crosby, G. A., Sansregret, J. L., *Inorg. Chem.* (1972) **11**, 1474.
29. Watts, R. J., Crosby, G. A., *Chem. Phys. Lett.* (1972) **13**, 619.
30. Baker, D. C., Crosby, G. A., "Abstracts of PaPpers," 166th National Meeting, ACS, Aug. 1973, PHYS 82.
31. Hipps, K. W., unpublished data.
32. Ballardini, R., Varani, G., Maggi, L., Balzani, V., Olson, K. R., Scandola, F., Hoffman, M. Z., *J. Am. Chem. Soc.* (1975) **97**, 728.
33. Tokel-Takvoryan, N. E., Hemingway, R. E., Bard, A. J., *J. Am. Chem. Soc.* (1974) **95**, 6582.
34. Elfring, Jr., W. H., Crosby, G. A., "Abstracts of Papers," 29th Annual Northwest Regional Meeting, ACS, June 1974.

RECEIVED January 24, 1975. Work supported by grant AFOSR-72-2207 from the United States Air Force Office of Scientific Research to Washington State University.

14

The Use of Quenching and Sensitization Techniques for the Study of Excited-State Properties of Transition Metal Complexes

V. BALZANI, L. MOGGI, F. BOLLETTA, and M. F. MANFRIN

Istituto Chimico "G. Ciamician" dell'Università, Bologna, Italy

The chemistry of electronically excited states may be considered a new dimension in chemistry, and its systematic exploration is very promising both theoretically and practically. Quenching and sensitization techniques constitute a powerful tool for obtaining detailed information on excited-state behavior. Sensitization and quenching by electronic energy transfer, chemical reactions of excited states, and exciplex formation are discussed, and examples with transition metal complexes are cited. The roles of the various excited states of Cr(III) complexes, the factors affecting the electronic energy transfer, the reducing properties of $(^3CT)Ru(dipy)_3^{2+}$, and the exciplex formation between $(^3CT)cis\text{-}Ir(phen)_2Cl_2^+$ and naphthalene are discussed in detail.

It is well known that for each molecule there is the so-called ground state and there are many electronically excited states. In general, the ground state is the one that is responsible for the ordinary chemistry while the electronically excited states are those responsible for the photochemical reactions. Each electronically excited state is virtually a new molecule with respect to the corresponding ground state.

A classic example is formaldehyde (*1*) which in its ground state is a planar molecule with no unpaired electrons (singlet state), a double CO bond, and a dipole moment of 2.3 D. In its lowest excited state, this molecule has 76 kcal/mole more energy than the ground state, it is not planar but pyramidal, and it has two unpaired electrons (triplet state), an essentially single CO bond, and a dipole moment of 1.3 D. Obviously, the two states exhibit completely different reactivity.

Similar changes upon electronic excitation may occur with other molecules including transition metal complexes. For example, excitation of a square planar complex can lead to tetrahedral excited states (*2, 3, 4*) thereby offering a path for intramolecular photoisomerization (*5*). Drastic geometrical changes can also take place upon excitation of octahedral complexes (*6*) although precise data on such molecules are difficult to obtain (*7*). The different chemical reactivities of ground and excited states of transition metal complexes have been observed in a number of cases (*5, 8*). Some new, outstanding examples of this type of behavior are discussed in this paper.

Sensitization and Quenching Processes

The events of a photochemical process can be discussed on the basis of the simple example schematized in Figure 1. Light excitation of a

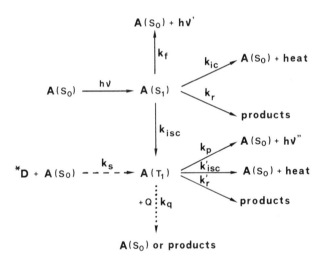

Figure 1. Schematic representation of a photochemical system including sensitization by energy transfer and quenching

ground state molecule, which is supposed to be a singlet state, leads to a spin-allowed excited state (S_1) which can deactivate by fluorescence, internal conversion to the ground state, chemical reaction, or intersystem crossing to a lower lying spin-forbidden excited state (T_1). T_1 in turn, can deactivate by phosphorescence, intersystem crossing to the ground state, or chemical reaction. As was mentioned above, each electronically excited state is characterized by its particular energy content, electronic configuration, spin value, shape, size, dipole moment, etc. These factors determine the values of the rate constants (k's) of the radiative and non-radiative deactivations as well as those of the chemical reactions (Figure 1). With direct photochemistry experiments, it is usually impossible to evaluate these rate constants, and in most cases it is even impossible to establish which excited state is responsible for the experimentally observed phenomenon. For example, when a photoproduct is obtained, it is difficult to establish whether it originated from the excited state obtained directly by irradiation, S_1, or from the spin-forbidden excited state populated *via* intersystem crossing, T_1.

Sensitization and quenching processes can make an important contribution in solving these problems (9, 10). For example, electronic energy transfer from a suitable sensitizer ($*D$ in Figure 1) to the molecule of interest can populate a specific excited state in a selective way. In particular, it is possible to populate those states which cannot be reached directly by irradiation, as $A(T_1)$ in Figure 1. The behavior of a specific state can thus be studied individually, and this of course helps greatly in understanding the roles of the various excited states of the molecule. On the other hand, the quenching of an excited state by known concentrations of a suitable quencher (Q in Figure 1) allows a quantitative evaluation of the rates of the competing unimolecular deactivation steps, and it can also be used to suppress completely the manifestations of a specific excited state so as to make easier the study of other processes.

In general, sensitization (by electronic energy transfer) and quenching processes can be used (9, 10) (a) to identify the reactive state of a molecule, (b) to increase the quantum yield of a process, (c) to increase the range of useful exciting wavelengths, (d) to quench undesired photoreactions and (e) to study the intimate mechanism of bimolecular processes. They can also provide information on excited-state lifetimes, intersystem crossing efficiencies, emission and reaction efficiencies, and association and ion-pairing constants. (For reviews on these topics, *see* Refs. 9 and 10.)

The bimolecular quenching of an excited-state molecule can occur by several distinct mechanisms (10), the most important of which are expected to be: (a) electronic energy transfer, (b) chemical reaction, (c) exciplex (or excimer) formation, (d) spin-catalyzed deactivation, and (e) external heavy atom effect. Whereas it is possible to demonstrate that a mechanism is important, it is generally very difficult, if not impossible, to establish whether a mechanism is completely ineffective. The most recent findings, however, reveal that mechanisms d and e listed above are normally not efficient enough to compete with the others. Therefore, we shall discuss only those examples that involve electronic energy transfer, chemical reaction, and exciplex formation.

Role of the Excited States in the Photochemistry of Cr(III) Complexes

One of the crucial points in the recent developments of coordination compound photochemistry has been the debate concerning the identity of the excited state(s) responsible for the photosolvation reactions that are obtained by irradiating Cr(III) complexes in their ligand field bands (5, 8). Direct photolysis experiments revealed that the most likely candidates (*see e.g.*, Figures 2 and 3) are the lowest spin-allowed excited state ($^4T_{2g}$ in octahedral symmetry) and the lowest spin-forbidden excited state (^2E_g). Such experiments, however, did not warrant any definite conclusion about the actual role of each of these two states (5).

The problem of the reactive state is strictly related to the other problem of the importance of the intersystem crossing ($^4T_{2g} \rightsquigarrow {}^2E_g$) and back intersystem crossing ($^2E_g \rightsquigarrow {}^4T_{2g}$) steps. Sensitization and quenching have given a determinant contribution to the solving of both these problems. There is much evidence, including the sensitization of the phosphorescence emission (*11, 12, 13, 14, 15*), that energy transfer is the most important (if not the unique) sensitization and quenching mechanism for these complexes.

The best studied case is probably that of $Cr(CN)_6^{3-}$. In dimethylformamide (DMF) solution at room temperature, the direct excitation of this complex to its lowest quartet excited state causes a photosolvation reaction and the phosphorescence emission (*16*). Sensitization with high energy donors (*e.g.*, triplet xanthone), which can populate both the $^4T_{2g}$ and the ^2E_g excited states by energy transfer, causes both the solvation reaction and the phosphorescence emission (*13, 14*). However, sensitization with low energy donors, such as $Ru(dipy)_3^{2+}$, which are only able to transfer energy to the ^2E_g state, causes the phosphorescence emission, but not the solvation reaction (*13, 14*). This demonstrates that the state responsible for the reaction must be the quartet. The same conclusion was reached by quenching experiments. It was found that the phosphorescence emission in DMF is strongly quenched by oxygen or water whereas the photoreaction is not quenched at all; therefore, it must originate from a state different from ^2E_g (*16*). By the energy transfer technique, it was also possible (*17*) to establish that the efficiency of the $^4T_{2g} \rightsquigarrow {}^2E_g$ intersystem crossing step of $Cr(CN)_6^{3-}$ is about 0.5. This finding, together with the other available data for $Cr(CN)_6^{3-}$ in DMF at 20°C, allows us to elucidate the role of the $^4T_{2g}$ and ^2E_g states in determining the photochemical and photophysical behavior of this complex (Figure 2). Of the $^4T_{2g}$ molecules, 10% undergo a permanent solvation reaction, practically none emits, 50% undergo intersystem crossing to the doublet, and the remaining 40% undergo deactivation to the ground state through an internal conversion step. This last 40% may include deactivation by a

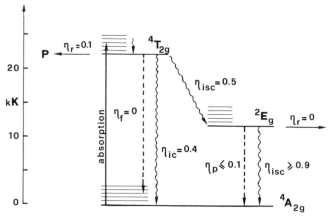

*Figure 2.   Excited-state behavior of $Cr(CN)_6{}^{3-}$ in DMF at
20°C*

reversible photodissociation which would not be possible in the solid
state where the intersystem crossing efficiency is in fact very near one
(*18*). Of the molecules which pass into the doublet state, practically
none reacts, less than 10% undergo a radiative transition (*i.e.* the observed phosphorescence), and none can give back intersystem crossing
to the excited quartet since the energy gap is too high (*19*); therefore,
the remaining molecules must be deactivated to the ground state through
an intersystem crossing step. In conclusion, for $Cr(CN)_6{}^{3-}$ it was possible
to identify the excited state responsible for the photochemical reaction
as well as to establish the relative importance of the various processes
which occur after light absorption.

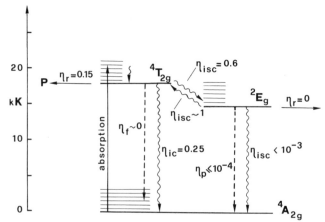

*Figure 3.   Excited-state behavior of $Cr(en)_3{}^{3+}$ in water at
20°C*

For other Cr(III) complexes, however, neither sensitization nor quenching can fully discriminate between the $^4T_{2g}$ and ^2E_g excited states because these are very close in energy and therefore back intersystem crossing from ^2E_g to $^4T_{2g}$ can also occur. The data obtained from quenching experiments, however, revealed (*20, 21, 22, 23, 24*) that the photoreaction is less quenched than the phosphorescence emission. This indicates that at least part of the photoreaction originates directly from the quartet state.

Detailed analysis of the quenching data and comparison with the data obtained from direct photolysis and luminescence experiments have led to the conclusions about Cr(en)$_3$$^{3+}$ (*23*) that are summarized in Figure 3. For this complex, 15% of the molecules that are excited to the quartet state undergo a prompt reaction, practically none emits, 25% undergo internal conversion to the ground state, and 60% pass into the doublet state. Since the energy separation between the two states is small, at room temperature the back intersystem crossing step is much faster than each one of the other possible deactivation paths of the doublet. Therefore, under these conditions, the doublet state plays the role of a parking place in which the excited molecules spend some time before coming back to the quartet where they react, deactivate, or pass again into the doublet state, as we have seen above. Thus, in this complex, there is a delayed as well as a prompt photoreaction. At low temperature, of course, the situation changes since the back intersystem crossing from ^2E_g to $^4T_{2g}$, which needs an activation energy of about 8 kcal/mole, slows down and the most important depopulation paths for the doublet are then the phosphorescent emission and the intersystem crossing deactivation to the ground state.

Results qualitatively similar to those described above for Cr(en)$_3$$^{3+}$ were also obtained for *trans*-Cr(NCS)$_4$(NH$_3$)$_2$$^-$ [which was actually the first case investigated (*20*)], Cr(NH$_3$)$_6$$^{3+}$ (*21*), and Cr(phen)$_3$$^{3+}$ (*22*).

Factors Affecting the Energy Transfer Efficiency

In a quenching experiment, one can measure the experimental bimolecular quenching constant k_q of the excited state *D by A. If it is assumed that the energy transfer is the only bimolecular quenching mode, then the kinetic scheme is simply:

$$*D + A \xrightarrow{k_q} D + *A \qquad (1)$$

If we consider this process in more detail, we must subdivide it into at least two steps (Reaction 2), *i.e.*, the diffusion together of the excited state

$$*D + A \underset{k_{-d}}{\overset{k_d}{\rightleftarrows}} [*D \ . \ . \ A] \overset{k_{et}}{\longrightarrow} D + *A \tag{2}$$

and the quencher to give the so-called encounter complex, and then the true energy transfer step which competes with the diffusion away of the two molecules. Thus, we can define the energy transfer efficiency, α_{et}, that is the fraction of encounters which results in the energy transfer from $*D$ to A, as

$$k_q = k_d \left(\frac{k_{et}}{k_{-d} + k_{et}} \right) = k_d \alpha_{et} \tag{3}$$

$$\alpha_{et} = \frac{k_q}{k_d} \tag{4}$$

In dealing with organic molecules in fluid solution, it is usually found (9) that, when energy and spin requirements are satisfied, the energy transfer process is diffusion controlled or nearly so, i.e., the energy transfer efficiency equals or is near one. This, however, is not generally true when transition metal complexes are involved (see Table I). The

Table I.   Selected Examples of Energy Transfer Efficiency[a]

Donor	Acceptor	k_q,[b] M^{-1} sec^{-1}	k_d,[b] M^{-1} sec^{-1}	$\alpha = \dfrac{k_q}{k_d}$[c]
$Ru(dipy)_3^{2+}$	$Mo(CN)_8^{4-}$	9×10^9	9×10^9	1
$Ru(dipy)_2(CN)_2$	O_2	5×10^9	7×10^9	0.7
Acridine	$Ni(NH_3)_6^{2+}$	3×10^8	7×10^9	0.04
$Ru(dipy)_3^{2+}$	$Ni(gly)_2$	4×10^7	7×10^9	0.006
Acridine	$Ni(H_2O)_6^{2+}$	3×10^7	7×10^9	0.004
$Ru(dipy)_3^{2+}$	$t\text{-}Cr(en)_2F_2^+$	3×10^6	3×10^9	0.001
$Ru(dipy)_3^{2+}$	$Cr(en)_3^{3+}$	$\leq 1 \times 10^6$	4×10^8	≤ 0.002
$Cr(en)_3^{3+}$	$Co(H_2O)_6^{2+}$	2×10^5	$> 4 \times 10^8$	< 0.0005

[a] Aqueous solutions at $\sim 20°C$. In each case, spin and energy requirements for electronic energy transfer are satisfied. For some of these systems, experimental evidence indicates that evergy transfer is the most important (if not the unique) quenching mechanism (10).
[b] From Ref. 10.
[c] Since the bimolecular quenching constant k_q is an upper limiting value for the rate constant that should be used for calculating the energy transfer efficiency (10), the α values are upper limiting values for α_{et}.

energy transfer efficiency can be as low as 10^{-3} when one of the species involved is a transition metal complex, and even lower when both the donor and the acceptor are transition metal complexes. Evidently, in these cases there are factors involved which reduce the quenching efficiency during the encounter. These factors can be (a) the nature of the ligands, (b) the nature of the metal, (c) the ionic charge, (d) the orbital nature of the excited states involved, (e) the geometry of the complex, (f) the coordination number, and (g) the nature of the solvent. Unfortunately, it is impossible to assess the role of each one of these factors because of the few available data.

A systematic investigation (15) of the quenching of excited $Ru(dipy)_3^{2+}$ by *cis*- and *trans*-$Cr(en)_2XY^+$ complexes revealed that the nature of the ligands is an important factor indeed. The quenching ability increases in the ligand series $F^- < Cl^- < NCS^- < Br^-$, which coincides neither with the nephelauxetic series (25) nor with the reduction ability of the free ligands (26). The findings also demonstrated that the quenching ability of the cis isomers is greater than that of the corresponding trans isomers, which indicates that geometric factors play some role in determining the energy transfer efficiency. The quenching of organic triplets by Ni and Pd chelates (27, 28) revealed that the energy transfer is more efficient when it can lead to ligand-centered rather than to metal-centered excited states, and that, when the latter type of excited state is involved, the shielding of the metal by a larger number of ligands (*e.g.*, octahedral vs. tetrahedral coordination) decreases the quenching ability of the complex. Many more systematic investigations are necessary in order to assess the roles of the various factors.

Chemical Reactions of the Excited States

Chemical reaction is an important quenching mechanism of electronically excited states. Because of the short lifetime (generally less than 1 μsec) of excited states in fluid solution at room temperature, quenching by chemical reaction must be very fast if it is to occur. We shall consider here only outer-sphere electron transfer reactions of excited states since these reactions are certainly fast enough to compete with the other deactivation modes.

First of all, it should be recalled that electronic excitation lowers the ionization potential and increases the electron affinity of a molecule so that each excited state is potentially both a better reductant and a better oxidant than the ground state molecule. The actual occurrence of reduction or oxidation reactions of an excited state is obviously related to favorable kinetic factors which are expected to depend strongly on the orbital nature of the excited state (*vide infra*).

It should also be noted that it is difficult to demonstrate conclusively an electron transfer quenching mechanism. In several cases, both energy and electron transfer are allowed, and both processes may lead to the same final products. A classical example is the quenching of the (^3CT)Ru-$(dipy)_3^{2+}$ excited state by Fe^{3+} ions (Figure 4). (The validity of the orbital (29) and spin (30) labels of the excited states of metal chelates of this type has recently been questioned.) In continuous irradiation experiments (31), the quenching of $(^3CT)Ru(dipy)_3^{2+}$ occurred without any permanent chemical change in the system, so that one cannot say whether the quenching occurs *via* electron or energy transfer. In flash photolysis experiments (32), the formation and the subsequent decay of $Ru(dipy)_3^{3+}$ were observed, which indicates that the process involves an electron transfer mechanism. It may be interesting to note that in principle there is another deactivation path which involves first the energy transfer from $(^3CT)Ru(dipy)_3^{2+}$ to Fe^{3+} and then an electron transfer reaction between $*Fe^{3+}$ and $Ru(dipy)_3^{2+}$ (this reaction is not included in Figure 4) that is followed by the back electron transfer reaction. Although in this specific case the lifetime of $*Fe^{3+}$ is too short and the concentration of $Ru(dipy)_3^{2+}$ too small [less than $10^{-5}M$ (32)] to make this process important, the possibility of such a deactivation path should be considered in other cases.

In order to demonstrate unequivocally the occurrence of electron transfer from $(^3CT)Ru(dipy)_3^{2+}$, a quencher must be chosen to which energy transfer is not allowed. Tl_{aq}^{3+} ions satisfy this criterion since their lowest excited state is much higher than ~ 50 kcal/mole, which is the energy content of the excited ruthenium complex. The study of the $Ru(dipy)_3^{2+}$–Tl_{aq}^{3+} system has revealed (31) that Reaction 5 is about 10^9

$$(^3CT)Ru(dipy)_3^{2+} + Tl_{aq}^{3+} \xrightarrow{k'} Ru(dipy)_3^{3+} + Tl_{aq}^{2+} \qquad (5)$$

$$k' = 1 \times 10^{+8}M^{-1}sec^{-1}$$

times faster than the analogous reaction (Reaction 6) involving the ground state Ru complex. This is a clear demonstration that the chemical

$$Ru(dipy)_3^{2+} + Tl_{aq}^{3+} \xrightarrow{k} Ru(dipy)_3^{3+} + Tl_{aq}^{2+} \qquad (6)$$

$$k = 2 \times 10^{-1}M^{-1}sec^{-1}$$

properties of the excited states can be profoundly different from those of the corresponding ground state molecules.

The extraordinary reducing ability of $(^3CT)Ru(dipy)_3^{2+}$, which was first discussed by Gafney and Adamson (33), has also been very well documented by Bock et al. (32) and by Navon and Sutin (34). A very recent work (35) also demonstrated that as much as 97% of the spectroscopically estimated excitation energy of this molecule can be used in energy conversion by electron transfer. This is in agreement with the fact that the $Ru(dipy)_3^{2+}$ excited state is metal-to-ligand charge transfer

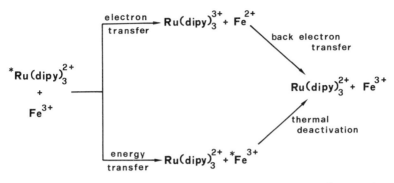

Figure 4. Paths of electron transfer and energy transfer deactivation of (^3CT)Ru(dipy)$_3^{2+}$ by Fe^{3+}

in nature, which is expected to favor the potential electron donor more than the potential electron acceptor properties of the excited state. The reverse should be true for ligand-to-metal charge transfer excited states (*see*, for example, Refs. *36* and *37*), but there has not yet been a systematic investigation of this matter. Similarly, the redox properties of metal-centered excited states have scarcely been investigated (*24, 38*). In conclusion, it seems evident that the study of electron transfer processes involving the excited states of transition metal complexes opens new perspectives for inorganic chemistry.

Exciplexes of Transition Metal Complexes

When two molecules in their ground state approach each other, no ground state complex can be obtained if the corresponding potential energy curve does not have any minimum. However, it may happen that, when one of the two molecules is excited, the approach between the excited molecule and the other molecule in its ground state leads to a minimum in the potential energy curve. In such a case, an exciplex is formed (Reactions 7 and 8). The exciplex formation results in the

$$A + h\nu \longrightarrow {}^*A \tag{7}$$

$$\begin{array}{c} {}^*A + Q \rightleftarrows {}^*(AQ) \\ \downarrow \\ A + h\nu' \end{array} \begin{array}{c} A + Q + h\nu'' \\ \longrightarrow \text{products} \\ A + Q + \text{heat} \end{array} \tag{8}$$

quenching of the originally excited molecule since the exciplex is subjected to its own deactivation processes, which can include luminescence and chemical reaction. The best proof of exciplex formation is a broad, structureless luminescence emission that is red-shifted with respect to the emission of the original excited state (39). In the absence of luminescence, exciplex formation is very difficult to demonstrate.

Exciplex formation is a rather common phenomenon with organic molecules (39). With transition metal complexes, there is only one well documented case, i.e. the exciplex formed by (^{3}CT)cis-Ir(phen)$_2$Cl$_2^+$ with ground state naphthalene (40). The enthalpy of association is about −4 kcal/mole. The exciplex lifetime is more than 1 μsec in DMF at 20°C, i.e. longer than that of (^{3}CT)cis-Ir(phen)$_2$Cl$_2^+$ ($\sim$ 0.3 μsec). The structure and the type of bonding of the exciplex are not known. However, since the excited state of the complex is iridium-to-phenanthroline charge transfer in nature, with a consequent increase in the electron density on the phenanthroline ligands, a charge transfer interaction between a phenanthroline ligand and naphthalene seems likely.

The different behavior of ground and excited cis-Ir(phen)$_2$Cl$_2^+$ toward naphthalene demonstrates once again that the ground and excited states of a molecule have different chemical properties. Since preliminary experiments (40) have revealed that exciplexes can also be obtained with other transition metal complexes, it seems easy to forecast that exciplex formation will be an important research topic in the next few years.

Conclusion

In this review, we have tried to demonstrate that quenching and sensitization are very powerful techniques for evidencing unsual chemical properties of the excited states of transition metal complexes. Generally speaking, each electronically excited state must be considered as a new chemical species with respect to the corresponding ground state molecule. This means that photochemistry offers a new dimension to chemistry (41), the excited-state dimension, which, although mostly unexplored, is very promising for the progress of chemical research.

Literature Cited

1. Turro, N. J., Dalton, J. C., Dawes, K., Farrington, G., Hautala, R., Morton, D., Niemczyk, M., Schore, N., Acc. Chem. Res. (1972) 5, 92.
2. Ballhausen, C. J., Bjerrum, N., Dingle, R., Eriks, K., Hare, C. R., Inorg. Chem. (1965) 4, 514.
3. Martin, D. S., Jr., Tucker, M. A., Kassman, A. J., Inorg. Chem. (1965) 4, 1682.
4. Ibid. (1966) 5, 1298.
5. Balzani, V., Carassiti, V., "Photochemistry of Coordination Compounds," Academic, London, 1970.
6. Pearson, R. G., Chem. Phys. Lett. (1971) 10, 31.
7. Hipps, K. W., Crosby, G. A., Inorg. Chem. (1974) 13, 1543.
8. Fleischauer, P. D., Adamson, A. W., Sartori, G., Prog. Inorg. Chem. (1972) 17, 1.

9. Lamola, A. A., Turro, N. J., "Energy Transfer and Organic Photochemistry," Interscience, New York, 1969.
10. Balzani, V., Moggi, L., Manfrin, M. F., Bolletta, F., Laurence, G. S., *Coord. Chem. Rev.* (1975) **15**, 321.
11. Binet, D. J., Goldberg, E. L., Forster, L. S., *J. Phys. Chem.* (1968) **72**, 3017.
12. Chen, S. N., Porter, G. B., *J. Am. Chem. Soc.* (1970) **92**, 3196.
13. Sabbatini, N., Balzani, V., *J. Am. Chem. Soc.* (1972) **94**, 7587.
14. Sabbatini, N., Scandola, M. A., Carassiti, V., *J. Phys. Chem.* (1973) **77**, 1307.
15. Bolletta, F., Maestri, M., Moggi, L., Balzani, V., *J. Am. Chem. Soc.* (1973) **95**, 7864.
16. Wasgestian, H. F., *J. Phys. Chem.* (1972) **76**, 1947.
17. Sabbatini, N., Scandola, M. A., Balzani, V., *J. Phys. Chem.* (1974) **78**, 541.
18. Castelli, F., Forster, L. S., *J. Phys. Chem.* (1974) **78**, 2122.
19. Diomedi Camassei, F., Forster, L. S., *J. Chem. Phys.* (1969) **50**, 2603.
20. Chen, S. N., Porter, G. B., *Chem. Phys. Lett.* (1970) **6**, 41.
21. Langford, C. H., Tipping, L., *Can. J. Chem.* (1972) **50**, 887.
22. Kane-Maguire, N. A. P., Langford, C. H., *J. Am. Chem. Soc.* (1972) **94**, 2125.
23. Ballardini, R., Varani, G., Wasgestian, H. F., Moggi, L., Balzani, V., *J. Phys. Chem.* (1973) **77**, 2947.
24. Maestri, M., unpublished data.
25. Cotton, F. A., Wilkinson, G., "Advanced Inorganic Chemistry," 3rd ed., Interscience, New York, 1972.
26. Latimer, W. M., "Oxidation Potentials," 2nd ed., Prentice-Hall, Englewood Cliffs, N. J., 1962.
27. Adamczyk, A., Wilkinson, F., *J. Chem. Soc. Faraday Trans. 2* (1972) **68**, 2031.
28. Allsopp, S. R., Wilkinson, F., *Chem. Phys. Lett.* (1973) **19**, 535.
29. DeArmond, M. K., Hillis, J. E., *J. Chem. Phys.* (1971) **54**, 2247.
30. Crosby, G. A., Hipps, K. W., Elfring, W. H., Jr., *J. Am. Chem. Soc.* (1974) **96**, 629.
31. Laurence, G. S., Balzani, V., *Inorg. Chem.* (1974) **13**, 2976.
32. Bock, C. R., Meyer, T. J., Whitten, D. G., *J. Am. Chem. Soc.* (1974) **96**, 4710.
33. Gafney, H., Adamson, A. W., *J. Am. Chem. Soc.* (1972) **94**, 8238.
34. Navon, G., Sutin, N., *Inorg. Chem.* (1974) **13**, 2159.
35. Bock, C. R., Meyer, T. J., Whitten, D. G., *J. Am. Chem. Soc.* (1975) **97**, 2909.
36. Symons, M. C. R., West, D. X., Wilkinson, J. G., *J. Phys. Chem.* (1974) **78**, 1335.
37. Balzani, V., Carassiti, V., *J. Phys. Chem.* (1968) **72**, 383.
38. Kane-Maguire, N. A. P., Langford, C. H., *J. Chem. Soc. Chem. Commun.* (1973) 351.
39. Stevens, B., *Adv. Photochem.* (1971) **8**, 161.
40. Ballardini, R., Varani, G., Moggi, L., Balzani, V., *J. Am. Chem. Soc.* (1974) **96**, 7123.
41. Quinkert, G., *Angew. Chem. Int. Ed. Engl.* (1972) **11**, 1072.

RECEIVED January 24, 1975. Work supported by the Italian National Research Council.

15

Environmental Effects on Intra- and Intermolecular Photophysical Processes in Cr^{3+} Complexes

LESLIE S. FORSTER

University of Arizona, Tucson, Ariz. 85721

The effect of environment on the nonradiative processes $^4T_2 \rightsquigarrow ^4A_2$, $^4T_2 \rightsquigarrow ^2E$, and $^2E \rightsquigarrow ^4A_2$ is reviewed. The rate constants for $^2E \rightsquigarrow ^4A_2$ are insensitive to environment more often at low temperatures than at room temperature. The relatively few data on the $^4T_2 \rightsquigarrow ^4A_2$ relaxation rates indicate a value of $\leq 10^5$ sec^{-1} for most complexes, but notable exceptions have been found. Some evidence indicates a variation in intersystem crossing ($^4T_2 \rightsquigarrow ^2E$) efficiency with environment. Concentration quenching by energy transfer, where important, affects the ^2E state rather than the $^4T_2$ state. In general, the environmental variations in Φ_p occur mainly after the complexes have reached the ^2E state. The need to collect photophysical and photochemical data under identical conditions is emphasized.

With the exception of molecules in the gas phase at low pressures, the properties of any species are mediated to some degree by the environment. In particular, the photophysical processes within a transition metal ion complex are often markedly dependent on the surroundings in which the complex is embedded. Interest in transition metal ion photophysics has been high not only for the intrinsic importance (*e.g.*, phosphors and lasers) but also because photochemistry and photophysics are intimately interrelated. It is this connection between photochemistry and photophysics that will be emphasized in this discussion.

The most striking contrast between solids, either glassy or crystalline, and fluids is the inhibition of diffusional processes in rigid media. However, photophysical processes are sensitive to nondiffusional perturbations as well, and the effect of environmental factors in this latter category is the subject of this review. With few exceptions, photochemistry has been studied in fluid media, while most photophysical measurements have been made when the complex is in a solid environment. Three types of crystalline environments are discussed: (a) undiluted crystals, (b) the guest species diluted in an isostructural host, and (c) double salts. The noncrystalline solid hosts include alcohol–water glasses and a plastic, poly(methyl methacrylate).

It is useful to distinguish ionic complexes (*e.g.*, $Cr^{3+}:Al_2O_3$) from molecular complexes (*e.g.*, $Cr(NH_3)_6^{3+}$) which persist in fluids as well as in solids. In the molecular complexes, the ligands are coupled much more strongly to the central metal ion than to the surroundings, and it is meaningful to treat the system as a molecule with the environment acting as a perturbant on intramolecular processes. Moreover, in crystals of molecular complexes, the bulky ligands prevent the close approach of metal ions that is possible with ionic complexes. Consequently, intermolecular exci-

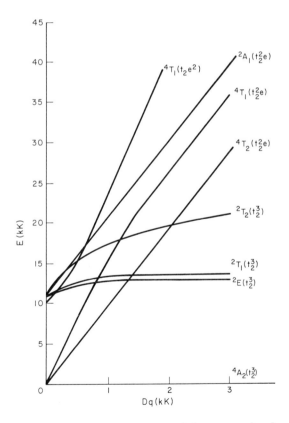

Figure 1. The variation in Cr^{3+} energy levels with ligand field (adapted from Ref. 49)

tation energy transfer may be less efficient in solids containing molecular complexes.

Because of the extensive literature on the photochemistry and photophysics of hexacoordinated Cr^{3+} complexes $(1, 2, 3, 4, 5)$, this group is very suitable for illustrating the effect of environment on the rates of intra- and intermolecular processes. The $^4T_2$ energy, relative to the ground $^4A_2$ state, is sensitive to Dq (Figure 1), and consequently the position of the $^4T_2 \leftarrow {}^4A_2$ absorption band varies markedly with the ligand. The ^2E energy is, in the crystal field approximation, relatively independent of ligand (6), but the ^2E energies actually span a range of ~ 3000 cm^{-1} (4).

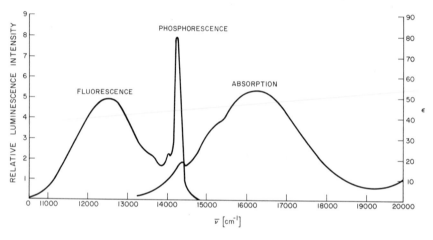

Figure 2. *The absorption and emission spectra of* $Cr(urea)_6^{3+}$ *(adapted from Ref. 50)*

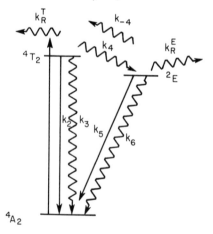

Figure 3. *Radiative* $(\rightarrow)$ *and non-radiative* $(\leadsto)$ *relaxation processes*

The differential shifting of $^4T_2$ and ^2E with ligand means that $^4T_2$ can be above or below ^2E. The absorption and emission spectra of $Cr(urea)_6^{3+}$ in Figure 2 are illustrative of the case where $^4T_2$ and ^2E are nearly equi-energetic. Although there are some energy variations from one environment to another, the effect of environment on the $^4T_2 \leftarrow {}^4A_2$ absorption and the $^2E \rightarrow {}^4A_2$ emission energies of a given molecular complex is relatively small ($\leq 500 \text{ cm}^{-1}$).

In contrast, radiative and especially nonradiative transition probabilities (Figure 3) can be very sensitive to environment. The evaluation of k_2–k_6 as a function of temperature and environment is the ultimate purpose of photophysical studies. The directly measurable quantities are the luminescence quantum yields (Φ) and lifetimes (τ) for fluorescence ($^4T_2 \rightarrow {}^4A_2$) and for phosphorescence ($^2E \rightarrow {}^4A_2$).

In the absence of thermally activated $^2E \rightsquigarrow {}^4T_2$ back transfer, the pertinent equations are:

$$\Phi_F = \frac{k_2}{k_2 + k_3 + k_4 + k_R^T} \tag{1}$$

$$\tau_F^{-1} = k_2 + k_3 + k_4 + k_R^T \tag{2}$$

$$\Phi_p = \frac{k_4}{k_2 + k_3 + k_4 + k_R^T} \cdot \frac{k_5}{k_5 + k_6 + k_R^E} \tag{3}$$

$$= \Phi_{2E} \frac{k_5}{k_5 + k_6 + k_R^E}$$

$$\tau_p^{-1} = k_5 + k_6 + k_R^E \tag{4}$$

$$\Phi_p/\tau_p = \Phi_{2E} k_5 \tag{5}$$

When back transfer is significant, it is necessary to distinguish the prompt fluorescence described by Equations 1 and 2 from delayed fluorescence which will exhibit the same lifetime as phosphorescence.

The inclusion of back transfer leads to rather complicated general expressions, but two limiting cases are important. The modified equations corresponding to these limits (7, 8) are (a) for the steady state limit $[4k_4k_{-4} \ll (k_{-4} + k_5 + k_6 + k_R^E - k_2 - k_3 - k_4 - k_R^T)^2]$:

$$\Phi_p = \frac{k_4 k_5}{(k_2 + k_3 + k_4 + k_R^T)(k_5 + k_6 + k_{-4} + k_R^E) - k_4 k_{-4}} \tag{3'}$$

$$\tau_p^{-1} = k_5 + k_6 + (1 - \Phi_{2E})k_{-4} \tag{4'}$$

$$\Phi_p/\tau_p = \Phi_{2E} k_5 \tag{5'}$$

and (b) for the equilibrium limit ($[^4T_2]/[^2E] = 3e^{-\Delta E/kT}$, where ΔE is the energy difference between the no-phonon levels of $^4T_2$ and ^2E):

$$\Phi_p = \frac{k_5}{(k_5 + k_6 + k_R{}^E) + (k_2 + k_3 + k_R{}^T)\,(3e^{-\Delta E/kT})} \qquad (3'')$$

$$\tau_p{}^{-1} = \frac{k_5 + k_6 + k_R{}^E + (k_2 + k_3 + k_R{}^T)\,(3e^{-\Delta E/kT})}{1 + 3e^{-\Delta E/kT}} \qquad (4'')$$

$$\Phi_p/\tau_p = \frac{k_5}{(1 + 3e^{-\Delta E/kT})} \qquad (5'')$$

Unless $\Delta E < 1000$ cm^{-1}, in the equilibrium limit, $\Phi_p/\tau_p \simeq k_5$ at ambient temperatures. Since the equilibrium limit can apply only when $\Phi_{2E} \simeq 1$, i.e. when $k_2 + k_3 + k_R{}^T \ll k_4$, at moderate or large ΔE Equation 5″ will lead to essentially the same temperature dependence as Equation 5′ if Φ_{2E} is constant. In principle, determination of $\Phi_p/\tau_p k_5$ as a function of temperature can be used to evaluate thermally induced changes in Φ_{2E}, but k_5 is not necessarily temperature-invariant. When $\Phi_p\tau_p$ is constant over a wide range of temperature, it is likely that both Φ_{2E} and k_5 are also constant.

In the following discussion, the effect of environment on the intramolecular relaxation rates k_3, k_6, and k_4 in Cr^{3+} complexes is reviewed. In addition, the variation in intermolecular transfer efficiency with environment is also discussed.

k$_6$

With few exceptions (e.g., Cr(urea)$_6{}^{3+}$ and Cr(H$_2$O)$_6{}^{3+}$), τ_p reaches the low temperature-limiting value ($\tau_p{}^\circ$) at temperatures above 77°K. In Table I are listed a representative selection of low temperature lifetimes. Since k_5 is seldom known accurately, it is necessary to estimate this quantity in order to evaluate k_6. For the symmetry-allowed $^2E \rightarrow {}^4A_2$ transition in Cr^{3+}:Al$_2$O$_3$, $k_5 = 260$ sec^{-1} (19) and $\tau_p{}^\circ$ in Cr(ox)$_3{}^{3-}$ is $\sim 10^{-3}$ sec (Table I). Consequently, $k_5{}^\circ \sim 10^2$–10^3 sec^{-1} in noncentrosymmetric complexes. In contrast, for symmetry-forbidden transitions in centrosymmetric complexes (e.g., Cr(CN)$_6{}^{3-}$), $k_5{}^\circ \simeq 10$ sec^{-1}. These values vary with the extent of vibronic coupling but the $k_6{}^\circ$ quantities in Table I were evaluated by assuming $k_5{}^\circ = 10^3$ and 10 sec^{-1} for noncentrosymmetric and centrosymmetric complexes, respectively.

In some complexes (viz., dilute solid solutions of Cr(acac)$_3$) $k_6{}^\circ$ is quite insensitive to environment, while in other complexes (e.g., Cr(D$_2$O)$_6{}^{3+}$ and Cr(CN)$_6{}^{3-}$) the lattice is the dominant influence (Table I). In other complexes (e.g., Cr(ox)$_3{}^{3-}$) $k_6{}^\circ$ may be relatively constant in several hosts yet increase markedly in others. It is significant that with

Table I.   Low Temperature Limiting $\tau_P{}^\circ$ and $k_6{}^\circ$

Complex[a]	Host[a]	$\tau_P{}^\circ$, μ sec	$k_6{}^\circ$, sec^{-1}	Ref.
$Cr(acac)_3$	$Al(acac)_3$	475	2,000	9
	$Cr(acac)_3$	115	8,600	9
		17	58,700	10
	alcohol–water glass	400	2,400	9
	poly(methyl methacrylate) plastic	460	2,100	11
$Cr(ox)_3{}^{3-}$	$NaMgAl(ox)_3 \cdot 9H_2O$	910	1,000	12
	$NaMgCr(ox)_3 \cdot 9H_2O$	960	1,000	13
	alcohol–water glass	900	1,000	12
	$K_3Al(ox)_3 \cdot 3H_2O$	100	9,900	14
	$K_3Cr(ox)_3 \cdot 3H_2O$	180	5,500	10
$Cr(D_2O)_6{}^{3+}$	$C(NH_2)_3Al(SO_4)_2 \cdot 6D_2O$	1,500	650	12
	$AlCl_3 \cdot 6D_2O$	350	2,850	12
	$KAl(SO_4)_2 \cdot 12D_2O$	750	1,300	15
	alcohol–water glass	130	7,700	12
$Cr(CN)_6{}^{3-}$	$K_3Co(CN)_6$	120,000	∼0	12
	$K_3Cr(CN)_6$	10	100,000	16
	$[Cr(en)_3][Cr(CN)_6]$	62	16,000	16
	$[Cr(NH_3)_6][Cr(CN)_6]$	3	330,000	16
	alcohol–water glass	3,300	300	17
$Cr(urea)_6{}^{3+}$	$[Cr(urea)_6]I_3$	160	6,200	18
	$[Cr(urea)_6]Br_3$	130	7,600	18
	$[Cr(urea)_6]Cl_3$	240	4,100	18
	$[Cr(urea)_6](ClO_4)_3$	100	10,000	18
	$[Cr(urea)_6](NO_3)_3$	240	4,100	18
$Cr(en)_3{}^{3+}$	$[Cr(en)_3](ClO_4)_3$	53	18,800	16
	$[Cr(en)_3]Cl_3$	40	25,000	16
	$[Cr(en)_3]Br_3 \cdot 4H_2O$	26	38,400	16
	$[Cr(en)_3]I_3$	80	12,400	16
	alcohol–water glass	100	10,000	17

[a] Acac: acetylacetonate; en: ethylenediamine; ox: oxalate.

one exception ($Cr(CN)_6{}^{3-}$ in $K_3Co(CN)_6$) $k_6{}^\circ \gg k_5{}^\circ$ for molecular complexes. Any reduction in Φ_p below one in these systems results mainly from nonradiative reactivation of the ^2E state.

A rather different type of environmental effect was detected in glassy solutions of $Cr(CN)_6{}^{3-}$ where excitation on the red edge of the $^4T_2 \leftarrow {}^4A_2$ absorption band leads to a multiexponential decay (20). The red-edge species are shorter lived.

The invariance of $k_6{}^\circ$ to environment does not imply a similar constancy of k_6 at higher temperatures. The validity of this assertion is clearly evident in Figure 4 where the thermal quenching of the $Cr(acac)_3$ τ_p begins at a lower temperature in a plastic host than in $Cr^{3+}:Al(acac)_3$ (11).

With $Cr(acac)_3$, both high viscosity and low temperature are necessary in order to reach k_6° (9). The microviscosity in plastics is not directly related to the macroscopic rigidity (21) and the microviscosity of poly-(methyl methacrylate) may decrease with temperature. Sometimes an apparently minor change—e.g., the removal of one water molecule from $Cr^{3+}:NaMgAl(ox)_3 \cdot 9 H_2O$—induces a large change in k_6 (12). An analogous situation prevails in other complexes, e.g., $Cr(en)_3^{3+}$ and $Cr(NH_3)_6^{3+}$ (16). These findings do not encourage the intermixing of photophysical and photochemical data recorded under different conditions of temperature and environment.

k_3

In contrast to k_6, relatively little data exist for k_3. This dearth can be attributed directly to the absence of prompt fluorescence when $^4T_2$ is above ^2E, the most common disposition of the energy levels in Cr^{3+} complexes (22). However, in those few systems where $^4T_2$ is below ^2E, intense fluorescence with $\tau_F \simeq 10^{-5}$ sec (23, 24, 25, 26) is observed at low temperatures. Since this value is close to the natural lifetime of $^4T_2 \to {}^4A_2$, $k_3^\circ \lesssim k_2$ and consequently k_3° is not much larger than 10^5 sec^{-1} in these cases.

Even though prompt fluorescence is not observed when $^4T_2$ is above ^2E, (20, 24, 27) τ_F can still be determined by monitoring the risetime of the $^2E \to {}^4A_2$ emission. In the systems studied so far—$Cr^{3+}:Al_2O_3$, $Cr(acac)_3$, $Cr(CN)_6^{3-}$, $Cr(en)_3^{3+}$, and $Cr(ox)_3^{3-}$—only an upper limit (10^{-9} sec) can be estimated for τ_F from risetime measurements (23, 24, 28). Since $k_4 \gg k_3$ in these complexes (vide infra), this τ_F limit is not helpful in estimating k_3.

The central question is: how does k_3 vary with Dq? Even though the $^4T_2$–$^4A_2$ separation is proportional to Dq, an energy gap dependence analogous to that encountered in aromatic compounds (7) cannot be assumed. The equilibrium geometries in the ground and excited states differ little in the aromatics and correspond to the so-called weak coupling limit in the theory of nonradiative transitions (29). In contrast, the potential surfaces of $^4T_2$ and $^4A_2$ are displaced and a different situation prevails. A change in Dq of some 100–200 cm^{-1} is often sufficient to invert the order of ^2E and $^4T_2$. It is reasonable to suppose that k_3 would not be markedly altered by such a small change in Dq. Indeed, delayed fluorescence is observed in the weak field complexes $Cr(D_2O)_6^{3+}$ (12), $Cr(urea)_6^{3+}$ (27, 30), and $Cr^{3+}:MgO$ (24) which indicates that k_3 is still $\leq 10^5$ sec^{-1}.

Despite the difficulties in determining k_3 quantitatively, changes in k_3 with environment have been detected. Evidence for the effect of

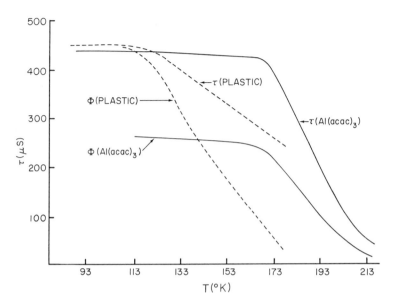

Figure 4. Φ_p *(relative) and* τ_p *for* $Cr(acac)_3$ *in* $Al(acac)_3$ (——) *and in poly(methyl methacrylate)* (– – – –) *as a function of temperature. Only the variation in* Φ_p *in a given host is meaningful.*

environment on k_3 is found in the derivatives of $Cr(acac)_3$. For $Cr(3$-bromoacac$)_3$ dissolved in $Al(3$-bromoacac$)_3$, Φ_p is reduced 100-fold when $^4T_2$ is excited rather than higher electronic states (*9*). Apparently, in this host $k_3 \gg k_4$, but intersystem crossing to the doublet manifold is still very efficient from higher energy states. No comparable variation is found for glassy solutions of the complex. When $Cr(acac)_3$ is dissolved in a crystalline host, Φ_p/τ_p is independent of temperature (which is consistent with Equation 5), but Φ_p/τ_p decreases with temperature when the host is a plastic (Figure 4). A similar decrease in Φ_p/τ_p is observed when $Cr(acac)_3$ is dissolved in absolute ethanol, but the effect is not significant until the temperature reaches $\sim 150°K$ (*28*). Since $k_3 + k_4 > 10^9$ sec^{-1}, these results indicate an enhanced k_3 ($> 10^9$ sec^{-1}) in the noncrystalline media at higher temperatures.

Φ_{2E}

A quantity of considerable importance for photochemical considerations is the intersystem crossing efficiency (Equation 6). The variation

$$\Phi_{2E} = \frac{k_4}{k_2 + k_3 + k_4 + k_R{}^T} \qquad (6)$$

in Φ_{2E} may be much smaller than that in the individual rate constants if $k_2 + k_3 \ll k_4$. Since $\tau_F^{-1} = k_2 + k_3 + k_4$ is larger than 10^9 sec^{-1}, it is likely that in many cases $k_3 \ll k_4$, at least at low temperatures (*vide supra*), and therefore:

$$\Phi_{2E} = \frac{k_4}{k_4 + k_R^T} \tag{7}$$

In the absence of appreciable photochemistry ($k_R^T = 0$), $\Phi_{2E} = 1$ for all values of k_4.

For powdered samples of $Cr^{3+}:K_3Co(CN)_6$, $\Phi_p = 1$ at room temperature (*31*). Consequently in this system, $\Phi_{2E} = 1$ (Equation 3) and $k_3 + k_R^T \ll k_4$. However, when the same complex, $Cr(CN)_6^{3-}$, is dissolved in fluid dimethylformamide (DMF), $k_3/k_4 \geq 0.8$ (*31a*). Since $^2E \leadsto {}^4T_2$ back transfer is negligible in $Cr(CN)_6^{3-}$, Equation 5 is valid. For $Cr^{3+}:K_3Co(CN)_6$, Φ_p/τ_p increases two-fold from $77°-300°K$ (*12*). Since Φ_p is temperature-invariant and equals one in this crystal, all of this increase must be ascribed to the thermal enhancement of k_5. This is quite reasonable since $^2E \rightarrow {}^4A_2$ is symmetry-forbidden in this centrosymmetric environment (*32*). In contrast, in alcohol–water solutions of $Cr(CN)_6^{3-}$, Φ_p/τ_p is constant ($\pm 5\%$) as the temperature is changed from $145°$ to $225°K$, and it increases about 20% when the temperature is raised to $289°K$ (*28*). These data are consistent with a diminished Φ_{2E} in fluid solutions at room temperature, but, in the absence of quantitative information about the temperature effect upon k_5, no definitive statement is warranted.

Incidentally, the evaluation of Φ_{2E} from Equation 5 is fraught with uncertainty. Although τ_p is readily measured and reliable values of Φ_p were determined in some cases, the estimation of k_5 from absorption measurements is very difficult (*33*). Computation of k_5 by perturbation theory is unreliable, and values of Φ_{2E} that depend on such a calculation should be viewed with skepticism (*4*).

In alcohol–water solutions of $Cr(en)_3^{3+}$, Φ_p/τ_p is independent of temperature from $153°$ to $298°K$ (*28*). Since the $^2E \rightarrow {}^4A_2$ transition in $Cr(en)_3^{3+}$ is symmetry-allowed, k_5 should not be very temperature-dependent. If one assumes that k_5 is constant, the $Cr(en)_3^{3+}$ Φ_{2E} is the same in fluid and in rigid glass solutions. In a crystalline environment ($[Cr(en)_3][Cr(CN)_6] \cdot 2\,H_2O$) at $25°C$, $\Phi_{2E} \simeq 1$ for $Cr(en)_3^{3+}$ (*34*). Yet, in aqueous solution, $\Phi_{2E} = 0.63$ is inferred from the photoaquation data. Since the photoaquation yield for direct reaction in $^4T_2$ is 0.15, these findings imply some $^4T_2 \leadsto {}^4A_2$ deactivation, *i.e.* k_3 competes favorably with k_4. However, the photochemical evaluation of Φ_{2E} is critically dependent upon the magnitude of the photoaquation yield excited by direct $^2E \leftarrow {}^4A_2$ absorption. If this quantity is 0.3 rather than the value of

0.4 that was used, then $\Phi_{2_E} = 0.85$, *i.e.* $k_3 \ll k_4$, and the difference between the solid and fluid environments results solely from photolysis. It is of course possible that k_3 is actually much larger in a noncrystalline than in a crystalline environment. Alternatively, if photoreaction involves a dissociative mechanism as the primary step, geminate recombination could serve to reduce the overall photoaquation yield. In this way, the entire decrease in Φ_{2_E} to 0.63 would be attributable to $k_R{}^T$.

The striking environmental sensitivity of k_3 in Cr(3-bromoacac)$_3$ has already been described. The enhancement of k_3 in crystalline solids is sufficient to make Φ_{2_E} very small for $^4T_2 \rightsquigarrow {}^2E$. As a group, the chromium β-diketonates exhibit a singular behavior. Replacement of methyl groups by hydrogen atoms markedly reduces Φ_p/τ_p (35). Furthermore, although Φ_p/τ_p for Cr(acac)$_3$ is temperature-independent in a crystalline host, it decreases with temperature in noncrystalline media, in both rigid plastic (Figure 4) and absolute ethanol solutions (28). Under some circumstances k_3 apparently competes with k_4 in Cr(acac)$_3$.

Intermolecular Energy Transfer

If measurements of Φ_p and/or τ_p are to be used to evaluate intramolecular relaxation rates, the effect of intermolecular processes on these quantities must be assessed. In addition, the efficiency of excitation energy transfer is of intrinsic interest, especially in connection with solid state photochemical and photophysical processes. If excitation energy is transferred to an identical center, *i.e.* the same species in precisely the same environment, then all of the relaxation rates k_2–k_6 are unaffected, and no change in measureable quantities (except polarization) is expected. However, if transfer occurs between non-identical centers, then observable changes will occur. Two situations can be distinguished (36): (a) single step donor–acceptor transfer and (b) migration transfer.

In category a are the diffusion-controlled processes that prevail in fluid media [*e.g.*, benzil$\rightsquigarrow$ Cr(CN)$_6{}^{3-}$ (37) and Cr(NH$_3$)$_2$(NCS)$_4{}^-$ $\rightsquigarrow$ Cr(CN)$_6{}^{3-}$ (38)] and energy transfer between different species in solids [*e.g.*, Co(CN)$_6{}^{3-}$ $\rightsquigarrow$ Cr(CN)$_6{}^{3-}$ (31) and Cr(en)$_3{}^{3+}$ $\rightsquigarrow$ Cr(CN)$_6{}^{3-}$ (34)]. This type of transfer is characterized by the quenching of donor luminescence and in some cases by the sensitization of acceptor emission. Migration transfer (category b) can be visualized as a random walk process that is terminated either by de-excitation of the excited species or by transfer to a sink. In Cr^{3+}, migration energy transfer may take place in either $^4T_2$ or ^2E:

$$^4T_2 \rightsquigarrow {}^4T_2 \rightsquigarrow {}^4T_2 \rightsquigarrow \text{sink}$$

$$^2E \rightsquigarrow {}^2E \rightsquigarrow {}^2E \rightsquigarrow \text{sink}$$

The sink can be at lattice defect or another species such as a pair center. The number of Cr^{3+} sites traversed depends on the interaction energy and the lifetime of the excited state.

Two mechanisms have been suggested for energy transfer—multipole and exchange (39). Multipole transfer in the dipole approximation, often called Förster or resonance transfer, can take place over large distances (20–50 A). Exchange transfer requires orbital overlap between the interacting centers and is short range. The single step donor–acceptor transfers are short range as expected for an exchange mechanism, and they appear to constitute an important process whenever energetically possible if the species are in close proximity. When a comparison is made of energy transfer in crystalline and glassy environments, it should be noted that in Cr^{3+} systems doping at the 0.1 mole % level corresponds roughly to a $0.01M$ solution. Migration energy transfer in ionic solids, e.g. ruby, occurs in ^2E by an exchange mechanism (40). ^2E energy transfer in ruby is detectable in 0.1% crystals, yet neither τ_p nor Φ_p of $Cr(ox)_3^{3-}$ in NaMg-Al(ox)$_3 \cdot 9 H_2O$ crystals is much affected by Cr^{3+} concentration in the 0.1–100% range (41). The near constancy of τ_p and Φ_p does not necessarily rule out ^2E migration transfer since energy migration does not lead to quenching if the sink concentration is small. In fact, two-fold variations in τ_p are observed in 50% crystals and powders from one sample to another. The sinks in ruby are pair centers, and these increase rapidly with concentration. The 2nd–4th nearest neighbors distances in Al_2O_3 are less than 3.5 A, and superexchange interactions via O^{2-} are very important in this lattice. On the other hand, in oxalates the closest Cr^{3+}–Cr^{3+} approach is greater than 7 A, and superexchange is not very important in Cr^{3+}:NaMgAl(ox)$_3 \cdot 9 H_2O$ (41). Consequently, the rate of migration is slower in molecular crystals than the 10^5 sec^{-1} estimated from the ruby data (40).

Concentration quenching of ^2E is evident in Cr^{3+}:Al(acac)$_3$ (9) and especially in Cr^{3+}:K$_3$Co(CN)$_6$ (42). The relatively long ^2E lifetime in Cr^{3+}:K$_3$Co(CN)$_6$ favors concentration quenching. In addition, surface defects become more important with increasing Cr^{3+} concentration (31). Perhaps bulk defects also increase. A comparison of $\tau_p{}^\circ$ in glasses and undiluted crystals suggests, at most, a minor concentration quenching of $Cr(en)_3^{3+}$ ^2E in [Cr(en)$_3$]I$_3$, but a somewhat larger effect in other lattices. The near constancy of $\tau_p{}^\circ$ for $Cr(urea)_6^{3+}$ in several crystals (10) again indicates the relative unimportance of ^2E concentration quenching. The nephelauxetic shift of $^2E \rightarrow {}^4A_2$ in $Cr(acac)_3$ and $Cr(CN)_6^{3-}$ indicates π delocalization of the Cr^{3+} excitation onto the ligands, and the $^2E \longleftrightarrow {}^4A_2$ spectra of $Cr(acac)_3$ clearly show the splittings associated with intermolecular interactions (43, 44). In the absence of π bonding, the ligands serve as insulators and inhibit energy transfer. Although ^2E concentration quenching reduces τ_p and Φ_p, Φ_p/τ_p should be independent of concentration, which is true for Cr^{3+}:$NaMgAl(ox)_3 \cdot 9 H_2O$ (*41*) and Cr^{3+}:$NaCa$-$Al(ox)_3 \cdot 3 H_2O$ (*14*). On the other hand, quenching of $^4T_2$ will reduce Φ_p without changing τ_p. There is some evidence for a reduction in Φ_p/τ_p with increasing Cr^{3+} in Cr^{3+}:$K_3Co(CN)_6$ (*42*), but the decay becomes non-exponential as τ_p decreases. The rate of $^4T_2$ migration energy transfer by a multipole mechanism is likely to be greater than the rate of ^2E transfer by an exchange mechanism, but the $^4T_2$ lifetime is less than 10^{-9} sec and $^4T_2$ migration probably cannot compete effectively with k_4. Consequently, processes originating in the $^4T_2$ state are unaffected by intermolecular interactions.

Thexi States

In a crystalline host, the potential curves as drawn in Figure 5 describe the total energy, complex and environment. If vibrational relaxation within an electronic state is faster than other competing steps, then photophysical and photochemical processes occur in thermally equilibrated populations. Figure 5 is also applicable for a rigid, noncrystalline medium, but as the solvent melts and solvent relaxation takes place during the excited-state lifetime, a more complex representation is required.

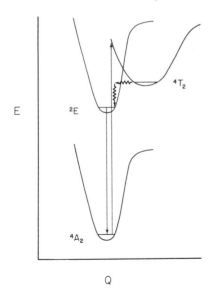

Figure 5. Potential curves for Cr^{3+} complexes in rigid media

In Cr^{3+}, the $^4A_2$ and ^2E states are both derived from the t_2^3 configuration, and they have nearly the same equilibrium geometries. Consequently the equilibrium solvent orientations are little changed in the two states. Since $^4T_2$ is derived from the t_2^2e configuration, the equilibrium position

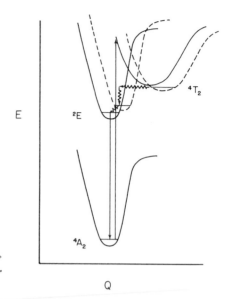

Figure 6. Potential curves for Cr^{3+}
*complexes in fluid media (see text for
key)*

is different, and the potential curves before and after solvent relaxation
are different. In Figure 6, the solid curves correspond to the solvent distribution appropriate to the ground state, while the dotted curves refer to
the solvent orientation that prevails in the relaxed $^4T_2$ state. Thermally
equilibrated states have been termed thexi states (*1*), and it has been suggested that thexi states are very much more distorted in fluid than in rigid
media. Possible consequences of this distortion include (a) increased τ_F
due to a reduced k_4, (b) smaller energy for $^2E \longrightarrow {}^4T_2$ back transfer,
and (c) alteration of the delayed fluorescence spectrum.

    Direct evidence for highly distorted thexi states is difficult to obtain.
Intersystem crossing might precede solvent reorientation and Φ_{2E} then
reflects only before-relaxation intersystem crossing. However, if a significant fraction of the excited molecules relax in $^4T_2$ and if Φ_{2E} is smaller in
the thexi state, then a reduction in Φ_p/τ_p would accompany the melting
of the rigid glass solvent. Within the precision of our measurements
(5%), no such change was observed for either $Cr(CN)_6^{3-}$ or $Cr(en)_3^{3+}$
(*28*).

    A definitive test for an increased lifetime in the relaxed state is to
monitor the excited-state absorption, with pulsed excitation, as a function
of medium rigidity. This experiment has not yet been reported for a
Cr^{3+} complex.

    The Stokes shift for delayed fluorescence is another measure of
excited-state distortion. Even in crystalline environments, the Stokes shift
can vary with the host as much as 1500 cm⁻¹, *e.g.* $Cr(urea)_6^{3+}$ (*30*).
There are few direct data on the dependence of the delayed fluorescence

spectrum on solvent rigidity, but in one instance, *viz.* $Cr(urea)_6^{3+}$ in methanol–DMF, no marked spectral change is observed at the glass point (45).

Conclusions

Although nonradiative relaxation of the $^4T_2$ state may be sensitive to environment, the major variations of Φ_p with environment originate in the ^2E state. This would result in the lack of correlation between any photochemistry that occurs in the $^4T_2$ state and $^2E \rightarrow {}^4A_2$ luminescence.

From the photophysical data amassed for Cr^{3+} complexes as a function of environment and temperature, one salient point emerges. If photophysical data are to be used to infer photochemical mechanisms, the photophysical determinations must include, but not be limited to, measurements made under precisely the same conditions as the photochemical determinations. This requirement has been fulfilled in a few instances— *e.g.* $Cr(CN)_6^{3-}$ in DMF (46), $Cr(en)_3^{3+}$ in H_2O (8), and $Cr(NH_3)_2$-$(NCS)_4^-$ in a H_2O–alcohol mixture (47)—but the widespread availability of pulsed excitation in the nanosecond domain now makes it possible to record the faster decays that prevail under photochemically meaningful conditions, *e.g.* $^2E \rightarrow {}^4A_2$ emission in aqueous solutions at room temperature (48). However, the ambiguities associated with back transfer require that photophysical data be recorded over a range of temperatures.

The determination of intersystem crossing efficiencies, *e.g.* Φ_{2E}, will continue to be difficult, but a comparison of results from direct and sensitized photolyses and measurements of Φ_p/τ_p will be very useful.

Literature Cited

1. Fleischauer, P. D., Adamson, A. W., Sartori, G., in "Inorganic Reaction Mechanisms," J. O. Edwards, Ed., Part II, p. 1, Interscience, New York, 1972.
2. Schlafer, H. L., Z. *Chem.* (1970) 10, 9.
3. Balzani, V., Carassiti, V., "Photochemistry of Coordination Compounds," Academic, London, 1970.
4. Forster, L. S., *Transition Met. Chem.* (1969) 5, 1.
5. Kirk, A. D., *Mol. Photochem.* (1973) 5, 127.
6. Tanabe, Y., Sugano, S., *J. Phys. Soc. Jpn.* (1954) 9, 753.
7. Porter, G. B., in "Concepts in Inorganic Photochemistry," A. Adamson and P. Fleischauer, Eds., chap. 2, Wiley, New York, 1975.
8. Ballardini, R., Varani, G., Wasgestian, H. F., Moggi, L., Balzani, V., *J. Phys. Chem.* (1973) 77, 2947.
9. Targos, W., Forster, L. S., *J. Chem. Phys.* (1966) 44, 4342.
10. Otto, H., Yersin, H., Gliemann, G., Z. *Phys. Chem.* (NF) (1974) 92, 193.
11. Camassei, F. D., Saldinger, J., unpublished data.
12. Camassei, F. D., Forster, L. S., *J. Chem. Phys.* (1969) 50, 2603.
13. Coleman, W. F., Forster, L. S., *J. Lumin.* (1971) 4, 429.
14. Coleman, W. F., *J. Lumin.* (1975) 10, 72.

15. Goldsmith, G. J., Shallcross, F. V., McClure, D. S., *J. Mol. Spectrosc.*
    (1965) **16**, 296.
16. Zander, H. U., Dissertation, Frankfurt/Main, 1969.
17. Chatterjee, K. K., Forster, L. S., *Spectrochim. Acta* (1965) **20**, 1603.
18. Yersin, H., Otto, H., Gliemann, G., *Theor. Chim. Acta* (1974) **33**, 63.
19. Nelson, D. F., Sturge, M. D., *Phys. Rev. A* (1965) **137**, 1117.
20. Castelli, F., Forster, L. S., *J. Am. Chem. Soc.* (1973) **95**, 7223.
21. Reich, S., Raziel, S., Michaeli, I., *J. Phys. Chem.* (1973) **77**, 1378.
22. Schläfer, H., Gausmann, H., Witzke, H., *J. Chem. Phys.* (1967) **46**, 1423.
23. Everett, P. N., *J. Appl. Phys.* (1971) **42**, 2106.
24. Castelli, F., Forster, L. S., *Phys. Rev. B* (1975) **11**, 920.
25. Glass, A. M., *J. Chem. Phys.* (1969) **50**, 1501.
26. Reynolds, M. L., Hagston, W. E., Garlick, G. F. J., *Phys. Status Solidi*
    (1968) **30**, 113.
27. Castelli, F., Forster, L. S., *J. Am. Chem. Soc.*, in press.
28. Castelli, F., unpublished data.
29. Englman, R., Jortner, J., *Mol. Phys.* (1970) **18**, 145.
30. Laver, J. L., Smith, P. W., *Aust. J. Chem.* (1971) **24**, 1807.
31. Castelli, F., Forster, L. S., *J. Phys. Chem.* (1974) **78**, 2122.
31a. Sabbatini, N., Scandola, M. A., Carassiti, V., *J. Phys. Chem.* (1973) **77**,
    1307.
32. Condrate, R., Forster, L. S., *J. Chem. Phys.* (1968) **48**, 1514.
33. Forster, L. S., in "Concepts in Inorganic Photochemistry," A. Adamson and
    P. Fleischauer, Eds., chap. 1, Wiley, New York, 1975.
34. Castelli, F., Forster, L. S., *Chem. Phys. Lett.* (1975) **30**, 465.
35. DeArmond, M. K., Forster, L. S., *Spectrochim. Acta* (1963) **19**, 1687.
36. Ganrud, W. B., Moos, H. W., *J. Chem. Phys.* (1968) **49**, 2170.
37. Binet, D. J., Goldberg, E. L., Forster, L. S., *J. Phys. Chem.* (1968) **72**,
    3017.
38. Chen, S., Porter, G. B., *J. Am. Chem. Soc.* (1970) **92**, 3196.
39. Dexter, D. L., *J. Chem. Phys.* (1953) **21**, 836.
40. Birgenau, R. J., *J. Chem. Phys.* (1969) **50**, 4282.
41. Castelli, F., Forster, L. S., *J. Lumin.* (1974) **8**, 252.
42. Kirk, A. D., Ludi, A., Schläfer, H. L., *Ber. Bunsenges. Phys. Chem.* (1969)
    73, 669.
43. Armendarez, P. X., Forster, L. S., *J. Chem. Phys.* (1964) **40**, 273.
44. Courtois, M., Forster, L. S., *J. Mol. Spectrosc.* (1965) **18**, 396.
45. Klassen, D. M., Schläfer, H. L., *Ber. Bunsenges. Phys. Chem.* (1968) **72**,
    663.
46. Wasgestian, H. F., *J. Phys. Chem.* (1972) **76**, 1947.
47. Adamson, A., *J. Phys. Chem.* (1967) **71**, 798.
48. Kane-Maguire, N. A. P., Langford, C. H., *Chem. Commun.* (1971) 895.
49. Liehr, A. D., *J. Phys. Chem.* (1963) **67**, 1314.
50. Porter, G. B., Schläfer, H. L., *Z. Phys. Chem.* (NF) (1963) **37**, 109.

RECEIVED February 6, 1975. Work supported by the National Science Foundation.

16

The Use of Ligand Substituents to Modify Photochemical Reactivities

Studies of Ruthenium(II) and Rhodium(III) Ammine Complexes

PETER C. FORD, G. MALOUF, J. D. PETERSEN, and V. A. DURANTE

University of California, Santa Barbara, Calif. 93106

Quantum yield data demonstrate that substituents can be used to modify the photoaquation reactivities of the complexes $Rh(NH_3)_5L^{3+}$ and $Ru(NH_3)_5L^{2+}$ (L = meta- or para-substituted pyridine). For the Rh(III) complexes, these modifications involve only minor shifts in ligand field excited-state energies, and the relatively small changes in the photolability of L follow an order inverse to the ligands' Brönsted basicities. For the Ru(II) complexes, these modifications involve dramatic shifts in the energy of metal-to-ligand charge transfer excited states with the result that photoaquation quantum yields vary by three orders of magnitude. It is concluded that with the unreactive complexes, the charge transfer state is lowest in energy whereas with the more reactive complexes, the ligand field excited state is lowest in energy.

Substituents at positions remote from reaction sites have been used extensively to study organic reaction mechanisms, but such effects have been examined to a lesser degree in transition metal systems. The advantages of substituent studies is readily apparent for certain systems. For π-unsaturated ligands (*e.g.*, pyridine and other aromatic heterocycles), substituents in meta- and para-ring positions can be used to generate relatively subtle electronic perturbations that include changes in ligand basicity and in the energies of π symmetry molecular orbitals. In these positions, little or no steric interaction with the coordination site results.

The electronic perturbations appear to be ideally suited for use in the modification of photochemical properties.

Previously reported work demonstrated that substituents can be used to tune the energies of excited states responsible for the emission spectra of certain group VIII metal complexes (1) and to modify significantly the absorption spectra of complexes displaying metal-to-ligand charge transfer (MLCT) bands (2). In this presentation, we summarize some recent attempts to use ligand substituents in our studies of transition metal complex photochemical reaction mechanisms. The particular subjects of interest are the metal ammine complexes $M(NH_3)_5L^{n+}$, where M is Rh(III) or Ru(II) and L is a meta- or para-substituted pyridine.

Spectral Properties

Table I lists the spectral similarities and differences of analogous ruthenium(II) and rhodium(III) complexes. The hexaammine complexes each have the two ligand field transitions ($A_1 \rightarrow T_1$ and $A_1 \rightarrow T_2$) expected for these low spin $4d^6$ octahedral complexes. For the pyridine complexes, the higher energy absorption in each case is a ligand-centered $\pi_L-\pi_L^*$ band analogous to that of free pyridine; however, the lower energy bands differ in character. For the rhodium(III) complex, the longest wavelength band has an energy and an intensity comparable to those of the lower energy band of the hexaammine and is assigned to a ligand field transition (3). In contrast, the lower energy band of aqueous $Ru(NH_3)_5py^{2+}$ has a very high extinction coefficient and, because of the sensitvity of the band to ring substituents (*vide infra*) and solvent, the band was designated MLCT in character (2).

Table I.   Spectral Data from Some Ru(II) and Rh(III) Complexes in Aqueous Solution

Complex	λ_{max}		Assignment
	nm	Log ε	
Rh(III)			
$Rh(NH_3)_6^{3+}$	305	2.13[a]	LF
	255	2.00	LF
$Rh(NH_3)_5py^{3+}$	302	2.20	LF
	259	3.44	$\pi_L-\pi_L^*$
Ru(II)			
$Ru(NH_3)_6^{2+}$	385	1.59[b]	LF
	275	2.83	LF + CTTS (?)
$Ru(NH_3)_5py^{2+}$	407	3.89[b]	MLCT
	244	3.66	$\pi_L-\pi_L^*$

[a] From Ref. *4*.
[b] From Ref. *2*.

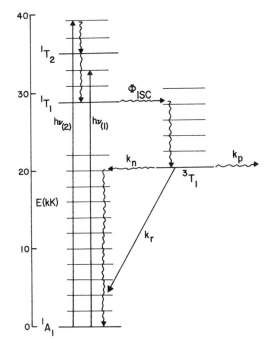

Figure 1.   Energy level diagram for photo-reaction of $Rh(NH_3)_6{}^{3+}$. k_p: aquation of re-active excited state to give photoproducts; k_n and k_r: nonradiative and radiative deacti-vation to ground state, respectively.

Photochemistry of Rhodium (III) Complexes

Photolysis in aqueous solution of the rhodium(III) pentaammine complexes $Rh(NH_3)_5X^{2+}$ or $Rh(NH_3)_5L^{3+}$ (where X^- is a halogen anion and L is an uncharged base such as NH_3, H_2O, or pyridine) always leads to photoaquation pathways when the lower energy LF bands are irradiated (*4, 5*). The ligand field excitation was interpreted as resulting in an efficient intersystem crossing/internal conversion to a lowest energy LF state followed by nonradiative deactivation to ground state, by aquation to products (in ambient temperature fluid solution), or by radiative deactivation (in low temperature glasses) (*6*). These processes are illustrated in Figure 1 for the hexaammine complexes where the lowest excited state is rpresented as the triplet $^3T_{1g}$. (It should be noted, however, that in several recent articles (*7, 8, 9*) it was emphasized that spin orbit coupling constants of the heavier metals are of such magnitude that spin multiplicities may not be a meaningful way to label the reactive low-energy state.) If the lowest state is presumed to be responsible for the photoreactions, the quantum yield for aquation is then represented by Equation 1 where

$$\Phi_p = \Phi_{ic} \frac{k_p}{k_p + k_n + k_r} \tag{1}$$

Φ_{ic} is the efficiency of interconversion of higher states to the reactive state (it was shown to have a value of one in previous studies with the halopentaammines (5)).

For the pyridine complex, aquation of the unique ligand is the predominant photoreaction pathway for 313-nm irradiation (Reaction 2).

$$\mathrm{Rh(NH_3)_5py^{3+}} + \mathrm{H_2O} \xrightarrow{h\nu \ (313 \ nm)} \mathrm{Rh(NH_3)_5H_2O^{3+}} + \mathrm{py} \tag{2}$$

A quantum yield of 0.14 mole/einstein was measured for Equation 2 by a spectral technique while the quantum yield of competitive NH_3 photoaquation can be estimated from pH changes as having an upper limit of 0.02. Similar or smaller limits for Φ_{NH_3} can be estimated for the other Rh(III) pyridine complexes discussed here.

Photoaquation quantum yields for several pyridine complexes are summarized with absorption and emission spectral data in Table II. These data indicate the following. The lowest energy absorption band, ligand field in character, is little affected by a 4-methyl or 3-chloro substituent, and it appears at the same wavelength for all three pyridine

Table II.   Photochemical and Photophysical Properties
of $[\mathrm{Rh(NH_3)_5L}][\mathrm{ClO_4}]_3$

L	$p\mathrm{K_a}$[a]	λ_{max},[b] nm	Φ_L,[c] moles/ einstein	E_{em},[d] kK
NH_3	9.25	305	0.075	16.9
![pyridine with CH₃]—CH$_3$	6.02	302	0.090	16.6
![pyridine]	5.25	302	0.14	16.9
![pyridine with Cl] Cl	2.84	302	0.34	16.5

[a] From Ref. 13; measured in aqueous solution at 25°C for the free ligand.
[b] Maximum for the lowest absorption band $(A_1 \rightarrow E)$ in aqueous solution.
[c] Photoaquation yields measured at 296°K in pH 3 aqueous solution.
[d] Energy of the maxima of the emission spectral band as measured for the solid salt at 77°K.

complexes. Interestingly, this wavelength is very close to that of the analogous band of the hexaammine complex. Similarly, the low-temperature emission spectra indicate that the emitting state, designated ligand field in character (6), is not strongly perturbed by either substituting pyridine for NH_3 or by the pyridine substituents examined. The photoaquation quantum yields are perturbed to a greater extent by the substituents and Φ_L values follow an inverse order to ligand basicities (according to free ligand pK_a values). The differences are not dramatic, but the pyridine substituents have a relatively minor effect on the rhodium(III) ligand field excited-state energies and a somewhat greater effect on the labilities of the reactive excited states.

Photochemistry of Ruthenium(II) Complexes

Visible region photolysis of $Ru(NH_3)_5py^{2+}$ initially involves excitation of the MLCT band in Table I. The overwhelmingly dominant photoreaction observed is substitutional in character with both NH_3 and py being labilized competitively (10) (see Reaction 3). The MLCT excited

$$Ru(NH_3)_5py^{2+} + H_2O \xrightarrow[(405\ nm)]{h\nu} \begin{cases} \longrightarrow Ru(NH_3)_5H_2O^{2+} + py \qquad (3) \\ \\ \longrightarrow Ru(NH_3)_4(H_2O)py^{2+} + NH_3 \end{cases}$$

state can be conceptualized as having an oxidized metal ion and a radical ion ligand in a coordination complex, *e.g.* **MLCT***. It appears unlikely

MLCT*

that **MLCT*** would be labile; in particular, it appears unlikely that it would be labile toward NH_3-aquation. The electron promoted from the low spin $Ru(II)$ has π symmetry with respect to the metal–ligand bond, and the σ metal ammonia bonds and the σ component of the Ru–py bond should be little affected or perhaps even enhanced by the more positive nature of the ruthenium in the excited state. The π component of the Ru–py bond no doubt is less important in **MLCT***. However, since π backbonding is relatively insignificant in the stability of $Ru(III)$ complexes, it is improbable that **MLCT*** would be noticeably more reactive than the corresponding Ru^{III} compound which is not labile under photolysis conditions.

On the basis of these considerations, one might conclude that if one excited state is responsible for the aquation of both NH_3 and py from

Ru(NH$_3$)$_5$py^{2+}, it should be a ligand field state. This state would be analogous to the reactive LF state responsible for photoaquation of Ru(NH$_3$)$_5$CH$_3$CN^{2+} (where the lowest absorption band is LF in character (9)) and of Rh(NH$_3$)$_5$L^{3+} (*vide supra*). Estimates of the energies of the lowest MLCT and LF states suggest that for Ru(NH$_3$)$_5$py^{2+} the lowest state may be LF in character (10).

The above discussion can be summarized as suggesting that the lowest LF excited state is labile toward aquation while MLCT states are not. Therefore the reactivity of Ru(NH$_3$)$_5$py^{2+} is attributed to an energy level scheme such as that depicted in Figure 2 where the initially populated MLCT states interconvert to the reactive LF state which is also the lowest energy state. Present data do not discriminate whether nonradiaive deactivation directly from MLCT states also occurs. Figure 2, however, does suggest the following experimental evaluation of this interpretation. The data for the Rh(III) systems indicate that LF excited-state energies are not wildly perturbed by pyridine ligand substituents. At the same time, however, it is known that pyridine substituents have dramatic effects on the energies of MLCT absorption band maxima (2) and presumably similar effects on the energies of all MLCT states. Therefore appropriate use of substituents would tune the MLCT energy to a point where the lowest state is MLCT in character. If initial excitation is followed by efficient interconversion to the lowest energy excited state, the

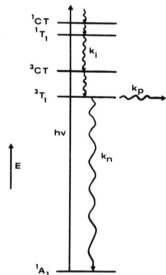

Figure 2. Excited state diagram for proposed mechanism for photoaquation of Ru(NH$_3$)$_5$py^{2+}. k$_p$: reaction leading to photoproducts.

observed photoreactivity should reflect the reactivity of the lowest MLCT state, not that of the lowest LF state.

With regard to photoaquation, these predictions appear to be borne out by the data in Table III. Irradiation at the MLCT λ_{max} of a variety of py–X complexes in pH 3 aqueous solution gives Φ_L values of 0.02–0.05 mole/einstein for $\lambda_{max} < \sim 460$ nm and to values less than 0.0003 when $\lambda_{max} > \sim 490$. For discussion purposes, we shall refer to the former species as reactive complexes and to the latter as unreactive complexes. (The nicotinamide complex, the sole exception to this pattern, was discussed previously (11).) An overall range of three orders of magnitude is observed in the measurable Φ_L values. These observations strongly suggest that modification of the MLCT energy with appropriate electron-withdrawing substituents leads to a reversal in the excited state order to give a substitution-unreactive charge transfer state with the lowest energy.

Several questions arise regarding the conclusions drawn from the data in Table III. Among these is the concern that the different λ_{irr} used with the different py–X complexes may affect the quantum yields since, unless there is a rapid realaxation to the lowest energy state in its thermally equilibrated form, the excitation energy may reflect factors other than the intrinsic reactivity of the lowest state. Table IV illustrates the dependence of Φ_L on the excitation wavelength for two cases. The 3,5-dichloropyridine complex (λ_{max} 447 nm) is classified according to previous discussion as reactive. Over the wavelength range 405–500 nm, Φ_L is essentially wavelength-independent for this complex. In contrast, the isonicotinamide complex shows a definite wavelength dependence in its Φ_L values. With a MLCT λ_{max} of 479 nm and a Φ_L value of 1.1×10^{-3} when irradiated at that wavelength, it appears to fall in the intermediate reactivity category. The effect of changing λ_{irr} is a roughly 12-fold variation in Φ_L over the wavelength range of 405–500 nm; however, the Φ_L values are essentially wavelength-independent for $\lambda_{irr} \geq 500$ nm. Clearly, the patterns of Φ_L wavelength dependence are different for the two complexes.

Figure 3 illustrates a rationalization of the observed wavelength dependences. Figure 3a represents the excited states of the reactive type complexes, *i.e.* those with lowest energy LF excited states. It is proposed that initial excitation at any wavelength results in relatively efficient interconversion to the lowest excited state, a substitution-reactive LF state. Figure 3b represents the nonreactive complexes, *i.e.* those with an MLCT state as the lowest energy excited state. The wavelength-dependent reactivity toward photoaquation is interpreted as indicating that LF states intermediate between the states initially populated and the lowest excited state can undergo reaction at a rate competitive with interconversion to the lowest excited state. The wavelength dependence may also

Table III.   Spectroscopic Quantum Yields for the Photoaquations of $Ru(NH_3)_5L^{2+}$ in Aqueous Solution[a]

$$Ru(NH_3)_5L^{2+} + H_2O \xrightarrow{h\nu} Ru(NH_3)_5H_2O^{2+} + L$$

L	λ_{max} (CT), nm	ν_{max} (CT), kK	λ_{irr}, nm	Φ_L, moles/einstein $\times 10^3$
4-methylpyridine (N⟨⟩—CH₃)	398	25.1	405	$37 \pm 3\ (2)$[b]
pyridine (N⟨⟩)	408	24.5	405	$45 \pm 2\ (3)$[b]
3-chloropyridine (N⟨⟩—Cl)	426	23.5	436	$48 \pm 2\ (2)$[b]
nicotinamide (N⟨⟩—C(=O)—NH₂)	427	23.4	433	$8.5 \pm .2\ (4)$
3,5-dichloropyridine (Cl,N⟨⟩,Cl)	447	22.4	450	$42 \pm 2\ (4)$
4-trifluoromethylpyridine (N⟨⟩—CF₃)	454	22.0	455	$22 \pm 5\ (2)$[c]
isonicotinate (N⟨⟩—CO₂⁻)	457	21.9	460	$26 \pm 3\ (4)$[d]
pyrazine (N⟨⟩N)	472	21.2	475	$1.4 \pm 0.1\ (4)$[e]

Complex				
N⟨benzene⟩—C(=O)—NH$_2$	479	20.9	480	1.07 ± 0.04 (4)
N⟨benzene⟩—C(=O)—O—CH$_3$	495	20.2	500	0.28 ± 0.04 (3) 0.25 ± 0.04 (4)^f
N⟨benzene⟩—C(=O)—CH$_3$	523	19.1	520	0.25 ± 0.06 (3)
N⟨benzene⟩NH$^+$	529	18.9		0.11 ± 0.02 (4) ^{g}
N⟨benzene⟩NCH$_3^+$	540	18.5	540	0.04 ± 0.01 (2)
N⟨benzene⟩—C(=O)—H	545	18.3	546	0.05 ± 0.01 (3)

^{a} Measured at 25°C in pH 3 aqueous 0.2M NaCl solution except where noted otherwise. New data reported here were obtained with a 150 W xenon short arc lamp as light source with interference filters for wavelength selection. *See* Ref. *10.*
^{b} From Ref. *3.*
^{c} From Ref. *11.*
^{d} At pH 10.
^{e} At pH 7.
^{f} In 1.0M HCl.
^{g} In 0.1M HCl.

indicate that the higher energy LF states are more reactive, or that the LF manifold is populated more efficiently from the higher MLCT states. Nonetheless, if the intrinsic reactivity of the lowest LF states is assumed to give a Φ_L value of 0.025–0.05, then the quantum yield (0.0045) measured for the isonicotinamide complex at 405-nm irradiation suggests that interconversion to the lowest state is at least 80–90% efficient.

For λ_{irr} greater than 500 nm, Φ_L for the isonicotinamide complex becomes approximately wavelength-independent with a limiting value of $\sim 3 \times 10^{-4}$ mole/einstein. This may represent the intrinsic reactivity of the MLCT state; however, several complexes with lower energy λ_{max} values for the MLCT are significantly less reactive. Thus an alternative

Table IV.   Wavelength Dependence of Quantum Yields[a]

$$\text{for Ru(NH}_3)_5\text{L}^{2+} + \text{H}_2\text{O} \xrightarrow{h\nu} \text{Ru(NH}_3)_5\text{H}_2\text{O}^{2+} + \text{L}$$

$$\Phi_L \times 10^3$$

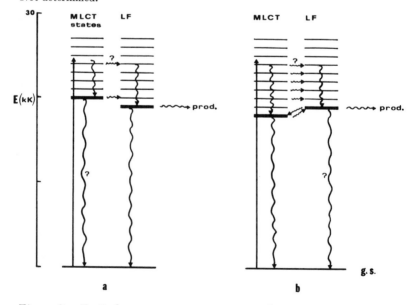

λ_{irr}, nm		
405	38. ± 1.	4.5 ± 0.1
449	42. ± 2.	1.5 ± 0.1
480	43. ± 1.	1.1 ± 0.1
500	48. ± 2.	0.37 ± 0.02
520	[b]	0.35 ± 0.02
546	[b]	0.30 ± 0.02

[a] Spectroscopic quantum yields were measured at 296°K in pH 3 aqueous solution; $\mu = 0.2M$ (NaCl).
[b] Not determined.

Figure 3. Excited-state representations revised to account for temperature and λ_{irr} dependence of aquation quantum yields for Ru(NH₃)₅L²⁺ quantum yields in aqueous solution. (a) Diagram for photoreactive complex, e.g. Ru(NH₃)₅(3,5-dichloropyridine)²⁺; and (b) diagram for a relatively unreactive complex, Ru(NH₃)₅(isonicotinamide)²⁺.

Table V. Effect of Temperature on Quantum Yields[a] of
L Aquation from Ru(NH$_3$)$_5$L^{2+}

L	T, °K	λ_{irr}, nm	Φ_L, moles/ einstein $\times 10^3$	E$_a$(app), kcal/mole
	296 328	500 500	0.37 1.3	7.6 ± 0.4
	296 328	449 449	42. 82.	4.1 ± 0.3

[a] Measured in pH 3 aqueous 0.2M NaCl solution.

explanation is attractive, namely that even the small reactivity noted for λ_{irr} greater than 500 nm with the isonicotinamide complex is the result of thermal back population of the lowest LF state. Evidence in support of this concept comes from examining the temperature-dependent Φ_L values (Table V). Photolysis of the isonicotinamide complex at 500 nm displays an apparent activation energy (7.6 kcal) that is substantially greater than that observed for photolysis of the 3,5-dichloropyridine complex at 449 nm (4.1 kcal). This difference (3.5 kcal) could easily represent the difference in energy between the lowest LF* and the lowest MLCT* for the isonicotinamide complex.

Flash Photolysis Studies

A previous flash photolysis study (12) of aqueous Ru(NH$_3$)$_5$py^{2+} reported transient bleaching of the MLCT band followed by relatively slow decay to substrate and aquation products at a rate, in part, inversely proportional to [H$^+$]. Since the transient bleaching was too long-lived to represent the behavior of an excited state, this observation was interpreted

$$(4)$$

T   TH$^+$

in terms of an intermediate having a free radical coordinated to a Ru(III) center with the pyridine nitrogen assuming an insulating tetrahedral configuration capable of reversible protonation (12) (see Reaction 4). The effect of [H⁺] on the transient bleaching parallels the pH effect on the quantum yield of photoaquation reported (10) previously for continuous photolysis studies. However, the proposed mechanism for formation of **T** involves reaction of a charge transfer state such as **MLCT*** and implies that a charge transfer state is precursor to the photoaquation of pyridine (12). The continuous photolysis experiments using

Table VI. Preliminary Results for the Flash Photolysis of the Complexes Ru(NH₃)₅L²⁺ in Deaerated Aqueous Solutions

L	λ_{max}, nm	Φ_L, moles/ einstein $\times 10^3$	Flash Photolysis Observations[a]
N⟨pyridine⟩	408	45.	transient bleaching; pH-dependent decay to starting material and products[b]
N⟨pyridine-Cl⟩	426	48.	transient bleaching; pH-dependent decay[b]
N⟨pyridine-Cl,Cl⟩	447	42.	transient blecahing; pH-dependent decay[c]
N⟨pyridine⟩—C(=O)—CH₃	523	0.25	no transient bleaching[c]
N⟨pyridine⟩N—CH₃⁺	540	0.04	no transient bleaching[c]

[a] $\mu = 0.2M$ (NaCl/HCl), 23 $\pm$ 2°C, substrate concentration $\sim 10^{-5}M$.
[b] λ(flash) > 409 nm.
[c] λ(flash) > 320 nm.

various substituted pyridines (*vide supra*) clearly demonstrate that an MLCT state is not the precursor to the photoaquation of L in the pH 3 aqueous solutions used in those studies. In addition, no enhancement of Φ_L was observed in 1M HCl for the photo-unreactive methylisonicotinate complex (Table III) (*11*). We consider a ligand field the more likely precursor to transients in this system and have suggested that the pyridine ring could turn 90° with respect to the normal Ru^{II}–N band axis and enter into a weak π complex with $Ru(NH_3)_5{}^{2+}$. In this configuration, the pyridine nitrogen would be free to undergo protonation.

We are examining the flash photolysis of several substituted pyridine complexes, and preliminary results are summarized in Table VI. Our studies confirm that transient bleaching is indeed observed when $Ru(NH_3)_5py^{2+}$ is flashed and that the kinetics of this transient's decay have a pH dependence similar to that reported (*12*). Each complex for which transient bleaching is observed also undergoes photoaquation. In no case does a complex which is relatively unreactive because of the electron-withdrawing substituent display a flash-induced transient or transient bleaching. Since the latter complexes are considered to have an MLCT state as the lowest energy excited state, these qualitative observations provide strong support for the argument that a ligand field excited state, not a charge transfer state, is the precursor of both the transient bleaching and the photoaquation pathways.

Concluding Remarks

In this presentation we have demonstrated how ligand substituents can be used to modify the photochemical reactivity of transition metal complexes. For the rhodium(III) complexes, where a ligand field state is lowest in energy, the substituents for the series $Rh(NH_3)_5pyX^{3+}$ have relatively little effect on the energies of the lowest state. Modifications of quantum yields were more substantial but still relatively modest. On the basis of an assumed dissociative mechanism, the increase in Φ_L with decreasing ligand basicity argues for the substituent effect being reflected most strongly in the reactivity of the excited state toward aquation. However, in the absence of photochemical and photophysical data under identical conditions, this conclusion remains somewhat speculative.

The use of substituents with the Ru(II) system made possible more dramatic modification of quantum yields. In this case, however, the modifications can be attributed to changes in the order of excited states, specifically creating a situation where a substitution unreactive MLCT state becomes the lowest energy excited state. For these systems, it would appear likely that some new reaction pattern might prove characteristic of the MLCT state.

Experimental Section

Continuous photolyses were carried out using apparatus and techniques described previously (*4, 10, 11*). All solutions were thermostated. Dark reactions with solutions prepared under identical conditions were run parallel to all photolysis studies. Quantum yields were corrected in each case for dark reaction, and the reported data represent initial yields.

Flash photolyses were carried out on a kinetic flash photolysis apparatus based on a Xenon Corp. model A micropulser and a model C trigger system. Filtered flash pulses of 500 J were used. Optical density *vs.* time traces were analyzed using a Xenon Corp. model G high stability kinetic analyzing light source, appropriate filters, and monochromators; they were recorded with an EMI 6256B photomultiplier tube and a Tektronic model 564 storage oscilloscope. The solution cell was 12 cm long and 1 cm in diameter, and all substrate solutions ($\sim 10^{-5}M$ in substrate, $0.2M$ NaCl/HCl) were rigorously deoxygenated. Luminescence spectra and emission lifetimes were measured on apparatus constructed by R. J. Watts of this department.

Literature Cited

1. Watts, R. J., Crosby, G. A., *J. Am. Chem. Soc.* (1972) **94**, 2606.
2. Ford, P. C., *Coord. Chem. Rev.* (1970) **5**, 75.
3. Foust, R. D., Ford, P. C., *Inorg. Chem.* (1972) **11**, 899.
4. Petersen, J. D., Ford, P. C., *J. Phys. Chem.* (1974) **78**, 1144.
5. Kelly, T. L., Endicott, J. F., *J. Phys. Chem.* (1972) **76**, 1937.
6. Thomas, T. R., Crosby, G. A., *J. Mol. Spectrosc.* (1971) **38**, 118.
7. Crosby, G. A., Hipps, K. W., Elfring, W. H., *J. Am. Chem. Soc.* (1974) **96**, 629.
8. Konig, E., Kremer, S., *J. Phys. Chem.* (1974) **78**, 56.
9. Ford, P. C., Petersen, J. D., Hintze, R. E., *Coord. Chem. Rev.* (1974) **14**, 67.
10. Chaisson, D. A., Hintze, R. E., Stuermer, D. H., Petersen, J. D., McDonald, D. P., Ford, P. C., *J. Am. Chem Soc.* (1972) **94**, 6665.
11. Malouf, G., Ford, P. C., *J. Am. Chem. Soc.* (1974) **96**, 601.
12. Nataragan, P., Endicott, J. F., *J. Am. Chem. Soc.* (1972) **94**, 5909.
13. Schofield, K., "Hetero Aromatic Nitrogen Compounds," Plenum, New York, 1967.

RECEIVED January 24, 1975. Work supported by the National Science Foundation. P. C. Ford was the recipient of a Camille and Henry Dreyfus Foundation Teacher–Scholar Grant.

A Comparative Study of Some Luminescence Properties of Homo- and Hetero-Bischelated Complexes of Iridium(III)

RICHARD J. WATTS, BOULDEN GRIFFITH, and
JOHN S. HARRINGTON

University of California, Santa Barbara, Calif. 93106

Ligand substituent alterations affect the luminescence decay of two hetero-bischelated complexes of iridium(III). Luminescence decay curves of cis-dichloro-1,10-phenanthroline-4,7-dimethyl-1,10-phenanthroline iridium(III) chloride com-plex A) *and of* cis-dichloro-1,10-phenanthroline-5,6-dimethyl-1,10-phenanthroline iridium(III) chloride (complex B) *reveal strikingly dissimilar nonexponential behavior. While both complexes exhibit an emission wavelength dependence of the nonexponential decays, analysis of complex A on the basis of a two-level emission model yields two exponential components having lifetimes of 8.8 and 22 μsec, with the short-lived level lying 300–400 cm⁻¹ lower in energy. This contrasts with complex B whose short- and long-lived com-ponents are both longer lived than the corresponding levels of complex A and whose short-lived level lies higher in energy.*

Numerous studies of the emission of visible and ultraviolet radiation by large molecules after they have been promoted to an excited state have resulted in several broad generalizations. In light-atom molecules, such as hydrocarbons, as many as two thermally non-equilibrated excited states may give rise to light emissions. These emissions are classified as fluorescence and phosphorescence and are associated with transitions in which the spin quantum number remains the same (fluorescence) or changes by one unit (phosphorescence). Phosphorescence is generally caused by a transition from a spin triplet to a spin singlet in molecules with an even number of electrons; fluorescence is generally the result of a singlet-singlet transition. Thus, there are four emitting levels in these molecules when the combined effects of electron spin and orbital momenta are taken into consideration. Under the common conditions at which emission measurements of large molecules are made (77°K, glassy medium), the three triplet sublevels are in thermal equilibrium with each other whereas thermal equilibration of the singlet and triplet manifolds does not occur.

When heavy atoms are introduced into the framework of a molecule, marked changes in the light emission characteristics occur. In particular, for complex molecules which contain metal ions from the second and third transition series, only a single emission from a set of thermally equilibrated excited levels is generally observed (1–7). This behavior is apparently associated with a spin-orbit coupling effect, and, indeed, it has been suggested that this effect is so large as to render spin labels inappropriate in second- and third-row metal complexes (8).

Although the number of emitting levels which are thermally populated at 77°K in heavy metal complexes is not generally known, several studies of complexes of the d^6 metal ions, Ru(II) and Ir(III), have concentrated on this question (4–7). These studies, which use measurements of the temperature dependence of luminescence lifetimes, indicate that complexes with charge-transfer emissions luminesce from four thermally equilibrated levels at 77°K. These levels lie in an energy region of 100–200 cm^{-1} from each other. The nature of these levels has been researched in considerable detail in the ruthenium trisbipyridyl ion and in related ions of D_3 symmetry. In these complex ions, two of the four emitting levels are degenerate although this degeneracy should be raised in complexes with reduced symmetry.

The contrasting luminescence behavior of light-atom and heavy-atom molecules indicates the vital role of spin-orbit coupling in determining the mechanisms by which eletcronic energy is converted to vibrational energy in the molecular framework. It is well known however that spin-orbit coupling alone will not ensure rapid energy transfer (9–11). Other molecular parameters, such as energy gaps (12, 13), vibrational frequencies (14), and the orbital nature of the excited states (15, 16), are important in determining energy transfer rates in light atom molecules and must at least have complementary roles to spin-orbit coupling in heavy-atom complexes (17). Unfortunately, the roles of these other parameters appear, in general, to be submerged by the dominant role of spin-orbit coupling, which causes the rapid energy transfer processes leading to a single emission from a set of thermally equilibrated levels in heavy-metal complexes.

Several rare exceptions to the general luminescence behavior of heavy-atom molecules described above have recently been reported. It has been found that hetero-trischelated complexes of Rh(III) with 2,2'-

bipyridine (bipy) and 1,10-phenanthroline (phen) have non-exponential luminescence decay curves (*18*). Although this is characteristic of light emission from several thermally non-equilibrated states, the individual luminescence spectra of the non-equilibrated levels cannot be resolved by conventional emission spectroscopy. This is apparently the result of a near degeneracy of the emitting levels. In these complexes, the non-equilibrated luminescent levels are thought to arise from excited states which have their energy localized in either the 2,2'-bipyridine or 1,10-phenanthroline ligand (*18*).

Several hetero-bischelated complexes of Ir(III) with 1,10-phenanthroline and substituted 1,10-phenanthroline have also been reported to have non-exponential luminescence decay curves (*19*). Although the individual emission spectra of the non-equilibrated levels of these complexes are again too close to resolve by conventional emission spectroscopy, partial resolution has been accomplished by time-resolved emission spectroscopy *via* box-car averaging techniques (*20*). Complete resolution has been accomplished by computer analysis of luminescence decay curves as a function of emission wavelength (*20*). In these complexes, the luminescent levels appear to arise from both ligand-localized ($\pi\pi^*$) states and charge-transfer ($d\pi^*$) states.

In this paper we first review the experimental data which characterizes the luminescence of several complexes formed by the binding of two chloride ions and two bidentate ligands to Ir(III). On the basis of this experimental information, we present a simple molecular orbital model which describes the orbital parentage of the luminescent states of these complexes. This model is used to interpret the non-exponential luminescence of hetero-bischelated complexes of Ir(III).

Review of Experimental Results

The luminescence of *cis*-dichlorobis-1,10-phenanthroline iridium(III) chloride, [IrCl$_2$(phen)$_2$]Cl, has been studied in several laboratories (*21*, *22*). Although there is general agreement that a single emission from a set of thermally equilibrated levels occurs upon excitation of the complex at 77°K in glassy media, there is some dispute concerning the orbital parentage of the emitting levels. Whereas Crosby and Carstens have assigned the emitting levels to a $d\pi^*$ orbital parentage (*21*), DeArmond and Hillis (*22*) have suggested that the emitting levels contain significant contribution from both $d\pi^*$ and $\pi\pi^*$ configurations. [In the partial study by Crosbey and Carstens (*21*) the *cis*-dichlorobis(1,10-phenanthroline)-iridium(III) ion was incorrectly identified as the tris species.] Homo-bischelated complexes of Ir(III) with several substituted phenanthroline ligands have been studied to determine the magnitude of this configuration interaction, and the matrix element responsible for the interaction has been estimated to be about 280 cm^{-1} (23). Based upon this estimate, the emitting levels of the unsubstituted complex would contain only a very small contribution from the $\pi\pi^*$ configuration, and we concur with the assignment of these levels to a $d\pi^*$ orbital parentage.

The luminescence lifetime of this complex at 77°K in glassy medium has been found to be 6.92 μsec, and the luminescence quantum yield is 0.496 (24). From these values, the intrinsic lifetime for the radiative decay of these levels at 77°K is estimated to be 14.0 μsec. Studies of the temperature dependence of the luminscence lifetime of similar complexes between 2° ond 100°K indicate that the emission is caused by four thermally equilibrated levels which lie within roughly 100 cm^{-1} of one another (4). The lifetime and quantum yield data reported at 77°K are therefore representative of the Boltzmann weighted average values of these quantities for a manifold of four thermally equilibrated levels.

The luminescence of cis-dichlorobis-5,6-dimethyl-1,10-phenanthroline iridium(III) chloride, [IrCl$_2$(5,6-mephen)$_2$]Cl, has also been studied in glassy media at 77°K (23). This emission has been assigned to a transition from a $\pi\pi^*$ excited level to the ground state in polar media such as methanol–water (25). In less polar solvents, increasing contributions from $d\pi^*$ configurations occur, as evidenced by a shortening of the luminescence lifetime. The lifetimes for this complex at 77°K range from 448 to 66 μsec as the solvent polarity is decreased (23). A luminescence quantum yield of 0.77 has been determined in ethanol–methanol glass at 77°K, and the intrinsic radiative lifetime is 86 μsec under these conditions. The number of emitting levels which are in thermal equilibrium at 77°K has not been determined for this complex.

The complex ion, cis-dichlorobis-4,7-dimethyl-1,10-phenanthroline iridium(III) chloride [IrCl$_2$(4,7-mephen)$_2$]Cl, has luminescence properties which are intermediate between the aforementioned homo-bischelated complexes (23). The emitting levels of this complex are best classified as nearly equal admixtures of $d\pi^*$ and $\pi\pi^*$ configurations. The luminescence lifetime at 77°K ranges from 208 to 22 μsec as the solvent polarity is decreased. The quantum yield of 0.62 in ethanol–methanol glass at 77°K indicates an intrinsic radiative lifetime of 35 μsec under these conditions. The number of equilibrated levels responsible for the emission of this complex as in the previous case, has not been determined.

The hetero-bischelated complex, [IrCl$_2$,phen)(5,6-mephen)]Cl, displays a non-exponentital luminescence decay curve when excited at 337 nm in ethanol–methanol glass at 77°K (19). Analysis of the decay curves of this complex by a non-linear least squares fit to a function representing the sum of two exponentials indicates that the emission is caused by levels with lifetimes of 65 and 9.5 μsec (20). Both time-resolved spectroscopy and analysis of decay curves as a function of emission wavelength indicate that the non-equilibrated manifolds of emitting levels are separated by 200–300 cm^{-1}, with the 65-μsec manifold lowest in energy. Time-resolved emission spectroscopy shows that the 65-μsec manifold has an emission spectrum which is insensitive to solvent polarity whereas the 9.5-μsec manifold moves to higher energy in more polar solvents. This data suggests that the long-lived level has $\pi\pi^*$ orbital parentage, and the short-lived level has $d\pi^*$ parentage.

The complex [IrCl$_2$(phen)(4,7-mephen)]Cl also displays a non-exponential luminescence decay when excited at 337 nm in ethanol–methanol glass at 77°K (20). Analysis of the luminescence decay curves of this complex by a two-state model indicates that the lifetimes of the two manifolds are 22 and 8.8 μsec. In contrast to the previous complex, it is the short-lived manifold of levels which lie lowest in this molecule. The splitting of the two manifolds of levels is about 300 cm^{-1} in this case.

In both hetero-bischelated complexes mentioned above, there is some evidence that there may be three sets of non-equilibrated levels responsible for the luminescence excited by 337-nm excitation (26). The determination of the lifetimes and energies of three levels by least squares analysis of decay curves as a function of emission wavelength is currently being studied and will be reported. For now, we will analyze these complexes based upon the two-level model which was used to interpret our data.

Orbital Parentage of Emitting Levels

Here we discuss a simple model for describing the orbital parentage of the luminescent levels of the complexes discussed above. We concentrate on a model which emphasizes the changes which occur in going from a homo- to a hetero-bischelated complex. The model is presented in this way to facilitate a description of the circumstances which lead to the unusual luminescence properties of the hetreo-bischelated complexes.

We begin with a consideration of the changes which occur in the d-orbitals associated with the central metal atom when the binding to two identical bidentate ligands in the homo-bischelated complexes is replaced by binding to two dissimliar ligands in the hetro-bischelated complexes. This change results in the elimination of the C_2 symmetry axis. There are, however, no degenerate representations in the C_2 symmetry group of the homo-bischelated complex, and no degeneracies in the d-orbitals are present to be split by the symmetry reduction brought about by formation of the hetero-bischelated complex. Futrhermore, no large changes in the average crystal field are expected to be caused by the slight alterations of the 1,10-phenanthroline ligand brought about by methyl substituents. Indeed, this has been shown to be true for the methyl-substituted derivatives of K[IrCl₄(phen)] (27). Thus, we conclude that no significant alterations of the d-orbitals need be considered in describing the luminescent states of these complexes.

We now turn to the π-orbitals of the ligands. Although the p-orbitals of the chloride ligands in these complexes undoubtedly make some contribution to the π-orbitals of the complex molecules, these ligands are common to all of the complexes we have studied. We therefore neglect these chloride orbitals and concentrate on the contributions of the phenanthroline molecular orbitals to the π-orbitals of the complex.

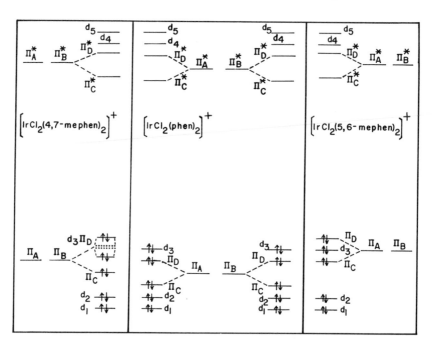

Figure 1. Molecular orbital energy level diagram for homo-bischelated complexes of iridium(III) with 1,10-phenanthroline, 4,7-dimethyl-1,10-phenanthroline, and 5,6-dimethyl-1,10-phenanthroline

We begin by considering the formation of complex π-orbitals from the π molecular orbitals on two identical 1,10-phenanthroline ligands. This process is pictured in the center section of Figure 1. Here, π_A and π_B represent the highest filled molecular orbitals of the two unbound phenanthroline ligands, and $\pi_A{}^*$ and $\pi_B{}^*$ represent the lowest unfilled molecular orbitals. Upon complex formation, the degenerate orbitals of the two identical ligands, A and B, will interact to form complex molecular orbitals. Since the phenanthroline orbitals are degenerate before the interaction, these resultant complex orbitals will contain equal contributions from the two ligand orbitals. Thus, π_C, π_D, $\pi_C{}^*$, and $\pi_D{}^*$ will be completely delocalized over the two bidentate ligands. On the basis of Figure 1, the lowest energy $\pi\pi^*$ state of the complex would arise from the promotion of an electron from π_D to $\pi_C{}^*$. This state should be at slightly lower energy than the lowest $\pi\pi^*$ state of the ligand. This effect is often observed in metal complexes with $\pi\pi^*$ luminescent states (28).

The other states which may arise from simple orbital promotions may be depicted by introducing a set of five non-degenerate d-orbitals. The lowest three of these stem from the degenerate t_{2g} set in octahedral symmetry, and these three contain the six paired d-electrons of Ir(III). Crystal field or d-d excited states of the complex arise from promotion of electrons from these filled d-orbitals to the two empty ones which stem from the e_g set in octahedral symmetry. Since no luminescence from a d-d excited level has been observed in the complexes we are considering, all of the d-d excited states are presumably at a higher energy than the $\pi\pi^*$ and $d\pi^*$ states. Charge-transfer or $d\pi^*$ states arise from promotion of an electron from the filled d-orbitals to the π^* complex orbitals.

The luminescent levels in the Ir(III) complexes we are considering arise from $d\pi^*$ or $\pi\pi^*$ orbital parentage, and the highest filled complex π orbital (π_D) and the highest filled d-orbital (d_3) are presumably close in energy. Since a $d\pi^*$ luminesecnce is observed for the [IrCl$_2$(phen)$_2$]$^+$ ion, d_3 has been placed slightly above π_D to represent this situation. The energy gap between the $d\pi^*$ emitting state of this complex and the lowest $\pi\pi^*$ state has been estimated to be about 1.1 kK (23). This presumably arises from a combination of the difference in orbital promotion energy and the difference in electron repulsion terms for the two types of states.

The observation of a $\pi\pi^*$ luminescence from the [IrCl$_2$(5,6-me-phen)$_2$]$^+$ ion is depicted on the right side of Figure 1. The lowest $\pi\pi^*$ state of 5,6-dimethyl-1,10-phenanthroline is at lower energy than that of 1,10-phenanthroline, and a smaller orbital promotion energy is depicted between π_A and $\pi_A{}^*$ or π_B and $\pi_B{}^*$. Since we feel that the d-orbital energies will be very nearly the same as before methyl substitution of the two ligands, the highest filled π-orbitals of 5,6-dimethyl-1,10-phenanthroline have been placed above those of 1,10-phenanthroline. Assuming an equivalent ligand–ligand interaction in the two homo-bischelated complexes, this leads to the placement of π_D above d_3 in the [IrCl$_2$(5,6-mephen)$_2$]$^+$ ion. Hence, the orbital promotion energy for the lowest $\pi\pi^*$ transition in this complex is less than the $d\pi^*$ promotion energy. The energy gap between these two states has been estimated to be about 0.6 kK (23), and is again presumably caused by a combination of differences in this promotion energy as well as electron repulsion terms.

On the left side of Figure 1 we compare the orbital energy levels of the [IrCl$_2$(phen)$_2$]$^+$ and [IrCl$_2$(4,7-mephen)$_2$]$^+$ complex ions. The low-

est $\pi\pi^*$ state of 4,7-dimethyl-1,10-phenanthroline is known to have an energy intermediate between that of 1,10-phenanthroline and 5,6-dimethyl-
1,10-phenanthroline. The homo-bischelated Ir(III) complex of this ligand
displays a mixed $\pi\pi^*$-$d\pi^*$ luminescent state. These facts are rationalized
by placing π_A and π_B for this ligand between the highest filled π-orbitals
of 1,10-phenanthroline and 5,6-dimethyl-1,10-phenanthroline. Again, assuming an equivalent ligand–ligand interaction in this complex, this would
make π_D and d_3 nearly degenerate. The interaction of π_D and d_3 would
then be partciularly large and would lead to complex molecular orbitals
delocalized over both the ligands and the central metal ion. This delocalization is depicted in Figure 1 by the formation of orbitals labelled π_D, d_3
from the interaction of the metal d-orbital and complex π-orbital. The
transition, $\pi_C^* \rightarrow \pi_D, d_3$ would then lead to the mixed $\pi\pi^*, d\pi^*$ luminescence of this complex.

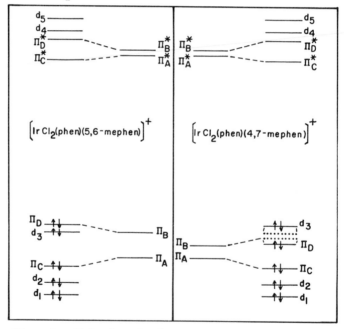

*Figure 2. Molecular orbital energy level diagram for hetero-
bischelated complexes of iridium(III) with 1,10-phenanthro-
line, 4,7-dimethyl-1,10-phenanthroline, and 5,6-dimethyl-1,10-
phenanthroline*

   In Figure 2 we depict the formation of complex molecular π-orbitals
from the π-orbitals of two non-identical phenanthroline ligands. On the
left side of the diagram, we treat the hetero-bischelated complex ion,
$[IrCl_2(phen)(5,6-mephen)]^+$. The ligand orbtials π_A and π_A^* represent
molecular orbitals on 1,10-phenanthroline, and the ligand orbitals π_B and

$\pi_B{}^*$ represent molecular orbitals on 5,6-dimethyl-1,10-phenanthroline. In this diagram, π_B has been placed above π_A in agreement with our placement of these levels in Figure 1. Since π_A and π_B are no longer degenerate in this complex, a much smaller ligand–ligand interaction will occur. Whatever interaction does occur leads to the formation of complex molecular orbitals, π_C and π_D. The most important point of our treatment is that π_C and π_D are not delocalized equally over the two bidentate ligands. Rather, π_C is localized primarily on 1,10-phenanthroline, and π_D is localized primarily on 5,6-dimethyl-1,10-phenanthroline. The lowest unfilled π orbitals on the two ligands are represented by $\pi_A{}^*$ and $\pi_B{}^*$. In our treatment we assume that the 5,6-dimethyl-1,10-phenanthroline orbital, $\pi_B{}^*$ lies slightly above $\pi_A{}^*$ on 1,10-phenanthroline. Since $\pi_A{}^*$ and $\pi_B{}^*$ are again not degenerate, they will interact less strongly than in the homo-bischelated complex to form complex molecular orbitals $\pi_C{}^*$ and $\pi_D{}^*$. Of these two, $\pi_C{}^*$ will be localized mainly on 1,10-phenanthroline and $\pi_D{}^*$ on 5,6-dimethyl-1,10-phenanthroline. From our diagram, the lowest promotional energy is seen to occur for the $\pi_D \rightarrow \pi_C{}^*$ transition. However, this transition would represent the transfer of charge from one bidentate ligand to the other, and we expect that this transition between weakly ineracting chromophores will lie at much higher energy than indicated by its orbital promotion energy. We believe it is the $\pi_D \rightarrow \pi_D{}^*$ transition which represents the lowest energy $\pi\pi^*$ state of the complex. Note that the luminescence which would arise from this transition would be classified as a $\pi\pi^*$ emission localized on 5,6-dimethyl-1,10-phenanthroline. Our diagram predicts that the next excited state of the complex would arise from the $d_3 \rightarrow \pi_C{}^*$ transition. This excited state may be classified as a $d\pi^*$ state localized in the part of the complex in the vicinity of the central metal and the 1,10-phenanthroline ligand. Thus, our treatment suggests that of the two lowest excited states in this hetero-bischelated complex, the lowest is localized on 5,6-dimethyl-1,10-phenanthroline, and the next is localized in the metal-1,10-phenanthroline part of the complex.

On the right side of Figure 2 we depict the formation of complex molecular orbitals in the $[IrCl_2(phen)(4,7\text{-mephen})]^+$ ion. As before, π_A and π_B represent molecular orbitals on 1,10-phenanthroline and 4,7-dimethyl-1,10-phenanthroline, respectively, as do $\pi_A{}^*$ and $\pi_B{}^*$. The 4,7-dimethyl-1,10-phenanthroline orbitals are again placed above those of 1,10-phenanthroline but not as far above as was done for 5,6-dimethyl-1,10-phenanthroline. This corresponds to our treatment of the homo-bischelated complexes in Figure 1. Interaction of π_A and π_B leads to the formation of complex molecular orbitals π_C and π_D localized primarily on 1,10-phenanthroline and 4,7-dimethyl-1,10-phenanthroline. Since this interaction is somewhat smaller than in the homo-bischelated complex of 4,7-dimethyl-1,10-phenanthroline, we indicate that π_D now lies close to d_3

but below it. Interaction of π_D and d_3 now leads to two complex orbitals, the higher of which contains more d_3 than π_D and the lower of which contains more π_D than d_3. The lowest excited π-orbitals consist of $\pi_C{}^*$, which is localized mainly on 1,10-phenanthroline and $\pi_D{}^*$, which is localized mainly on 4,7-dimethyl-1,10-phenanthroline. The lowest promotional energy corresponds to the $d_3 \rightarrow \pi_C{}^*$ transition. The excited state arising from this transition would be described as a $d\pi^*$ level, and the transition density is localized mainly in the part of the complex between the metal and the 1,10-phenanthroline ligand. A small change in the electron density on the 4,7-dimethyl-1,10-phenanthroline ligand is associated with this state because of the interaction of d_3 and π_D. Other low-lying states would arise from the $\pi_D \rightarrow \pi_C{}^*$ and $\pi_D \rightarrow \pi_D{}^*$ promotions. As before, $\pi_D \rightarrow \pi_C{}^*$ is expected to lie at somewhat higher energy than indicated by the promotional energy, and the state arising from the $\pi_D \rightarrow \pi_D{}^*$ promotion is expected to be the lowest $\pi\pi^*$ state. Thus, our treatment of this complex indicates that the lowest excited level is a charge-transfer primarily from Ir(III) to 1,10-phenanthroline with some $\pi\pi^*$ character mixed in. The next highest level is a $\pi\pi^*$ state localized on 4,7-dimethyl-1,10-phenanthroline with some $d\pi^*$ character mixed in.

A recent study of the photochemistry of $[IrCl_2(phen)_2]Cl$ and $[IrCl_2-(5,6-mephen)_2]Cl$ attributes the photochemical activity of these complexes to thermal population of a low-lying set of levels of d-d orbital parentage (29). We have now studied the time-resolved emission spectroscopy of these complexes between $-196°$ and $0°C$ in glycerol, and have concluded that the emission around $0°C$ is primarily d-d in character whereas the emission at $-196°C$ is mainly d-$\pi^*\{[IrCl_2(phen)_2]Cl\}$ or π-$\pi^*\{[IrCl_2(5,6-mephen)]Cl\}$ (30). We attribute the lack of d-d emission at $-196°C$ to a thermal barrier for the radiationless process which leads from the d-π^* or π-π^* levels to the d-d levels.

In view of these results, it follows that the placement of the orbital labelled d_4 in Figures 1 and 2 may have to be altered. This orbital may be, in fact, the lowest unfilled molecular orbital in these complexes. We are currently studying the time resolved spectroscopy of the hetero-bischelated complexes between $-196°$ and $0°C$ to establish the correct placement of the d-d levels.

Energy Transfer Processes

In the simple molecular orbital model for the lowest excited states of these molecules, the primary difference between the homo- and hetero-bischelated complexes is in the localization of the excitation energy in the latter. In this section we associate this localization of energy with the failure of the lowest excited states of the hetero-bischelated complexes to attain thermal equilibrium. On the other hand, the lowest excited states of the homo-bischelated complexes are thermally equilibrated, and this equilibration is associated with the delocalization of the excitation energy over the complex.

The description of the radiationless processes which lead to the transfer of excitation energy from one excited state to another in heavy metal complexes has been discussed by several authors (*11, 17, 22*). Spin-orbit coupling must be included in the description of the initial and final states for the radiationless process but may also contribute to the coupling terms responsible for this process (*11, 17*). There are also vibronic coupling terms which arise from breakdown of the Born-Oppenheimer approximation which presumably play some role in radiationless energy transfer processes in both heavy- and light-atom molecules (*11, 17*). Beyond these electronic terms, there are vibrational contributions to the rate constant for a radiationless process, and these contributions are generally expressed as Franck-Condon factors (*11*).

The essential meaning of these coupling terms is that the electron distribution for a given excited state of a molecule, with energy E_1, may be altered to one for another excited state, with energy E_2, if the energy difference, $E_1 - E_2$, is transferred into nuclear vibrational energy. In this transfer process, energy must flow into vibrations of nuclei in the region where the electronic distribution is changing from that of the initial state to that of the final state (*11*). In a sense, the nuclear vibrations may thus be viewed as a communication medium between the initial and final electron distributions. In the case where the two changing electron distributions are delocalized over the entire molcule, as in the homo-bischlated complexes, the nuclei can effectively communicate messages between the two distributions. However, where two localized changing distributions are encountered, as in the hetero-bischelated complexes, there may be no nuclei in common to the two. Alternatively, the nuclei which are in the region where changes in both occur may not participate in vibrations which have the appropriate frequency to provide the communication medium.

Since the non-equilibrated levels of the hetero-bischelated complexes are separated by only several hundred wavenumbers, there are a limited number of vibrations of the appropriate frequency to enable the molecule to undergo a transition from one of these levels to the other. It is our point of view that in these complexes, there are no appropriate vibrational frequencies of nuclei in the region where the two changing electron distributions overlap to bring about a transition from one to the other. Thus, thermal equilibration of these localized electron distributions does not occur, and emissions from several independent manifolds of different orbital parentage are observed.

Concluding Remarks

We have described the orbital parentage of the lowest excited states of homo- and hetero-bischelated complexes of Ir(III). This description stresses the localization of transition charge density in excited states of hetero-bischelated complexes in specific areas of the molecular framework. In the homo-bischelated complexes, delocalization of the transition charge density over large segments of the molecular framework is emphasized. We believe that these factors are important in determining the pathways for energy transfer processes in these complexes.

The hetero-bischelated complexes discussed represent key molecules for the further study of energy transfer processes in heavy-atom molecules. Since two or more non-equilibrated sets of excited manifolds of different orbital parentage are known to emit light in these complexes, it will now be possible to determine whether selective population of any of these manifolds from higher excited states occurs by studying the effect of excitation wavelength on the ratio of the emission intensity from the luminescent levels. In this way, we may be able to determine whether or not there are selective pathways for radiationless deactivation of the higher excited states of heavy-atom molecules such as those which are known to occur for light-atom molecules (15, 16).

Literature Cited

1. Demas, J. N., Crosby, G. A., *J. Amer. Chem. Soc.* (1970) **92**, 7262.
2. DeArmond, M. K., *Accounts Chem. Res.* (1974) **7**, 309.
3. Watts, R. J., Crosby, G. A., *J. Amer. Chem. Soc.* (1972) **94**, 2606.
4. Watts, R. J., Harrigan, R. W., Crosby, G. A., *Chem. Phys. Lett.* (1971) **8**, 49.
5. Harrigan, R. W., Crosby, G. A., *J. Chem. Phys.* (1973) **59**, 3468.
6. Hipps, K. W., Crosby, G. A., *Inorg. Chem.* (1974) **13**, 1544.
7. Harrigan, R. W., Hager, G. D., Crosby, G. A., *Chem. Phys. Lett.* (1971) **8**, 49.
8. Crosby, G. A., Hipps, K. W., Elfring, W. H., *J. Amer. Chem. Soc.* (1974) **96**, 629.
9. Robinson, G. W., Frosch, R. P., *J. Chem. Phys.* (1962) **37**, 1962.
10. *Ibid.*, (1963) **38**, 1187.
11. Robbins, D. J., Thomson, A. J., *Mol. Phys.* (1973) **25**, 1103.
12. Siebrand, W., *J. Chem. Phys.* (1967) **47**, 2411.
13. *Ibid.*, (1966) **44**, 4055.
14. *Ibid.*, (1967) **46**, 440.
15. El-Sayed, M., *J. Chem. Phys.* (1963) **38**, 2834.
16. El-Sayed, M., *Accounts Chem. Res.* (1968) **1**, 8.
17. Thomas, T. R., Watts, R. J., Crosby, G. A., *J. Chem. Phys.* (1973) **59**, 2123.
18. Halper, W., DeArmond, M. K., *J. Lumin.* (1972) **5**, 225.
19. Watts, R. J., *J. Amer. Chem. Soc.* (1974) **96**, 6186.
20. Watts, R. J., Brown, M., Griffith, B. G., Harrington, J. S., *J. Amer. Chem. Soc.* (1975) **97**, 6029.
21. Crosby, G. A., Carstens, D. H. W., in "Molecular Luminescence," E. C. Lim, Ed., p. 309, W. A. Benjamin, New York, 1969.

22. DeArmond, M. K., Hillis, J. E., *J. Chem. Phys.* (1971) **54**, 2247.
23. Watts, R. J., Crosby, G. A., Sansregret, J. L., *Inorg. Chem.* (1972) **11**, 1474.
24. Demas, J. N., Crosby, G. A., *J. Amer. Chem. Soc.* (1971) **93**, 2841.
25. Crosby, G. A., Watts, R. J., Carstens, D. H. W., *Science* (1970) **170**, 1195.
26. Watts, R. J., Griffith, B. G., Harrington, J. S., *J. Amer. Chem. Soc.*, in press.
27. Watts, R. J., Van Houten, J., *J. Amer. Chem. Soc.* (1974) **96**, 4334.
28. Carstens, D. H. W., Crosby, G. A., *J. Mol. Spectry.* (1970) **34**, 113.
29. Ballardini, R., Varani, G., Moggi, L., Valzani, V., Olson, K. R., Scandola, F., Hoffman, M. Z., *J. Amer. Chem. Soc.* (1975) **97**, 728.
30. Watts, R. J., White, T. P., Griffith, B. G., *J. Amer. Chem. Soc.* (1975) **97**, 6914.

RECEIVED January 24, 1975. Work supported in part by the Petroleum Research Fund and by the Committee on Research of the University of California, Santa Barbara.

18

Fluorescence Properties of Rare Earth Complexes Containing 1,8-Naphthyridines

M. NGSEE, H. W. LATZ, and D. G. HENDRICKER

Ohio University, Athens, Ohio 45701

The solid state fluorescence spectra of some Eu^{+3}, Sm^{+3}, Tb^{+3}, and Dy^{+3} nitrate, chloride, and acetylacetonate (acac) complexes of 1,8-naphthyridine (napy) and 2,7-dimethyl-1,8-naphthyridine (2,7-dmnapy) are recorded. The relative fluorescence intensities of analogous complexes of the same metal are in the order $NO_3^- > Cl^- > acac$ and 2,7-dmnapy > napy. The Stark splittings of the europium emission lines are consistent with site symmetries of C_2 for the 10-coordinate $Eu(napy)_2(NO_3)_3$, of C_s for the 8-coordinate $Eu(acac)_3$-(2,7-dmnapy), and of C_3 for the chloride arrangement in the 11-coordinate $EuCl_3(napy)_2(H_2O)$.

$\mathbf{F}$luorescence of a rare earth complex was first reported by Weissman in 1942 (*1*). By the early 1960's, a detailed understanding of the mechanism of fluorescence of rare earth chelates had been achieved that was provided in large part by the extensive studies of Crosby and co-workers (*2, 3, 4, 5, 6*). The discovery of laser action of an alcohol solution of europium benzoyl-acetonate by Lempicki and Samelson (*7*) stimulated interest in the preparation of numerous rare earth complexes and in the investigation of their potential as laser materials (*8, 9*). Of the many rare earth complexes which are known to fluoresce, the vast majority contain β-diketonate ligands. However, there are also a few which contain only nitrogen heterocyclic ligands (*9, 10, 11*).

The ligands 1,10-phenanthroline (phen) and 2,2-bipyridine (bipy) yield fluorescent complexes wherein five-membered chelate rings are formed (*9, 10, 11*). Recently, studies of the coordinating ability of the four-membered ring-chelating agents, 1,8-naphthyridine (napy) and 2,7-dimethyl-1,8-naphthyridine (2,7-dmnapy) with rare earth halides (*12*) and nitrates (*13, 14*) yielded complexes with proposed coordination numbers of 10–12. In addition, mixed ligand complexes of the type $M(\beta\text{-diketonate})_3(2,7\text{-dmnapy})$ where $M =$ rare earth were also prepared (*12*).

napy   2,7-dmnapy

The quantum yields for emission lines of these complexes are expected to increase as the coordination number about the rare earth becomes larger and as the energies of the ligand triplet state and ion resonance level approach one another. On the basis of these criteria, the naphthyridine complexes of Sm^{+3}, Eu^{+3}, Tb^{+3}, and Dy^{+3} might be expected to yield more intense emission spectra than similar phen and bipy complexes. Also, the number and splitting of emission lines observed for the europium complexes are indicative of the symmetry about the metal (*9, 11, 15*), and thus they may be used to delineate possible coordination geometries. For these reasons, we undertook an investigation of the fluorescence spectra of several naphthyridine complexes of Sm^{+3}, Eu^{+3}, Tb^{+3}, and Dy^{+3}.

Experimental

The rare earth complexes were prepared as previously reported (*12, 13, 14*), and their formulation was verified by elemental analysis. Solid state fluorescence spectra were recorded at room temperature with an Aminco–Bowman model 4-8202 model spectrophotofluorometer equipped with a 600 line/mm grating and a xenon lamp. The spectra were obtained by the front surface fluorescence method using the solid sample accessory C73-62140 and a quartz cell. A paste comprised of nitromethane and complex was prepared and placed on the surface of the cell. The nitromethane was then evaporated by a nitrogen stream, which left a film of complex. An excitation wavelength of 310–320 nm was used in all experiments. Spectra were observed using 0.5 mm slits and slit program number one. In this operating mode, resolution to ± 1 nm is expected. Phosphorescence decay times were determined by mechanically blocking the excitation radiation and recording the emission decay curve using an X–Y recorder driven at 2 in./sec. The lifetime values (τ) are accurate to ± 0.02 sec.

Results and Discussion

The details of the solid state fluorescence spectra are tabulated in Table I. The band energies for the samarium complexes are nearly

Table I. Details of Uncorrected Fluorescence

Compound

$Sm(NO_3)_3(2,7\text{-}dmnapy)_2$		$Sm(NO_3)_3(napy)_2$		$SmCl_3(napy)_2(H_2O)$	
λ, nm	RI[a]	λ, nm	RI	λ, nm	RI
465	vw	463	w	463	vw
506	w	506	w	505	w
562	2.5	558	2.0	559	.27
568	2.8	566	2.3		
595	3.1	593	2.5	595	.45

$Eu(NO_3)_3(2,7\text{-}dmnapy)_2$		$Eu(NO_3)_3(napy)_2$		$EuCl_3(napy)_2(H_2O)$	
λ, nm	RI	λ, nm	RI	λ, nm	RI
478	vw	470	vw	467	vw
529	1	525	w	535	w
548	w	552	w	553	w
573	2	572	2	580	1.5
586	18	586	6.5	591	5.5
608	52	608	44	614	5.5

$Tb(NO_3)_3(2,7\text{-}dmnapy)_2$		$Tb(NO_3)_3(napy)_2$		$TbCl_3(napy)_2(H_2O)$	
λ, nm	RI	λ, nm	RI	λ, nm	RI
488	1020	487	110	488	74
539	1420	541	148	542	121
584	60	583	8	581	14
617	5	620	w	616	w

$Dy(NO_3)_3(2,7\text{-}dmnapy)_2$		$Dy(NO_3)_3(napy)_2$		$DyCl_3(napy)_2(H_2O)$	
λ, nm	RI	λ, nm	RI	λ, nm	RI
458	1.7	456	w	456	vw
481	60	479	17	480	13
573	54	572	15	570	7

[a] Relative intensity (RI) is assigned arbitary units; vw = very weak, w = weak.
[b] Mol. phos. = molecular phosphorescence.

identical. The assignments of the transitions of the fluorescence emissions at 505 nm to $^4G_{7/2}$–$^6H_{5/2}$, those at 559 nm to $^4G_{5/2}$–$^6H_{5/2}$, those at 568 nm to $^4F_{3/2}$–$^6H_{7/2}$, and those at 595 nm to $^4G_{5/2}$–$^6H_{7/2}$ agree with assignments made by others (16, 17, 18). The samarium complexes have the weakest fluorescence of the four rare earths studied. The very weak band at 465 nm, which is also observed in the dysprosium and europium complexes, apparently results from ligand molecular phosphorescence. Strong molecular phosphorescence of napy was reported at 450–454 nm (19, 20). The lifetimes of this band of the 2,7-dmnapy complexes of samarium and dysprosium nitrate are $\tau = 0.14$ sec and 0.10 sec, respectively. Lifetimes of this magnitude are reasonable only for phosphorescence and not for

Spectra of Rare Earth Complexes

Compound		Assignment
$Sm(acac)_3(2,7\text{-}dmnapy)$		
λ, nm	RI	
		mol. phos.[6]
		$^4G_{7/2}-{}^6H_{5/2}$
560	w	$^4G_{5/2}-{}^6H_{5/2}$
		$^4F_{3/2}-{}^6H_{7/2}$
598	w	$^4G_{5/2}-{}^6H_{7/2}$

$Eu(acac)_3(2,7\text{-}dmnapy)$		
λ, nm	RI	
		mol. phos.
		$^5D_1-{}^7F_1$
		$^5D_1-{}^7F_2$
525	0.5	$^5D_0-{}^7F_0$
588	0.4	$^5D_0-{}^7F_1$
607	3.0	$^5D_0-{}^7F_2$

$Tb(acac)_3(2,7\text{-}dmnapy)$		
λ, nm	RI	
486	700	$^5D_4-{}^7F_6$
542	1320	$^5D_4-{}^7F_5$
579	44	$^5D_4-{}^7F_4$
615	4	$^5D_4-{}^7F_3$

$Dy(acac)_3(2,7\text{-}dmnapy)$		
λ, nm	RI	
458	0.9	mol. phos.
481	23	$^4F_{9/2}-{}^6H_{15/2}$
573	19	$^4F_{9/2}-{}^6H_{13/2}$

fluorescence emission (9, 10).

The band intensities of the dysprosium and europium complexes are fairly similar except for the acetylacetonate (acac) adducts. The assignments for the two ion emission bands observed for the dysprosium complexes are $^6F_{9/2}-{}^6H_{15/2}$ at 480 nm and $^4F_{9/2}-{}^6H_{13/2}$ at 572 nm (21). For many other dysprosium complexes, a band assigned to the $^4F_{9/2}-{}^6H_{11/2}$ transition was also reported at approximately 650 nm (5, 16). Second-order scatter radiation from the excitation source which occurs in this region prevented observation of this transition. The europium complexes have only two available resonance levels, $^5D_0$ and $^5D_1$ (8, 9, 22). For the appropriate assignments for the observed bands, *see* Table I. The $^5D_0-{}^7F_2$

transition is associated with the laser behavior of these complexes (7, 8, 9).

The terbium complexes have by far the most intense fluorescence. For terbium, the $^5D_4$ state at 20,500 cm^{-1} is the only ion resonance level below the 22,210 cm^{-1} triplet state of napy (8, 9, 22). The assignments for the five observed transitions agree with several reports (23, 24). No band attributable to ligand molecular phosphorescence was observed for any of these stronger fluorescent complexes.

The fluorescence intensity of the same ion resonance band of rare earth nitrate complexes of napy is always less than that of the corresponding 2,7-dmnapy analog. The increased basicity of 2,7-dmnapy relative to napy resulting from the electronic effects of the methyl groups ortho to the donor nitrogen sites results in a stronger metal–ligand bond and thus a more efficient energy transfer which enhances fluorescence. Similar effects were observed by Sinha (25) with europium and terbium complexes of bipy and methyl-substituted bipy and by Filipescu et al. in a study of methoxy-substituted β-diketonate rare earth complexes (26).

On a qualitative basis, the fluorescence of the napy complexes containing chloride is weaker than that of the corresponding nitrate compounds. Such variation can be ascribed to the difference in coordinating ability of the anion and/or to the amount of coordinated water. Conductivity studies of the nitrate and chloride complexes in nitromethane yield Λ values of 13–19, which indicates that the complexes are nonelectrolytes (12, 13, 14). The presence of bands attributable to νM–Cl and νM–OH$_2$ in the IR spectra of the chloride complexes substantiates the conclusion that all species are bound to the rare earth ion (12). Vibrational modes indicative of bidentate nitrate and the lack of bands suggesting monodentate nitrate are reported for the rare earth nitrate complexes of napy and 2,7-dmnapy (13, 14).

The ability of the anions to coordinate with rare earths decreases in the order NO$_3^-$ > Cl$^-$. The stronger covalent bond formed by nitrate than by chloride should provide better shielding of the metal ion and thus result in enhanced fluorescence intensity. Eu^{+3} in nitrate solution yields stronger fluorescence than in chloride solution (27). Intensities of emissions from hydrated complexes are significantly lower than those of the same metal complex in anhydrous form (6). The proposal that the decrease in fluorescence intensity is caused by a high energy vibration of the water that provides a radiationless de-excitation of the rare earth ion has been confirmed by deuterium substitution studies (28, 29). Because of the apparently greater coordination number of the chloride complex relative to the nitrate compound (vide infra), one would expect a greater fluorescence intensity for the former compound. Therefore, both the bonding ability of the anion and the presence of coordinated water contribute to the lower fluorescence intensity of the chloride complexes relative to the nitrate complexes. Since no anhydrous chloride complexes have yet been obtained, it is not now possible to evaluate the contribution of each effect to the total fluorescence intensity of the complex.

For 2,7-dmnapy complexes of the same rare earth, the recorded intensities of the nitrate complexes are always greater than those in which acac is the anion. Conductivity measurements on nitromethane solutions of $M(acac)_3(2,7$-dmnapy$)$ give Λ values of 6–9 which are typical of non-electrolytes (*11, 12*). IR spectra of the acac complexes have νCO absorptions at approximately 1600 and 915 cm^{-1} and νM–O bands at 405 and 313 cm^{-1}. These absorptions and the lack of a 1700-cm^{-1} band indicate that both oxygens of each of the acac units are coordinated to the metal (*11, 30*). The more intense fluorescence of the nitrate complex may result from the presence of a second 2,7-dmnapy ligand which would increase the coordination number of the rare earth from 8 to 10. The triplet state of acac is reported at 25,300 cm^{-1} (*31*). The triplet state of napy at 22,210 cm^{-1} is closer to the rare earth resonance levels and may contribute to a more efficient energy transfer which in turn would enhance fluorescence intensity.

The fluorescence spectra of the europium adducts also provide information about the geometry of the complex. The lower the local site symmetry about the europium ion, the more numerous the lines in the spectrum. This fine structure results from internal Stark splitting of the ionic levels. The maximum number of lines into which each transition will be split as a function of geometry has been tabulated by several authors (*11, 15, 32*). The observation of a $^5D_0$–$^7D_0$ transition for all the complexes reported indicates that none possesses a center of symmetry.

The fluorescence spectrum of Eu(napy)$_2$(NO$_3$)$_3$ has a quintuplet for the $^5D_0$–$^7F_2$ transition and three lines for the $^5D_0$–$^7F_1$ emission. This splitting pattern, in conjunction with the observation of the $^5D_0$–$^7F_0$ line, is consistent with site symmetries of C_n or C_{nv} where $n \leq 3$ (*15*). From the IR and conductivity data, a 10-coordinate europium ion is expected (*14*). The two most energetically favored polyhedra for coordination number 10 are the bicapped square antiprism and the bicapped dodecahedron (*33, 34*). In the structure of La(bipy)$_2$(NO$_3$)$_3$ as determined by x-ray methods, the ligands are so arranged as to give bicapped dodecahedral geometry with C_2 site symmetry for the lanthanum (*35*). A similar structure for the analogous napy complex would be completely consistent with the splitting pattern observed in the fluorescence spectrum and with the known preference of napy for the dodecahedral coordination polyhedron in eight-coordinate complexes (*36, 37*).

The splitting pattern in the fluorescence spectrum of the Eu(acac)$_3$-(2,7-dmnapy) complex is nearly identical to that reported for Eu(acac)$_3$-(phen) (*17, 38*). A singlet, triplet, and quintuplet are observed for the

$^5D_0-^5F_0$, $^5D_0-^5F_1$, and $^5D_0-^5F_2$ transitions respectively. This pattern limits the site symmetry of the metal to C_n or C_{nv} where $n \leq 3$. X-ray structure determination of the complex $Eu(acac)_3(phen)$ reveals that the europium ion is eight-coordinate (39). The ligands span the s edges of a distorted square antiprism about the europium whose site symmetry is C_s. The IR and conductivity data for the 2,7-dmnapy analog also support octacoordination (12). However, the smaller ligand bite of 2,7-dmnapy (~ 2.3 Å) compared with those of phen (2.74 Å) and acac (2.82 Å) suggests that a dodecahedral ligand arrangement which also yields C_s site symmetry for europium would be more reasonable for $Eu(acac)_3$-(2,7-dmnapy) (33, 34, 36).

The Stark splitting of the $^5D_0-^5F_1$ and $^5D_0-^5F_2$ lines into a doublet and quintuplet in the spectrum of $EuCl_3(napy)_2(H_2O)$ is identical to the splitting pattern reported for $EuCl_3(bipy)_2(H_2O)$ (40). The proposed structure for the bipy complex is an octahedron of bridging chlorides with the water molecule located on a C_3 axis of the octahedron and the two bipy molecules arranged so that their nitrogen atoms form a square about the same C_3 axis (40). Since the IR and conductivity data indicate that all species in $EuCl_3(napy)_2(H_2O)$ are bound to the metal, an 11-coordinate structure for the napy complex similar to that of the bipy analog might be expected (12). If all the chlorides of $EuCl_3(napy)_2(H_2O)$ were non-bridging, an 8-coordinate complex of symmetry C_s would be produced. As was observed for the 8-coordinate $Eu(acac)_3(2,7\text{-dmnapy})$ compound of similar symmetry, a triplet and quintriplet for the $^5D_0-^5F_1$ and $^5D_0-^5F_2$ lines should occur. Since such is not the case, the 11-coordinate model is presently favored.

Literature Cited

1. Weissman, S. I., *J. Chem. Phys.* (1942) 10, 214.
2. Crosby, G. A., *Mol. Cryst.* (1966) 1, 37.
3. Crosby, G. A., Whan, R. E., Freeman, J. J., *J. Phys. Chem.* (1962) 66, 2493.
4. Whan, R. E., Crosby, G. A., *J. Mol. Spectrosc.* (1962) 8, 315.
5. Freeman, J. J., Crosby, G. A., *J. Phys. Chem.* (1963) 67, 2717.
6. Freeman, J., Crosby, G., Lawson, K., *J. Mol. Spectrosc.* (1964) 13, 399.
7. Lempicki, A., Samelson, H., *Phys. Lett.* (1963) 4, 133.
8. Ross, D. L., Blanc, J., "Europium Chelates as Laser Materials," ADV. CHEM. SER. (1967) 71, 155.
9. Sinha, A. P. B., in "Spectroscopy in Inorganic Chemistry," C. N. R. Rao and J. R. Ferraro, Eds., Vol. 2, p. 255, Academic, New York, 1971.
10. Sinha, S. P., "Complexes of the Rare Earths," Pergamon, New York, 1966.
11. Forsberg, J. H., *Coord. Chem. Rev.* (1973) 10, 195.
12. NgSee, M., M.S. Thesis, Ohio University, 1974.
13. Hendricker, D. G., Foster, R. J., *J. Inorg. Nucl. Chem.* (1972) 34, 1949.
14. Foster, R. J., Hendricker, D. G., *Inorg. Chim. Acta* (1972) 6, 371.
15. Sinha, S. P., Butter, E., *Mol. Phys.* (1969) 16, 285.

16. Sinha, S. P., Z. *Naturforsch.* (1965) **20a**, 835.
17. Melby, L. R., Rose, N. J., Abramson, E., Caris, J. C., J. *Am. Chem. Soc.* (1964) **86**, 5117.
18. Mango, M., Dieke, G., J. *Chem. Phys.* (1962) **37**, 2354.
19. Nishimura, A., Vincent, J., *Chem. Phys. Lett.* (1971) **11**, 609.
20. Nishimura, A., Tinti, D., Vincent, J., *Chem. Phys. Lett.* (1971) **12**, 360.
21. Crosswhite, H., Dieke, G., J. *Chem. Phys.* (1966) **35**, 1535.
22. Dieke, G., "Spectra and Energy Levels of Rare Earth Ions in Crystals," Interscience, New York, 1968.
23. Van Uitert, L., Soden, R., J. *Chem. Phys.* (1960) **32**, 1161.
24. Thomas, S., Singh, S., Dieke, G., J. *Chem. Phys.* (1963) **38**, 2180.
25. Sinha, S. P., J. *Inorg. Nucl. Chem.* (1966) **28**, 189.
26. Filipescu, N., Sager, W., Serafin, F., J. *Phys. Chem.* (1964) **68**, 3324.
27. Gallagher, P., J. *Chem. Phys.* (1964) **41**, 3061.
28. Hutchinson, C., Magnum, B., J. *Chem. Phys.* (1960) **32**, 1261.
29. Wright, M., Frosch, R., Robinson, G., J. *Chem. Phys.* (1960) **33**, 934.
30. Joshi, K., Pathak, V., J. *Inorg. Nucl. Chem.* (1973) **35**, 3161.
31. Sager, W., Filipescu, N., Serafin, F., J. *Phys. Chem.* (1965) **69**, 1092.
32. Brecher, C., Samelson, H., Lempicki, A., J. *Chem. Phys.* (1965) **42**, 1081.
33. Muetterties, E. L., Wright, C. M., Q. *Rev. Chem. Soc.* (1967) **21**, 109.
34. Lippard, S. J., in "Progress in Inorganic Chemistry," F. A. Cotton, Ed., Vol. 8, p. 114, Interscience, New York, 1967.
35. Al-Karaghouli, A. R., Wood, J. S., J. *Am. Chem. Soc.* (1968) **90**, 6548.
36. Singh, P., Clearfield, A., Bernal, I., J. *Coord. Chem.* (1971) **1**, 29.
37. Epstein, J. M., Dewan, J. C., Kepert, D. L., White, A. H., J. *Chem. Soc. Dalton Trans.* (1974) 1949.
38. Bauer, H., Blanc, J., Ross, D., J. *Am. Chem. Soc.* (1964) **86**, 5125.
39. Watson, W. H., Williams, R. J., Stemple, N. R., J. *Inorg. Nucl. Chem.* (1972) **34**, 501.
40. Filipescu, N., Bjorklund, S., McAvoy, N., Dengnan, Jr., J. *Chem. Phys.* (1968) **48**, 2895.

RECEIVED February 28, 1975. Work supported in part by the Ohio University Research Committee (OURC-342).

19

Unusually Volatile and Soluble Metal Chelates: Lanthanide NMR Shift Reagents

R. E. SIEVERS,[1] J. J. BROOKS,[2] J. A. CUNNINGHAM,[3] and W. E. RHINE[3]

Aerospace Research Laboratories,[4] ARL/LJ, Wright–Patterson Air Force Base, Ohio 45433

Novel lanthanide β-diketonate complexes have been synthesized. Their properties include thermal, hydrolytic and oxidative stabilities, volatility, Lewis acidity, and unusually high solubility in nonpolar organic solvents. Various combinations of these properties make lanthanide complexes useful as NMR shift reagents and fuel antiknock additives and in other applications. NMR spectral studies revealed that the Pr(III), Yb(III), and Eu(III) complexes of 1,1,1,2,2,3,3,7,7,7-decafluoro-4,6-heptanedione have sufficient Lewis acidity to induce appreciable shifts in the proton resonances of weak Lewis bases such as anisole, acetonitrile, nitromethane, and p-nitrotoluene. Data from single-crystal structure determinations indicate that the NMR shift reagent–substrate complexes are not stereochemically rigid and that effective axial symmetry may exist by virtue of rapid intramolecular rearrangements.

Metal β-diketonate complexes display a variety of interesting and unusual properties. Among these are thermal, hydrolytic, and oxidative stabilities, volatility, Lewis acidity, and unusual solubility in nonpolar organic solvents. In general, it is a particular combination of these properties rather than any single one that makes possible the use of these complexes in a diverse range of applications. For example, be-

---

[1] Present address: Dept. of Chemistry, University of Colorado, Boulder, Colo. 80302.
[2] Present address: Monsanto Research Corp., Dayton Laboratory, Dayton, Ohio 45407.
[3] Present address: Air Force Materials Laboratory, Wright–Patterson Air Force Base, Ohio 45433.
[4] On June 30, 1975, The Aerospace Research Laboratories were abolished; consequently, all correspondence should be directed to the first author at his present address.

cause of the volatility and thermal stability of certain of these complexes, the application of gas chromatography to ultratrace metal analysis became feasible (*1*). Because of other combinations of properties, these compounds are used as NMR shift reagents (*2, 3, 4*), fuel antiknock additives (*5, 6*), homogeneous catalysts (*7*), model compounds for gas phase stereochemical studies (*8*), and selective gas chromatographic liquid phases (*9*).

Perhaps the most widely recognized use of lanthanide β-diketonates is as NMR shift reagents. This application takes advantage not only of the intrinsic paramagnetic nature of certain of the lanthanide ions, but also of the Lewis acidity, hydrolytic stability, and high solubility in nonpolar organic solvents of their complexes. This paper describes our recent studies of the use of these unusual chelates as NMR shift reagents.

Volatile, stable rare earth complexes were originally synthesized with the hope that differences in volatility would provide a means of separating and purifying the rare earths. Early claims that the lanthanide acetylacetonates were volatile were later shown to be incorrect. The tris complexes generally occur as hydrates, and they do not exhibit the necessary stability for fractional sublimation or gas chromatographic separation. The hydrated lanthanide acetylacetonates undergo self-hydrolysis on heating, and the reaction products are no longer appreciably volatile. The synthesis and characterization of the anhydrous, sterically hindered Ln(thd)$_3$ complexes represented a major advancement in the search for volatile, stable lanthanide complexes (*1*). Table I lists the ligands studied most extensively in our laboratory.

Table I. Structures and Abbreviations of β-Diketones

$$R^1 - \overset{O}{\underset{|}{C}} - \overset{\ominus}{CH} - \overset{O}{\underset{|}{C}} - R^2$$

R^1	R^2	Abbreviation
CH_3-	$-CH_3$	acac
$C(CH_3)_3-$	$-C(CH_3)_3$	thd
$CF_3CF_2CF_2-$	$-C(CH_3)_3$	fod
$CF_3CF_2CF_2-$	$-CF_3$	dfhd

An extension of this research led to the preparation of fluorinated β-diketones which form stable and even more volatile rare earth complexes. Detailed studies of the Ln(thd)$_3$ (*1*), Ln(fod)$_3$ (*12*), and Ln(dfhd)$_3$ (*13, 14*) complexes revealed that the volatility of the tris complex is directly proportional to the degree of fluorine substitution and inversely proportional to the ionic radius (15) of the metal atom for a given series of chelate complexes.

In 1969 (16) it was reported that addition of the bis-pyridine adduct of Eu(thd)$_3$ induced large shifts in the proton NMR spectrum of cholesterol. Subsequently, it was reported that the unsolvated Eu(thd)$_3$ was even more effective as an NMR shift reagent (17). Since these first reports appeared, more than 400 papers have been published (2, 3, 4) in this field on subjects ranging from spectral clarification to the selection of the best shift reagent for a given application. The reagents most widely used in the early studies were the tris-thd complexes of Eu(III), Pr(III), and Yb(III). In general, Eu(thd)$_3$ and Yb(thd)$_3$ induce downfield shifts whereas Pr(thd)$_3$ induces upfield shifts. Although the shifts induced by Yb(thd)$_3$ are usually greater than those induced by the Pr and Eu analogs, much of the fine structure in the NMR spectra is often lost because of signal broadening.

Despite the successes achieved with the Ln(thd)$_3$ complexes used as shift reagents, they have limited solubility in the usual NMR solvents (18, 19) such as chloroform and carbon tetrachloride, and they are nearly insoluble in solvents such as acetonitrile, nitromethane, and p-dioxane. In addition, the interaction between the Ln(thd)$_3$ chelates and weak nucleophiles is often not strong enough to result in complexes which exhibit large induced shifts.

During the course of our research on volatile rare earth complexes, we found that the fod chelates were more soluble in a wide range of solvents than either the acetylacetonates or the thd complexes. We postulated that the presence of the electronegative fluorine atoms increases the Lewis acidity of the metal which results in a stronger interaction with various nucleophiles (20). The greater Lewis acidity of Eu(fod)$_3$ relative to Eu(thd)$_3$ was demonstrated by the resolution of resonances in a mixture of isomeric azoxybenzenes (21). This phenomenon was also demonstrated independently by gas chromatography (GC) (9)—fluorinated β-diketonate complexes interact more strongly with organic nucleophiles than do nonfluorinated ones. A similar GC study that concentrated on Eu(fod)$_3$ (10) related the strengths of these interactions to the donating abilities of the organic nucleophiles and to steric constraints. The fod complexes are now the most widely used class of NMR shift reagents because of greater convenience in use and wider applicability to weak nucleophiles.

Further substitution of fluorine atoms in the β-diketone side chains has led to the synthesis and characterization of the Ln(dfhd)$_3$ complexes as NMR shift reagents. Although the hydrated Ln(fod)$_3$ complexes are more soluble in chloroform than the hydrated Ln(dfhd)$_3$ complexes, the lanthanide dfhd complexes have superior solubility in dioxane and aceto-

Table II. Solubility of Some Lanthanide β-Diketonate Shift Reagents[a]

	Solvent		
Complex	Acetonitrile	Dioxane	Chloroform
$Ln(dfhd)_3 \cdot xH_2O$[b]	>1	>1	0.08
$Ln(fod)_3 \cdot xH_2O$[b]	0.8	0.03	0.5
$Ln(thd)_3$	insoluble	0.03	0.03

[a] Solubility is given in g/g.
[b] Drying the shift reagent over P_4O_{10} increases its solubility in chloroform.

nitrile (Table II). It is important to have shift reagents that are soluble and that can function well in solvents such as acetonitrile and dioxane because many compounds of biological importance are soluble only in solvents such as these. The dfhd complexes appear to be very promising for such applications.

The high degree of solubility of the dfhd complexes in deuterated acetonitrile makes it possible to measure experimentally the degree of hydration of these complexes. The integrated intensity of the H_2O proton resonance compared with that of the methine proton of the shift reagent provides a measure of the relative amount of water present. The chemical shift of the water protons depends on the concentration, but it is observed downfield from the methine proton resonance of the shift reagent. Typical residual water after drying 3 days *in vacuo* over P_4O_{10} is 0.5 mole/mole europium chelate. This amount of water does not seriously interfere with the ability of the complex to function effectively as a shift reagent.

The increased Lewis acidity of the $Ln(dfhd)_3$ complexes relative to the thd and the fod compounds is demonstrated by a comparison of the induced shifts in $CDCl_3$ solutions of such weak bases as acetonitrile, nitromethane, and anisole (Table III). The induced shifts are for the methyl protons of each compound. Although $Pr(dfhd)_3$ and $Yb(dfhd)_3$ induce the largest shifts, line broadening is appreciable; in fact, in experi-

Table III. Comparison of Shifts Induced in the Spectra of Weak Lewis Bases[a]

	Anisole	Acetonitrile		Nitromethane	
Complex	0.1 R:S[b]	0.1 R:S	0.3 R:S	0.1 R:S	0.3 R:S
$Pr(dfhd)_3$	−1.58	−1.50	−3.45	−1.20	−2.27
$Yb(dfhd)_3$	2.03	2.28	5.67	1.50	3.00
$Eu(dfhd)_3$	0.68	0.77	1.92	0.45	0.88
$Eu(fod)_3$	0.22	0.40	0.85	0.20	0.23
$Eu(thd)_3$	0.35	0.38	0.75	0.20	0.23

[a] Shifts are given in ppm. Data obtained at 60 MHz with 10^{-4} mole shift reagent dissolved in 0.5 g $CDCl_3$.
[b] R:S—the mole ratio of shift reagent to substrate.

ments with tetrahydrofuran (THF), no fine structure was observed in the THF proton resonances. In similar experiments with Eu(dfhd)₃, there was no discernible broadening.

The greater Lewis acidity of the Ln(dfhd)₃ shift reagents is also demonstrated by the fact that adducts are formed and relatively large shifts are induced even with weak Lewis bases such as nitrobenzene derivatives. The induced shifts are perhaps best illustrated by the para-substituted derivatives such as *p*-chloronitrobenzene and *p*-nitrotoluene. The data for induced shifts with Eu(dfhd)₃ are summarized in Table IV.

Table IV.   NMR Shifts in *p*-Chloronitrobenzene and *p*-Nitrotoluene Induced by Addition of Eu(dfhd)₃[a]

	p-*Chloronitrobenzene*	p-*Nitrotoluene*
ortho-H	1.5	1.9
meta-H	0.5	1.2
methyl-H		0.33

[a] Shifts are given in ppm. Data obtained at 60 MHz with 10^{-4} mole complex dissolved in $CDCl_3$; mole ratio of shift reagent to substrate, 0.3.

The electron-withdrawing chloro group reduces the Lewis basicity of the nitro group, and therefore the induced shifts for *p*-chloronitrobenzene are smaller, as might be expected from consideration of the Hammett sigma function. Plots of chemical shift *vs.* mole ratio of shift reagent-to-substrate for each proton are linear over the range of 0.0–0.5 mole ratio. It is assumed that the slopes of these lines give the magnitudes of the induced shifts which contain information about the geometry of the complex.

In order to approximate the conformation of the complex in solution, a calculation of the type described by Willcott *et al.* (22) (assuming effective axial symmetry) was made on the *p*-nitrotoluene chemical shift data (*see* Ref. 2, p. 143). Because axial symmetry is assumed, the simplified McConnell–Robertson equation can be used. In several calculations the principal magnetic axis was varied so that it was directed from the europium atom to various points along the line bisecting the O–N–O angle. A broad minimum was obtained; therefore, within the above constraints, the orientation of the principal magnetic axis does not appear to affect seriously the calculated gross geometry of the complex. The best fit of the NMR data was obtained when the Eu atom was positioned 2.2 ± 0.2 A above the plane of the nitro group and coplanarity of the nitro group and the benzene ring was assumed. The calculation in which the principal magnetic axis was assumed to be colinear with the Eu–N vector is illustrated in Figure 1.

Relative to either the Ln(thd)₃ or the Ln(fod)₃ chelates, use of the Ln(dfhd)₃ complexes offers an additional advantage since the β-diketone

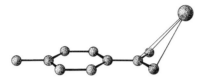

Figure 1. Calculated coordination geometry of the europium–p-nitrotoluene interaction

has only the methine proton resonance which can possibly coincide with the substrate proton resonances. In the dfhd complexes there are only 3 methine protons in the chelate shell whereas there are 3 methine and 27 *tert*-butyl protons in the Ln(fod)$_3$ complexes and 3 methine and 54 *tert*-butyl protons in the Ln(thd)$_3$ complexes. Furthermore, interference is very unlikely in Eu(dfhd)$_3$ complexes since the methine proton resonance experiences a shift in the opposite direction from that of the substrate resonances; for Eu(dfhd)$_3$ dried over P$_4$O$_{10}$, this resonance appears upfield from TMS. Consequently, one can accomplish the same objective of eliminating interfering proton peaks by substituting fluorine atoms for methyl groups in the fod ligand as was done previously by deuteration.

Many questions concerning the interpretation of NMR shift reagent data remain unanswered. Hopefully, single-crystal structural determinations of shift reagent–substrate complexes will allow basic structural principles to be deduced. The structures of Eu(thd)$_3$(DMF)$_2$, Eu(thd)$_3$-(DMSO), Eu(thd)$_3$(1,10-phenanthroline), and Yb(thd)$_3$(DMSO) were recently determined (23). These structures, in conjunction with structures determined in other laboratories (*see* Ref. 2, pp. 368–369), lead to several generalizations concerning the stereochemistry of shift reagent complexes.

In the single-crystal structure of Eu(thd)$_3$(DMSO), two conformations of the complex occupy the same unit cell. The two conformations have the same gross stereochemistry, but they differ significantly in detail. Therefore, this indicates that the NMR shift reagent complexes are not structurally rigid, but rather that the coordination polyhedron and the ligands can be easily distorted by packing effects. This is not an isolated example. The same phenomenon was also observed in crystalline Eu(thd)$_3$(DMF)$_2$ which also contains two non-equivalent molecules in the unit cell.

A comparison of Yb(thd)$_3$(DMSO) and Eu(thd)$_3$(DMSO) reveals that these compounds crystallize in different space groups and that there are significant differences in the stereochemistry of their solid state structures. The coordination geometry of the Yb complex can best be described as a trigonal base–tetragonal base polyhedron whereas the Eu complex can best be described as a distorted pentagonal bipyramid. Intramolecular and interligand contacts between *tert*-butyl groups indicate that there are no serious steric interactions. To date, the preferred stereochemistry of higher coordinate complexes has not been successfully correlated with factors such as crystal field stabilization, ligand–ligand interactions, and solvation energies, and the crystal lattice energy appears to be the most important factor that determines the solid state structure (24). We conclude that the observed differences between the solid state structures of Yb(thd)₃(DMSO) and Eu(thd)₃(DMSO) may be caused predominantly by packing considerations.

The bulky *tert*-butyl groups on the thd ligand do not appear to be the dominant factor in determining or limiting the coordination geometry to any one configuration. In the octacoordinate complex Eu(thd)₃(DMF)₂, the ligands form a distorted square antiprism with two of the thd moieties forming one square face and one thd and the two DMF ligands (cis to each other) forming the other face. The 1,10-phenanthroline adducts of Eu(thd)₃ and Eu(acac)₃ also have square antiprismatic solid state structures (25). Differences between the latter two structures cannot be attributed to the presence of bulky *tert*-butyl groups. In fact, it is the thd complex which has the most compact coordination polyhedron, and the β-diketonate ring in closest proximity to the 1,10-phenanthroline group has a smaller fold angle about the oxygen–oxygen vector. Perhaps the greater electron-donating ability of the thd ligand may account for the shortening of the Ln–O bond lengths in the thd complex. In contrast to the above pattern, the nucleophiles in Eu(thd)₃(pyridine)₂ (26) and Ho(thd)₃(picoline)₂ (27) are bonded to opposite faces of a square antiprism and are as far apart from each other as possible. Analyses of these structures imply that several arrangements of the ligands about the central metal atom are possible.

If one wishes to extract stereochemical information about substrates in solution from the data obtained in the NMR experiments, two physical–mathematical models are available. For shifts induced by a dipolar (pseudocontact) mechanism, the McConnell–Robertson equation (Equation 1) (28, 29)—where r_i, θ_i, and ϕ_i are the spherical polar coordinates of the ith resonating nucleus in the coordinate system of the principal magnetic axes—relates the direction and magnitude of the shift to the geometry of the substrate–chelate complex. If the substrate–chelate

$$\Delta H_i / H = K \, (3 \cos^2 \theta_i - 1)/r^3{}_i + K' \, (\sin^2 \theta_i \cos 2 \, \phi_i)/r^3{}_i \qquad (1)$$

complex has axial symmetry (a C_3 or higher axis of rotation), Equation 1 may be simplified to the more manageable form of Equation 2 were θ_i is the angle between the principal magnetic axis and the vector from the lanthanide ion to the ith resonating nucleus.

$$\Delta H_i / H = K \, (3 \cos^2 \theta_i - 1)/r^3{}_i \qquad (2)$$

Briggs *et al.* (30) have shown that an equation similar in form to Equation 2 can be derived and that it should be valid when the substrate ligand undergoes free rotation about an axis passing through the lanthanide ion as well as when the substrate–chelate complex forms three or more interconverting rotamers that are equally populated. Significantly, in this derivation no *a priori* assumptions are made concerning the symmetry of the complex. In the Briggs *et al.* equation, θ_i now denotes the angle between the rotation axis and the vector from the paramagnetic lanthanide ion to the *i*th resonating nucleus.

Although almost all the reported crystal structures of lanthanide NMR shift reagent complexes are devoid of any symmetry element greater than the trivial C_1 rotation axis, this need not eliminate the possibility of effective axial symmetry in solution. The coordination geometry exhibited by higher coordinate lanthanide ions in a crystalline arrangement apparently is affected by packing considerations. A direct comparison of solid state and solution structure as determined from NMR experiments may not be valid because of the great difference in the time scales of the two techniques. Averaging of several dissymetric arrays, such as those found in the crystal structure of $Eu(thd)_3(DMSO)$, leads to an equivalent and possibly an axially symmetric description for the lanthanide shift reagent–substrate complex. Low potential energy barriers between the idealized higher coordinate polyhedra might also permit the time-averaged solution configuration to be significantly different from that displayed in the crystal. This averaging process, whatever the details, must be rapid with respect to the NMR time scale since shift reagent studies at ambient temperatures always reveal a single NMR spectrum that is the average of free and complexed substrate and of all intermediate species in solution. Crystal structure determinations, however, indicated that, in the seven-coordinate substrate–lanthanide shift reagent complex, steric crowding is not as serious as was once believed, and that free rotation about the Ln–X bond is possible. Consequently there may be two processes operating in solution that allow successful application of the simplified equation: (a) free rotation of the substrate and (b) rapid interconversion of geometric isomers to produce an effective axially symmetrical complex.

Lanthanide complexes with optically active β-diketones have been used to determine the purity of optical isomers (*see* chap. 4, p. 87 of Ref. 2 for a review). The most widely used chiral shift reagents are based on 3-trifluoroacetyl-*d*-camphor, the anion of which is designated facam. The crystal structure determination of the DMF adduct of tris(3-trifluoroacetyl-*d*-camphorato)praseodymium(III), the first of a chiral shift reagent, has been completed (*31*). The asymmetric unit contains the dimer, $(facam)_3Pr(DMF)_3Pr(facam)_3$, with the DMF oxygen atoms forming bridges between the two $Pr(facam)_3$ moieties. Therefore, each $Pr(III)$ ion is nine-coordinate with a geometry best described as a capped

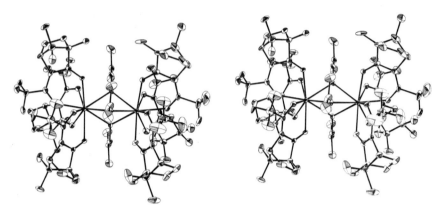

Figure 2.   A stereoscopic drawing of the dimer (facam)₃Pr(DMF)₃Pr(facam)₃. Average bond distances and angles: Pr–O(facam), 2.46(3) A; Pr–O(DMF), 2.60(2) A; Pr–Pr, 4.078(9) A; and Pr–O(DMF)–Pr, 103.6(1.8) A.

square antiprism (*see* Figure 2). By ignoring the ring backbones and focusing only on the Pr_2O_{15} core, one can see that there is a pseudo-mirror plane containing the three bridging DMF oxygen atoms. The presence of the asymmetric centers in the facam backbone prevents the entire dimer from having mirror symmetry. Even in this molecule, which contains large facam ligands, there is still sufficient space in the coordination sphere of Pr(III) to accommodate three DMF groups. One may conclude that the unsolvated complexes, particularly for the early members of the lanthanide series, should be able to bind large nucleophiles with little difficulty. Now that x-ray data are available, it is apparent that steric crowding is not as severe as was originally believed. It is also noteworthy that in all structures determined, the Ln–X bond lengths are in the expected ranges but that major reorganizations of the ligands about the metal atom are observed.

Literature Cited

1. Eisentraut, K. J., Sievers, R. E., *J. Am. Chem. Soc.* (1965) **87**, 5254.
2. Sievers, R. E., Ed., "Nuclear Magnetic Resonance Shift Reagents," Academic, New York, 1973.
3. Cockerill, A. F., Davies, G. L. O., Harden, R. C., Rackham, D. M., *Chem. Rev.* (1973) **73**, 553.
4. Reuben, J., *Prog. Nucl. Magn. Reson. Spectrosc.* (1973) **9**, 1.
5. Tischer, R. L., Eisentraut, K. J., Scheller, K., Sievers, R. E., Bausman, R. C., Blum, P. R., "New Rare Earth Antiknock Additives that Are Potential Substitutes for Tetraethyl Lead," Aerospace Res. Lab., Wright–Patterson Air Force Base, Ohio (1974) Rep. **ARL TR-74-0170** (Nat. Tech. Inf. Ser. Rep. **AD/A-006, 151/5 WP**).
6. Eisentraut, K. J., Tischer, R. L., Sievers, R. E., "Rare Earth β-Ketoenolate Antiknock Additives in Gasoline," U.S. Patent 3,794,473 (1974).
7. Sievers, R. E., *et al.*, unpublished data.

8. Kutal, C., Sievers, R. E., *Inorg. Chem.* (1974) **13**, 897.
9. Feibush, B., Richardson, M. F., Sievers, R. E., Springer, Jr., C. S., *J. Am. Chem. Soc.* (1972) **94**, 6717.
10. Brooks, J. J., Sievers, R. S., *J. Chromatogr. Sci.* (1973) **11**, 303.
11. Sicre, J. E., Dubois, J. T., Eisentraut, K. J., Sievers, R. E., *J. Am. Chem. Soc.* (1969) **91**, 3476.
12. Springer, C. S., Meek, D. W., Sievers, R. E., *Inorg. Chem.* (1967) **6**, 1105.
13. Richardson, M. F., Sievers, R. E., *Inorg. Chem.* (1971) **10**, 498.
14. Scribner, W. G., Smith, B. H., Moshier, R. W., Sievers, R. E., *J. Org. Chem.* (1970) **35**, 1969.
15. Sievers, R. E., in "Coordination Chemistry," S. Kirschner, Ed., p. 270, Plenum, New York, 1969.
16. Hinckley, C. C., *J. Am. Chem. Soc.* (1969) **91**, 5160.
17. Sanders, J. K. M., Williams, D. H., *Chem. Commun.* (1970) 422.
18. Demarco, P. V., Elzey, T. K., Lewis, R. B., Wenkert, E., *J. Am. Chem. Soc.* (1970) **92**, 5734.
19. *Ibid.* (1970) **92**, 5737.
20. Rondeau, R. E., Sievers, R. E., *J. Am. Chem. Soc.* (1971) **93**, 1522.
21. Rondeau, R. E., Sievers, R. E., *Anal. Chem.* (1973) **45**, 2145.
22. Willcott, M. R., Lenkinski, R. E., Davis, R. E., *J. Am. Chem. Soc.* (1972) **94**, 1742.
23. Cunningham, J. A., Sievers, R. E., *Proc. 10th Rare Earth Res. Conf., Carefree, Ariz., April–May 1973.*
24. Blight, D. G., Kepert, D. L., *Theor. Chim. Acta* (1968) **11**, 51.
25. Watson, W. H., Williams, R. J., Stemple, N. R., *J. Inorg. Nucl. Chem.* (1972) **34**, 501.
26. Cramer, R. E., Seff, K., *Chem. Commun.* (1972) 400.
27. Horrocks, Jr., W. De W., Sipe, J. P., Luber, J. R., *J. Am. Chem. Soc.* (1971) **93**, 5258.
28. McConnell, H. M., Robertson, R. E., *J. Chem. Phys.* (1958) **29**, 1361.
29. Horrocks, Jr., W. DeW., *J. Am. Chem. Soc.* (1974) **96**, 3022.
30. Briggs, J. M., Moss, G. P., Randall, E. W., Sales, K. P., *J. Chem. Soc. Chem. Commun.* (1972) 1180.
31. Cunningham, J. A., Sievers, R. E., *J. Am. Chem. Soc.* (1975) **97**, 1586.

RECEIVED February 26, 1975.

20

The Effect of Coordination Pattern and 5f Electron Configuration on Organoactinide Reactivity

TOBIN J. MARKS

Northwestern University, Evanston, Ill. 60201

This article discusses developments in two fields of organo-actinide chemistry: sigma-bonded actinide organometallics and actinide ions as templates in ligand cyclization reactions. The marked thermal stability of $(\eta^5\text{-}C_5H_5)_3U\text{-}R$ and $(\eta^5\text{-}C_5H_5)_3Th\text{-}R$ compounds results largely from coordinative saturation which thwarts decomposition via β-hydride elimination as well as other routes. Thermolysis of these compounds occurs by intramolecular elimination of R–H. The coordinatively unsaturated UR_4 and ThR_4 compounds are less thermally stable, and they decompose where possible by hydride elimination. The organometallic product of $(\eta^5\text{-}C_5H_5)_3ThR$ thermolysis is $(\eta^5\text{-}C_5H_5)_2Th(\eta^1\text{:}\eta^5\text{-}C_5H_4)_2Th(\eta^5\text{-}C_5H_5)_2$ which suggests the importance of an intermediate $(\eta^1\text{-}C_5H_4)Th(\eta^5\text{-}C_5H_5)_2$ carbene complex–ylid species. When phthalocyanine condensations are carried out in the presence of uranyl ion, a complex of the macrocyclic super-phthalocyanine ligand, a five-subunit, expanded analog of phthalocyanine, is obtained.

The great flowering in transition metal organometallic chemistry during the last 20 years has, until very recently, largely ignored the actinide elements. Full interest in actinide organometallic chemistry did not fully awaken until the recent synthesis of the novel compound uranocene. Since then it has become increasingly apparent that the actinides have a rich and diverse organometallic chemistry (1–7). Moreover, these metal ions have unique stereochemical (high coordination numbers, unusual coordination geometries) and electronic (5f valence orbitals) features that are unknown in the d transition series which suggest interesting prospects for their use as reagents and catalysts. This paper presents recent developments in two burgeoning fields: actinide-to-carbon sigma bonds and actinide ions as templates in cyclooligomerization reactions. The principal focus is on those coordinative as well as electronic features which control the rates and selectivities of the chemical transformations that are undergone when organic ligands are coordinated to actinide ions.

Actinide-to-Carbon Sigma Bonds

Prior to and during the Manhattan Project, considerable effort was directed toward the synthesis of volatile uranium alkyl compounds such as $U(C_2H_5)_4$ for use in isotope separation. These attempts were unsuccessful (8, 9) and it was assumed that the uranium alkyl bond was intrinsically unstable. Thus, for several decades the nature of the uranium-to-carbon linkage remained unexplored, and there was even some question as to whether such bonding could take place.

These preconceptions were dispelled by the synthesis of the remarkably thermally stable uranium organometallics of the formula $(\eta^5\text{-}C_5H_5)_3$-U–R *via* Reaction 1 (*10, 11, 12, 13, 14*). The molecular structure of the

$$(C_5H_5)_3UCl \xrightarrow[\text{RMgX}]{\text{RLi}} (C_5H_5)_3U–R \qquad (1)$$

$$
\begin{array}{lll}
R = CH_3 & = \text{neopentyl} & = p\text{-tolyl} \\
 = n\text{-}C_4H_9 & = tert\text{-}C_4H_9 & = \text{benzyl} \\
 = \text{allyl} & = C_6H_5 & = \text{-2-}cis\text{-2--butenyl} \\
 = i\text{-}C_3H_7 & = C_6F_5 & = \text{-2-}trans\text{-2-butenyl} \\
 = \text{vinyl} & = C_2C_6H_5 &
\end{array}
$$

R = phenylacetylide complex is depicted in Figure 1 (*15*). Although they are exceedingly air sensitive, these compounds generally have very high thermal stability. For example, the half-life of the R = *n*-butyl compound in toluene solution at 97°C exceeds 1000 hr.

That such monohapto organometallics are not unique to uranium was demonstarted by the high yield synthesis of the thorium(IV) analogs (Reaction 2) (*16, 17*). These complexes have even greater thermal stabil-

$$(C_5H_5)_3ThCl \xrightarrow[\text{RMgX}]{\text{RLi}} (C_5H_5)_3Th–R \qquad (2)$$

$$
\begin{array}{ll}
R = n\text{-}C_4H_9 & = \text{-2-}cis\text{-2-butenyl} \\
 = \text{allyl} & = \text{-2-}trans\text{-2-butenyl} \\
 = i\text{-}C_3H_7 & = \text{5-hexenyl} \\
 = \text{neopentyl} & = n\text{-}C_3H_7
\end{array}
$$

ity than the $(C_5H_5)_3UR$ compounds. For example, in toluene solution (sealed tube) at +170°C $(C_5H_5)_3Th(n\text{-butyl})$ has a half-life of more than 95 hr. The thorium complexes offer an intriguing opportunity to

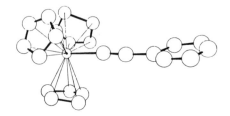

Figure 1. The structure of (η^5-C_5H_5)$_3$-
$UC_2C_6H_5$ (15)

examine the effects of two 5f electrons (Th(IV) is $5f^0$; U(IV) is $5f^2$) on the properties of actinide alkyls.

In view of the reported instability of uranium tetra-alkyls, we undertook an investigation of those factors which stabilize actinide-to-carbon sigma bonds (13, 17). The approach was to study the kinetic aspects of (C_5H_5)$_3$MR thermolysis in solution and to scrutinize the thermal decomposition products, including those in which deuterium and stereochemical labels were incorporated. Several significant patterns emerged. Thermolysis of both (C_5H_5)$_3$UR and (C_5H_5)$_3$ThR compounds in toluene solution does not occur *via* the commonly observed β-hydrogen elimination sequence (Reaction 3 sometimes followed by Reaction 4) (18–28). Rather,

$$\text{CH}_2\!\!=\!\!\text{CHR}$$
$$\text{M--CH}_2\text{CH}_3\text{R} \rightleftharpoons \overset{\downarrow}{\text{M--H}} \rightleftharpoons \text{M--H} + \text{CH}_2\!\!=\!\!\text{CHR} \qquad (3)$$
$$\mathbf{1}$$

$$\text{M--H} + \text{M--CH}_2\text{CH}_3\text{R} \rightarrow 2\text{M} + \text{CH}_3\text{CH}_2\text{R} \qquad (4)$$

for both M = Th and U, R–H is eliminated (13, 17). Deuterium-labelling studies—*e.g.* (C_5D_5)$_3$MR thermolyzed in toluene and (C_5H_5)$_3$MR thermolyzed in toluene-d_8—indicate that abstraction of a cyclopentadienyl ring hydrogen takes place. Crossover experiments (*e.g.* Reaction 5) indicate that abstraction is predominantly intramolecular. Some crossover

$$(C_5H_5)_3\text{MR}' \quad \text{R}'\text{--H}$$
$$\rightarrow \qquad\qquad (5)$$
$$(C_5D_5)_3\text{MR} \quad \text{R--D}$$

was observed in the products of the thorium experiment. However, mass spectra of starting material isolated from partially thermolyzed samples indicates that this scrambling occurs prior to thermal decomposition. Experiments were also designed in view of the fact that vinylic free radicals are known to invert at a rate which is competitive with diffusion out of solvent cages and interception by the most potent radical scavengers

$$\underset{\text{R}}{\overset{\text{R}}{\diagdown}}\text{C}\!\!=\!\!\text{C}\underset{\text{H}}{\overset{\text{R}}{\diagup}} \;\rightleftharpoons\; \underset{\text{R}}{\overset{\text{R}}{\diagup}}\text{C}\!\!=\!\!\text{C}\underset{\text{H}}{\overset{\text{R}}{\diagdown}} \qquad (6)$$

(Reaction 6) (*13, 29–37*). The -2-*cis*-2-butenyl and -2-*trans*-2-butenyl complexes of both uranium and thorium, thermoylze to produce pure *cis*-2-butene and *trans*-2-butene, respectively (*13, 17*). Thus, the hydrogen abstraction occurs with retention of configuration at the carbon atom bound initially to the actinide, and free radicals are apparently not present

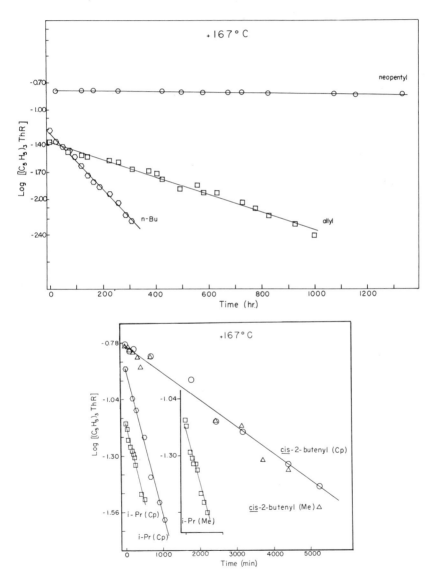

Figure 2. Kinetic plots for the thermal decomposition of various $(C_5H_5)_3ThR$ compounds in toluene solution. Functional groups in parentheses indicate the PMR resonances monitored for the data.

for any length of time along the reaction coordinate. For $(C_5D_5)_2(C_5H_5)$-U(n-butyl), the hydrogen abstraction exhibits a kinetic isotope effect (k_H/k_D) of approximately 8 ± 1 (38). As a comparison, some k_H/k_D values for β-hydride elimination in organometallic systems are rhodium 7.04 (39), copper 4.7 (40), lithium 2–3 (41), and iridium 2.28 (42). This suggests that C–H bond breaking is probably involved in the rate-determining step of $(C_5H_5)_3$UR thermolysis.

Kinetic investigations of $(C_5H_5)_3$UR and $(C_5H_5)_3$ThR thermal decomposition in solution indicate a unimolecular (good first-order kinetic plots) process (e.g., see Figure 2). Further support for unimolecularity is provided by the observation that, for both uranium and thorium systems, added $(C_5H_5)_3$M(i-propyl) does not affect the rate of decomposition of $(C_5H_5)_3$M(n-butyl) and vice versa. Analysis of PMR spectra, together with kinetic data for the rate of disappearance of resonances corresponding to various groups on the same $(C_5H_5)_3$MR molcule, demonstrates that facile β-hydride elimination prior to thermolysis (43, 44)—such as in the isomerization of Reaction 7 (see Ref. 45 and references cited therein)—does not occur to any appreciable extent. Data on ura-

$$
\begin{array}{ccc}
 & \overset{\displaystyle H}{\underset{\displaystyle |}{}} & \\
CH_3 & CH_2\!\!=\!\!C\text{--}CH_3 & \\
\underset{\displaystyle |}{|} & \downarrow & \\
M\text{--}CH & \rightleftharpoons \quad M\text{--}H & \rightleftharpoons \quad M\text{--}CH_2CH_2CH_3 \qquad (7) \\
\underset{\displaystyle CH_3}{|} & 1 &
\end{array}
$$

nium and thorium complexes are complied in Table I. In all cases, the thorium organometallics are more thermally stable.

In constructing any mechanistic scheme to describe the thermal decomposition of actinide alkyls, the first questions which arise are why β-hydride elimination does not occur in $(C_5H_5)_3$UR and $(C_5H_5)_3$ThR complexes and whether this behavior is a fundamental property of all actinide alkyl complexes. In order to answer these questions, we examined

Table I. Comparative Kinetic Data for
Thermolysis of $(C_5H_5)_3$MR Compounds in Toluene Solution[a]

R	$M = U$		$M = Th$	
	$\Delta G^{\ddagger}$, kcal/mole	$t_{\frac{1}{2}}$, hr	$\Delta G^{\ddagger}$, kcal/mole	$t_{\frac{1}{2}}$, hr
n-C$_4$H$_9$	33.3 (97°)	1130	37.6 (167°)	96
Neopentyl	32.2 (97°)	270	41.4 (167°)	7500
i-C$_3$H$_7$	29.8 (72°)	201	34.3 (167°)	7.2
Allyl	28.7 (72°)	40	38.6 (167°)	566
trans-2-Butenyl	34.6 (97°)	6730	36.8 (167°)	39

[a] Good first-order kinetic plots.

the thermally unstable products of Reaction 8 (*46, 47*). When R possesses

$$MCl_4 + 4RLi \rightarrow [MR_4] \rightarrow products \qquad (8)$$
$$M = Th, U$$

a β-hydrogen, the presumed actinide tetraalkyls decompose to yield large quantities of olefin. For example, for M $=$ U and R $=$ *n*-butyl, the organic product yields in hexane solution are 49% *n*-butane, 46% 1-butene, and 1% *n*-octane whereas for M $=$ Th and R $=$ *n*-butyl they are 38% 1-butene and 60% *n*-butane. In both cases, the actinide is isolated as the metal. Importantly, only traces of octane were detected. The products of a free radical homolytic decomposition (Reaction 9) would be predicted

$$M-R \rightarrow M \cdot \quad \cdot R \qquad (9)$$

(*48, 49, 50*) to be predominantly R–R from radical dimerization, together with smaller quantities of R–H from solvent hydrogen atom abstraction as well as R–H and the olefin R–H–(H_2) from radical disproportionation. Dimerization to disproportionation ratios are *ca.* 10:1 for primary alkyl radicals (*48, 49*). Thus, the products of the thermal decomposition of the unstable actinide tetraalkyls are those expected from β-hydride elimination (Reactions 3 and 4).

We believe that the differences in the thermolysis pathway and hence to a great extent in the stability of these compounds compared with that of the triscyclopentadienyls is primarily the result of the degree of coordinative saturation. It is well established (*24, 25, 26, 27, 28*) that β-hydride elimination (Reaction 3) proceeds *via* intermediate hydride olefin complexes (Structure 1). These configurations are usually attained by expansion of the metal coordination sphere or by dissociation of another ligand. In the coordinatively congested $(C_5H_5)_3MR$ complexes, apparently neither of these two possibilities can be achieved by a reasonable expenditure of free energy, and an alternative route is traversed. The exact degree of steric congestion is impossible to quantify; however, considerable crowding of groups is apparent from the x-ray data (the α alkyl carbon is less than 3.0 A from the ring carbon atoms) and various spectroscopic studies. For example, the barrier to rotation about the U–C sigma bond in $(C_5H_5)_3U$(*i*-propyl) is on the order of 10 kcal/mole (*13*) (Figure 3). On the other hand, the actinide tetraalkyls presumably have coordination numbers as low as four, and intermediates (or transition states) such as Structure 1 are readily achieved, thus facilitating β-hydride elimination. If no β-hydrogen is present on the R group in the MR_4 compounds, the major thermolysis product is R–H. For the 2-*cis*- and 2-*trans*-2-butenyl systems, stereochemical integrity at the α-carbon atom is maintained which indicates that free R $\cdot$ radicals are not involved. The thorium tetraalkyls are more thermally stable than those of uranium (*46*). Indeed,

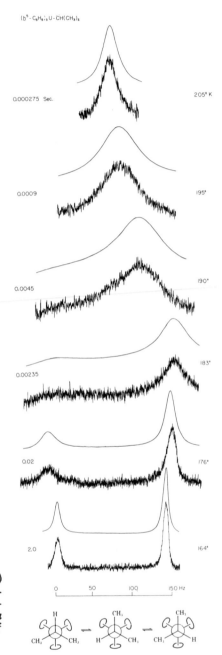

Figure 3.   PMR spectra (at 90 MHz) in the C_5H_5 region of $(C_5H_5)_3U(iso-propyl)$ as a solution in dimethyl ether-toluene.   Computer-generated spectra are for mean pre-exchange lifetimes of the low field resonance.

tetrabenzylthorium can be isolated as a colorless, crystalline solid which decomposes slowly at room temperature (51). The fact that none of the actinide tetraalkyls is as stable as the triscyclopentadienyls presumably indicates that coordinative saturation also plays a role in hindering thermolysis pathways other than β-hydride elimination.

No mechanistic study would be complete without an accounting of all reaction products. Though, as shown above, considerable information can be obtained from the organic products, the resultant actinide-containing thermolysis species are also of great interest. Solutions of $(C_5H_5)_3UR$ complexes deposit a pyrophoric brown precipitate upon thermal decomposition (*13*). This solid analyzes as $(C_5H_5UH_n, n = 2$ or 3; the IR spectrum clearly reveals the presence of an η^5-C_5H_5 functionality. Mass spectra reveal the presence of several uranium atoms. We have so far been unable to crystallize this compound. Thermolysis solutions of the $(C_5H_5)_3ThR$ series deposit a colorless solid. If thermolysis is carried out slowly (170°C), the product can be obtained as colorless needles (*17*) which analyze (C, H, and mass spectrum) as $[(C_5H_5)_2Th(C_5H_4)]_2$. Again the IR spectrum has bands that are attributable to η^5-C_5H_5 rings. The molecular structure of this complex was determined by x-ray diffraction (Figure 4) (*52*). The configuration of this remarkable molecule can be

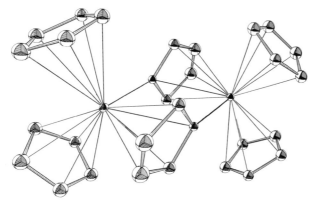

Figure 4. The molecular structure of $[(C_5H_5)_2(C_5H_4)$-Th]$_2$ (52)

viewed as that obtained by removing the elements R–H from a $(C_5H_5)_3$-ThR molecule and then dimerizing the resultant species. Hence, each thorium atom is bonded to two pentahaptocyclopentadienyl rings, and it shares a third *via* a double η^5:η^1-C_5H_4 bridge. In many ways the structure is similar to that of niobocene (*53*) except that there is no evidence for either a direct thorium–thorium bond or a metal hydride.

What pathways then are open for the thermal decomposition of the coordinatively saturated $(C_5H_5)_3UR$ and $(C_5H_5)_3ThR$ compounds? With the available data, we can greatly narrow the possibilities. Three descriptions of the primary event now seem most likely (Mechanisms 2, 3, and

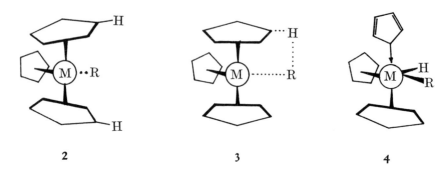

2   **3**   **4**

4) for the transfer of a hydrogen from a cyclopentadienyl ring to the R moiety. A homolytic free radical bond scission (Mechanism 2) followed by abstraction of a ring hydrogen is acceptable only if it occurs in a solvent cage (54, 55) which is sufficiently tight to ensure retention of configuration in RH. Considering the structure of the organoactinide in the ground state as well as the other spectroscopic evidence for coordinative congestion, such a situation cannot be ruled out since the α-carbon atom of R is within 3 A of several of the surrounding ring hydrogens. The stability trend as a function of R, primary > secondary > tertiary, also suggests radical character. However, the nearly complete retention of configuration we observe for the hydrogen abstraction was never observed previously for vinyl radical atom transfer reactions (29–37), nor is such stereospecificity commonly observed for caged radical pair recombination reactions in solution (54, 55, 56, 57). In addition, one should note that, if a caged radical pair is sufficiently constrained, it is not legitimate to consider the radicals free and the reaction verges upon a concerted process. Mechanism 3 is such a concerted, four-center process in which the hydrogen is transferred directly to R. It could also be viewed as an intramolecular proton transfer reaction (58). We have already observed that protonolysis of actinide–alkyl bonds is facile with alcohols (Reaction 10)

$$(C_5H_5)_3M\text{–}R \xrightarrow[\text{M = Th, U}]{\text{CH}_3\text{OH}} (C_5H_5)_3M\text{–}OCH_3 + RH \qquad (10)$$

(13, 17). Finally, Mechanism 4 is formally an oxidative addition reaction in which a ligand C–H bond is added to the metal center. In d transition metal chemistry, there is considerable precedent for this type of process (59, 60, 61) including examples in which the hydrogen is subsequently transferred to an alkyl group (62, 63) (formally a reductive elimination) as well as those in which a cyclopentadienyl ring provides the additive C–H bond, (64, 65).

The uncertainty in mechanistic pathways is essentially reduced to the question of whether the hydrogen is transferred directly to the R ligand, or whether it is transferred first to the metal and then subsequently to R. It is certain that these transfers must be in a tightly controlled, intramolecular fashion. Although the organic thermal decomposition products are identical, it is not certain that the uranium and thorium systems traverse exactly the same thermolysis reaction coordinate. In this connection, it is instructive to examine the relative abilities of uranium and thorium to shuttle between oxidation states (*66*). The uranium potentials in Reaction 11 and 12 are for aqueous solutions (*66*); data for thorium is esti-

$$\text{U(III)} \xleftarrow{\;-0.63\text{ V}\;} \text{U(IV)} \xrightarrow{\;+0.32\text{ V}\;} \text{U(VI)} \tag{11}$$

$$\text{Th(III)} \xleftarrow{\;-3.7\text{ V}\;} \text{Th(IV)} \longrightarrow \text{Th(VI)} \tag{12}$$

mated from spectroscopic correlations (*see* Ref. *67* and references cited therein). Uranium(IV) can be reduced or oxidized with relative ease. It is far more difficult to reduce thorium(IV) (few thorium(III) compounds are known—(C_5H_5)$_3$Th (*68*), ThOF (*69*), and ThBr$_3$ (*70*)), and it is impossible to exidize it without removal of low lying electrons from the closed $6p$ shell. To the extent that organometallic reaction mechanisms can be described in terms of formal changes in oxidation state, rate-determining processes involving either reduction (*e.g.* Mechanism **2**) or oxidation of the metal should be more facile for uranium(IV) than for thorium(IV). Hence the difference in rates of (C_5H_5)$_3$UR and (C_5H_5)$_3$ThR thermolysis may reflect either unequal energetic requirements for following the same reaction pathway or the following of different pathways.

The above discussion represents only a single facet of the thermal decomposition reaction: how H is transferred to R. However, once R–H expulsion occurs and by whatever mechanism, the actinide-containing species that remain must form the product which is isolated. For both uranium and thorium, the unimolecularity of R–H elimination implicates an intermediate such as Structure **5**. This is a carbene complex (*71, 72, 73*) of which diaryl (*74, 75*) and dialkyl (*76, 77*) examples are now

5   6   7

known. However, no d transition metal cyclopentadienylidene complexes have yet been isolated (*78, 79*) which, in view of the very electrophilic character of this carbene (*80 81, 82, 83*), is not surprising. The resonance hybrid (Structure **5**) has an antiaromatic four-electron π system. However, a very electron-rich metal (one actually capable of undergoing a formal two-electron oxidation) such as a divalent actinide might stabilize the aromatic ylid hybrid (Structure **7**). Thus an intermediate such as Structure **7** could be reasonbaly stable, and when M = thorium it apparently persists until dimerization occurs to yield the isolated product [(C$_5$H$_5$)$_2$Th(C$_5$H$_4$)]$_2$ in Figure 4. Further rationale for the existence of Structure **7** is that triphenylphosphine forms a very stable complex with cyclopentadienylidene, *i.e.* triphenylphosphoniumcyclopentadienylide (Structure **8**) (*84, 85*). It is highly probable that carbene–ylid (η^1-

8

C$_5$H$_4$)M species are involved in a large number of thermolysis and group transfer reactions of both d and f transition metal (η^5-C$_5$H$_5$)M compounds.

It is thus clear that coordinative saturation and presumably also immobilization imparted by other ligands plays an important role in stabilizing actinide-to-carbon sigma bonds. One way in which to examine the energetics of coordinative saturation is to study actinide allyl compounds. As is revealed by IR and laser Raman spectrascopy, the compounds (C$_5$H$_5$)$_3$U(allyl) (*13*) and (C$_5$H$_5$)$_3$Th(allyl (*17*) have mono-haptoallyl ground state geometries. In both cases, however, the room temperature PMR spectra have dynamic A$_4$X time-averaged spectra (Figures 5 and 6). With the uranium(IV) compound (*13*) isotropic shifts (attributable to the two unpaired 5f electrons) are sufficiently large (*i.e.* they impart sufficient time resolution) (*86, 87, 88*) so that the slow exchange limit spectrum can be observed at low temperature (Figure 5). This confirms the instantaneous monohapto geometry and allows calculation (from the spectral coalescence point) of the free energy of activation for the dynamic sigmatropic process in Reaction 13. The maximum in free energy by which the trihapto geometry can exceed the monohapto is the activation energy, $\Delta G^\ddagger = 8.0 \pm 1.0$ kcal/mole. Because of the dia-

(13)

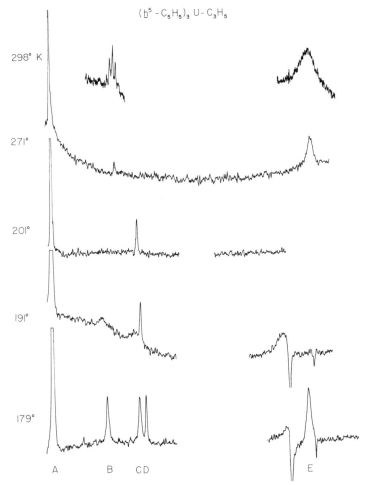

Figure 5.   Field sweep PMR spectra (at 90 MHz) of $(C_5H_5)_3U$-(allyl) as a solution in toluene-d_8. Resonance A is attributable to the cyclopentadienyl protons whereas resonances B, C, D, and E are attributable to the allyl protons. Dispersion bands are an artifact of 15 KHz field modulation.

magnetism of the thorium analog, the frequency separation between its exchanging sites at 90 MHz is insufficient for observation of spectra below the spectral coalescence point (Figure 6) before the sample solution freezes. However, by using reasonable estimates for the resonance positions (89, 90) and assuming the coalescence point to be $-100° \pm 10°C$, we calculate $\Delta G^{\ddagger} = 7.6$–$8.7$ kcal/mole. This is surprisingly close to the value found for the uranium system, and it suggests that the potential energy surface for the tautomeric process represented by Reaction 13 changes little when thorium is substituted for uranium.

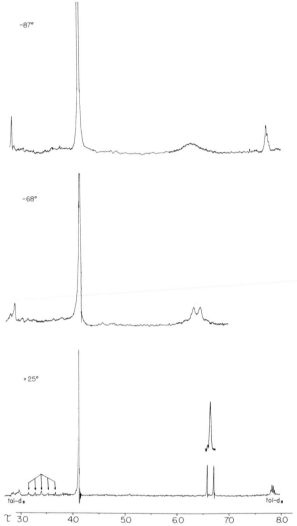

Figure 6. Variable temperature PMR spectra (at 90 MHz) of (C₅H₅)₃Th(allyl) in toluene-d₈. Inset resonance at bottom resulted from applying a decoupling frequency to the center of the methine multiplet (denoted by arrows).

These findings, along with the previous discussion of thermolysis, suggest that the organometallic chemistry of uranium and thorium that involves tautomeric processes and valence isomerizations (as well as certain catalytic processes) probably has rather similar products and activation energetics. However, any chemical transformations of uranium and thorium organometallics that involve changes in oxidation state may have quite different products and rates. In this way, the 5*f* electrons should have a significant effect on the chemistry.

The tetraallyls U(allyl)$_4$ (*91*) and Th(allyl$_4$ (*92*) are also known. They possess ground state trihaptoallyl structures. At temperatures above ambient, the thorium compound exhibits fluxional behavior. This is best represented (*93, 94*) by Reaction 14 where the potential energy surface

(14)

is an approximately inverted form of that for the monohaptoallyl. The inversion no doubt arises from a decrease in coordinative saturation. Unfortunately, the thermal instability of the corresponding uranium tetraallyl hinders efforts to observe the fluxional process.

The triscyclopentadienyl alkyls and the tetraalkyls represent two extremes in actinide alkyl chemistry. The former are thermally stable and relatively nonlabile whereas the latter are thermally unstable and labile. It is of interest to determine whether some intermediate compromise in these characteristics can be achivede by suitable choice of supporting ligands.

One promising approach to the adjustment of coordinative saturation is to alter the number of cyclopentadienyl rings. Since the great bulk of titanium, zirconium, and hafnium organometallic chemistry is that of (C$_5$H$_5$)$_2$MX$_2$ systems, compounds such as the recently reported (C$_5$H$_5$)$_2$-UCl$_2$ (*95*) would seem to be ideal precursors. We (*96*) as well as others (*97*) have found this to be an incorrect formulation of the reaction product of UCl$_4$ and 2TlC$_5$H$_5$. When the reaction is carried out in dimethoxyethane (DME) we find (*96*) by NMR that the product is actually a mixture of (C$_5$H$_5$)$_3$UCl and (C$_5$H$_5$)UCl$_3$ · DME (*98*).

Another approach is to tie the cyclopentadienyl rings together (*96*) (*e.g.* Reaction 15). For X = CH$_2$, the dark red crystalline product is

X = CH$_2$, CH$_2$CH$_2$CH$_2$, (CH$_3$)$_2$Si, etc.   (15)

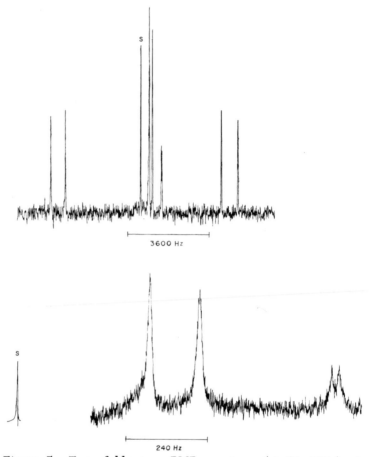

Figure 7. Top: field-sweep PMR spectrum (at 60 MHz) of
$[CH_2(\eta^5\text{-}C_5H_4)_2]_2U_2Cl_5^-Li^+ \cdot 2THF;$ *the peak marked S is attrib-*
utable to $C_6D_5H.$ *Bottom: expansion of spectrum to show coordi-*
nated THF and CH_2 *resonances.*

obtained as a tetrahydrofuran adduct, $[CH_2(C_5H_4)_2]_2U_2Cl_5^-Li^+ \cdot 2THF$.
The proton NMR spectrum (Figure 7) indicates that the molecule has
only one plane of symmetry as in Structure **9**. The coordinated THF

cannot be removed *in vacuo* without destruction of the compound; however, it rapidly exchanges with added free THF in solution (Figure 8). Since the rate of exchange is roughly a function of added THF, and since the exchange is accompanied by evolution of a time-averaged symmetry plane, rapid equilibration with a monomeric species such as Structure 10

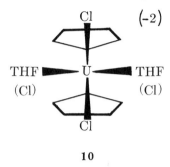

10

is suggested. These results indicate that it may be possible to tune the degree of coordinative saturation at uranium by simple manipulation of the size of the bridge between the cyclopentadienyl rings.

Preliminary studies indicate that these new dihalides can be converted to alkyl, aryl, and tetrahydroborate derivatives of varying stability (*96, 99*). Another useful precursor for some types of biscyclopentadienyl derivatives of uranium(IV) is the amido compound $(C_5H_5)_2U[N(C_2H_5)_2]_2$ (*100*).

Actinide Ions as Templates

Template reactions (*101, 102, 103, 104, 105*) constitute an extensive class of chemical transformations (both stoichiometric and catalytic) in which a metal ion serves as a framework for the coordinative cyclization of organic ligands. An intriguing question is whether, by increasing the ionic radius of the template (*e.g.* by using an actinide ion), it is possible to produce an expanded macrocyclic ligand. The pronounced tendency of the uranyl ion to achieve a pentagonal bipyramidal (Structure **11**) or hexagonal bipyramidal (Structure **12**) coordination geometry led us to

11   **12**

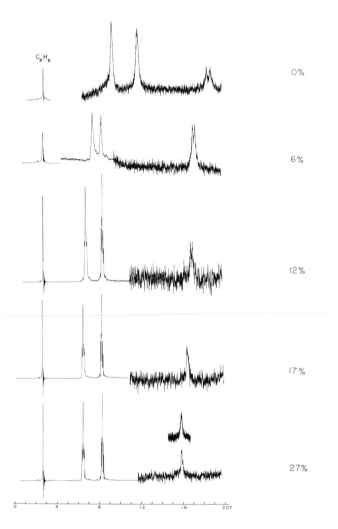

Figure 8. Proton NMR spectra (at 60 MHz) of [CH₂-
$(\eta^5\text{-}C_5H_4)_2]_2U_2Cl_5^-Li^+ \cdot 2THF$ *as a function of added*
THF (measured as a percentage of solution volume)

$$4 \quad \text{(}o\text{-}C_6H_4(CN)_2) + MCl_2 \longrightarrow \text{(phthalocyanine-M)} \qquad (16)$$

$M = Fe, Co, Ni, Zn$

investigate the variation in the phthalocyanine (*106, 107*) condensation (Equation 16) in which the uranyl ion serves as a template (*108, 109, 110, 111, 112*).

X-ray diffraction (*113*) reveals that the so-called uranyl phthalocyanine is in reality a complex of the superphthalocyanine ligand, an expanded five-subunit analog of phthalocyanine (*112, 113*). (The rigorous *Chemical Abstracts* name of this complex is 5,35:14,19-diimino-7,12:21,-26:28,33-trinitrilopentabenzo[c,h,m,r,w,][1,6,11,16,21]pentaazacyclopentacosinatodioxouranium(VI). The molecular structure of this unique complex is depicted in Figures 9 and 10. The coordinative preferences of the uranyl ion can dramatically alter the normal course of the cyclization reaction (Equation 17).

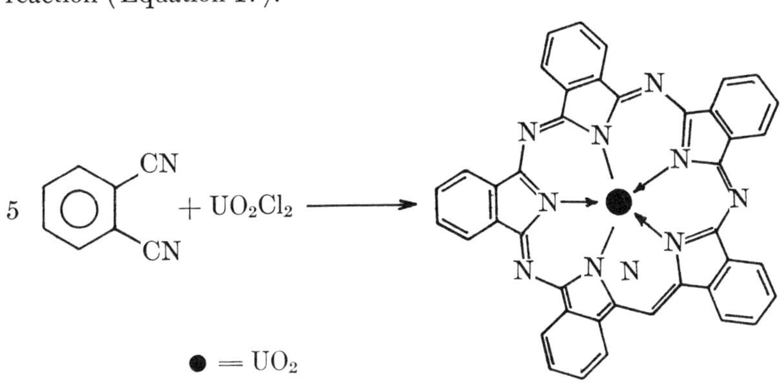

$$5 \quad \text{(o-dicyanobenzene)} + UO_2Cl_2 \longrightarrow \text{(superphthalocyanine complex)} \qquad (17)$$

$$\bullet = UO_2$$

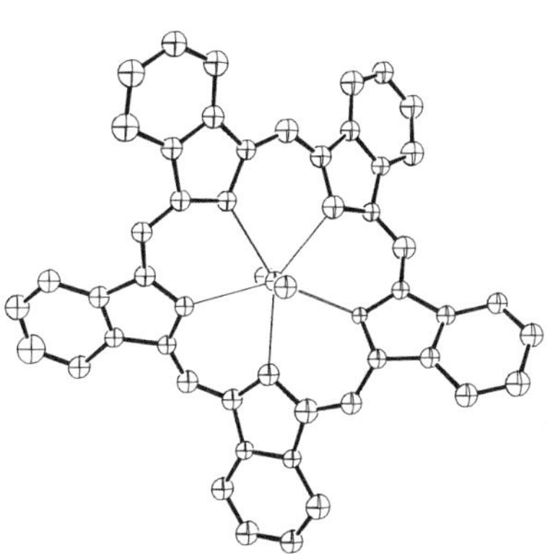

Figure 9.   Model in perspective of the uranyl superphthalocyanine molecule.   The view is approximately perpendicular to the mean plane of the macrocycle.

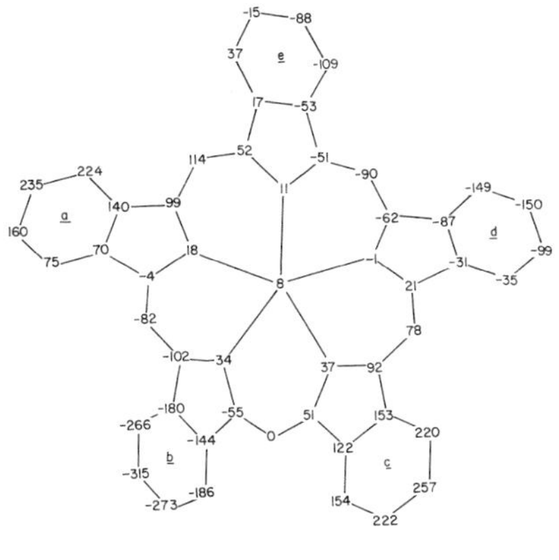

Figure 10.   Diagram derived from Figure 9 to illustrate the nonplanarity of the uranyl superphthalocyanine macrocyclic ligand. The perpendicular displacement of the atoms, in units of 0.01 A, is given with respect to an arbitrary mean plane.

It is of great interest to determine the degree to which the structural and chemical stabilities of the macrocyclic ligand (and by inference its formation and selectivity) are imparted by the unusual coordinative properties of the uranyl ion. Under metal ion displacement conditions, a surprising ligand contraction process (equation 18) occurs (*114*) which, to our knowledge, is unprecendented in macrocycle coordination chemistry. Products were confirmed by IR, mass, and visible spectra. An example of this reaction, monitored spectrophotometrically, is presented in Figure 11. Under identical conditions, metalation of the free phthalocyanine ligand is far slower. Also, reaction with trivalent metal salts such as lanthanide trihalides yields the corresponding PcMX compounds. Reaction with finely dispersed active metals (*e.g.* Na, K in refluxing mesitylene) produces the corresponding metal phthalocyanines. Treatment of uranyl superphthalocyanine with aqueous and nonaqueous acids results in demetalation accompanied by ring contraction.

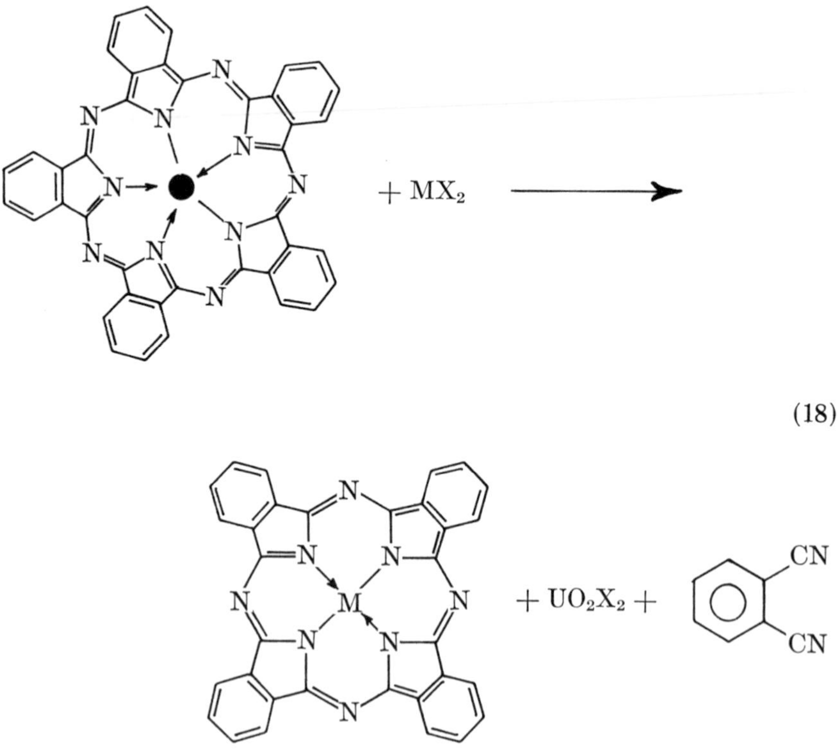

$$+ \; MX_2 \longrightarrow$$

(18)

$$+ \; UO_2X_2 \; +$$

● = UO$_2$
M = Co, Ni, Cu, Zn, Sn, Pb
X$^-$ = halide

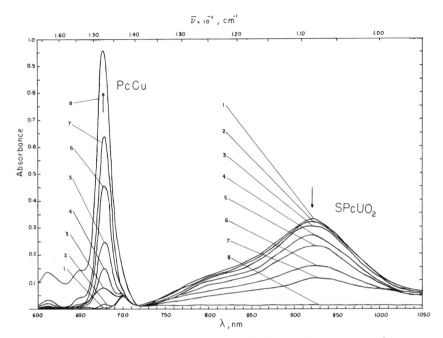

Figure 11. Spectrophotometric record of the reaction of uranyl super-phthalocyanine (SPcUO$_2$) with CuCl$_2$ in 200/1 1-chloronaphthalene/DMF at 75°C (PcCu: cupric phthalocyanine)

The propensity which the superphthalocyanine ligand exhibits for ring contraction indicates that the favorable pentagonal bipyramidal uranyl coordination geometry and the optimum U–N bond distance of *ca.* 2.5 Å (*113*) are important in stabilizing the expanded macrocyclic ligand system. Though stannous phthalocyanine is severely distorted (*115*), with the relatively large Sn(II) ion displaced 1.1 Å from the ring coordination plane (*115*), tin appears to fail the above criteria on both counts (Sn–N = 2.25 Å) (*115*) and contraction still occurs (Equation 18). In addition to these criteria, it is likely that steric strain in the five-subunit ligand (*113*) may be an important driving force for contraction. To a first approximation, electronic factors within the ligand do not appear to be a factor. Hückel molecular orbital calculations (*116*) on the free four- and five-subunit free ligands (assumed planar) minus the fused benzene rings indicate that the π energy of superphthalocyanine is not significantly different from that of phthalocyanine (*117, 118, 119*) plus phthalonitrile.

The marked tendency of the uranyl ion to form pentagonal bipyramidal seven-coordinate complexes with U–N bond lengths of *ca.* 2.5–2.6 Å suggests that this ion (as a template) is ideally suited for constructing pentadentate macrocyclic ligands such as phthalocyanine and porphyrin analogs. On a more general level, the present findings indicate that actinide ions may be capable of altering the normal course of a number of organic cyclooligomerization reactions *via* unusual requirements of coordination geometry and ionic radius. Further efforts to explore the scope of this reaction pattern are in progress.

Acknowledgments

It is a pleasure to acknowledge the collaboration of my able coworkers, William J. Kennelly, John R. Kolb, Afif M. Seyam, Djordje R. Stojakovic, and William A. Wachter. I thank A .L. Allred for the use of the HMO program (*116*).

Literature Cited

1. Cernia, E., Mazzei, A., *Inorg. Chim. Acta* (1974) **10**, 239.
2. Marks, T. J., in "Prospects in Organotransition-Metal Chemistry," Y. Ishii and M. Tsutsui, Eds., p. 81, Plenum, New York, 1975.
3. Marks, T. J., *J. Organomet. Chem.* (1974) **79**, 181.
4. Seaborg, G. T., *Pure Appl. Chem.* (1972) **30**, 539.
5. Kanellakopulos, B., Bagnall, K. W., in "M.T.P. International Review of Science, Inorganic Chemistry," H. J. Emeleus and K. W. Bagnall, Eds., Ser. 1, Vol. 7, p. 299, University Park Press, Baltimore, 1972.
6. Hayes, R. G., Thomas, J. L., *Organomet. Chem. Rev. Sect. A* (1971) **7**, 1.
7. Gysling, H., Tsutsui, M., *Adv. Organomet. Chem.* (1970) **9**, 361.
8. Gilman, H., Jones, R. G., Bindschadler, E., Blume, D., Karmas, G., Martin, Jr., G. A., Nobis, J. F., Thirtle, J. R., Yale, H. L., Yoeman, F. A., *J. Am. Chem. Soc.* (1956) **78**, 2790.
9. Katz, J. J., private communication.
10. Marks, T. J., Seyam, A. M., *J. Am. Chem. Soc.* (1972) **94**, 6545.
11. Gebala, A. E., Tsutsui, M., *Chem. Lett.* (1972) 775.
12. Gebala, A. E., Tsutsui, M., *J. Am. Chem. Soc.* (1973) **95**, 91.
13. Marks, T. J., Seyam, A. M., Kolb, J. R., *J. Am. Chem. Soc.* (1973) **95**, 5529.
14. Brandi, G., Brunelli, M., Lugli, G., Mazzei, A., *Inorg. Chim. Acta* (1973) **7**, 319.
15. Atwood, J. L., Haines, Jr., C. F., Tsutsui, M., Gebala, A. E., *J. Chem. Soc. Chem. Commun.* (1973) 452.
16. Marks, T. J., Kolb, J. R., Seyam, A. M., Wachter, W. A., *Proc. Sixth Int. Conf. Organomet. Chem.*, Amherst, Mass., 1973, Abstract **114**.
17. Marks, T. J., Wachter, W. A., *J. Am. Chem. Soc.*, in press.
18. Davidson, P. J., Lappert, M. F., Pearce, R., *Acc. Chem. Res.* (1974) **7**, 209.
19. Braterman, P. S., Cross, R. J., *Chem. Soc. Rev.* (1973) **2**, 171.
20. Wilkinson, G., *Pure Appl. Chem.* (1972) **30**, 627.
21. Calderazzo, F., *Pure Appl. Chem.* (1973) **33**, 453.
22. Cotton, F. A., Wilkinson, G., "Advanced Inorganic Chemistry," 3rd ed., Chap. 23–26, Interscience, New York, 1972.
23. Parshall, G. W., Mrowca, J. J., *Adv. Organomet. Chem.* (1968) **7**, 157.
24. Clark, H. C., Wong, C. S., *J. Am. Chem. Soc.* (1974) **96**, 7213.
25. Clark, H. C., Jablonski, C. R., *Inorg. Chem.* (1974) **13**, 2213.
26. Whitesides, G. M., Stedronsky, E. R., Casey, C. P., San Filippo, Jr., J., *J. Am. Chem. Soc.* (1970) **92**, 1426.
27. Whitesides, G. M., Gaasch, J. F., Stedronsky, E. R., *J. Am. Chem. Soc.* (1972) **94**, 5258.

28. Sneeden, R. P. A., Zeiss, H. H., *J. Organomet. Chem.* (1970) **22**, 713.
29. Fessenden, R. W., Schuler, R. H., *J. Chem. Phys.* (1963) **39**, 2147.
30. Fessenden, R. W., *J. Phys. Chem.* (1967) **71**, 74.
31. Kasai, P. H., Whipple, E. B., *J. Am. Chem. Soc.* (1967) **89**, 1033.
32. Kopchik, R. M., Kampmeier, J. A., *J. Am. Chem. Soc.* (1968) **90**, 6733.
33. Frantazier, R. M., Kampmeier, J. A., *J. Am. Chem. Soc.* (1966) **88**, 1959.
34. *Ibid.* (1966) **88**, 5219.
35. Singer, L. A., Kong, N. P., *J. Am. Chem. Soc.* (1966) **88**, 5213.
36. Kampmeier, J. A., Chen, G., *J. Am. Chem. Soc.* (1965) **87**, 2608.
37. Neuman, Jr., R. C., Holmes, G. D., *J. Org. Chem.* (1968) **33**, 4317.
38. Seyam, A. M., unpublished data.
39. Stille, J. K., Huang, F., Regan, M. T., *J. Am. Chem. Soc.* (1974) **96**, 1518.
40. Stedronsky, E. R., Ph.D. Thesis, Massachusetts Institute of Technology, 1970 (quoted in Ref. 27).
41. Finnegan, R. A., Kutta, H. W., *J. Org. Chem.* (1965) **30**, 4138.
42. Evans, J., Schwartz, J., Urquhart, P. W., *J. Organomet. Chem.* (1974) **81**, C37.
43. Tamaki, A., Kochi, J. K., *J. Chem. Soc. Chem Commun.* (1973 423.
44. Tamaki, A., Magennis, S. A., Kochi, J. K., *J. Am. Chem. Soc.* (1973) **95**, 6487.
45. Lehmkuhle, H., Olbrysch, O., *Ann.* (1973) 715.
46. Marks, T. J., Seyam, A. M., Wachter, W. A., unpublished data.
47. Marks, T. J., Seyam, A. M., *J. Organomet. Chem.* (1974) **67**, 61.
48. Ingold, K. U., in "Free Radicals," J. K. Kochi, Ed., Vol. I, Chap. 2, Wiley, New York, 1973.
49. Gibian, M. J., Corley, R. C., *Chem. Rev.* (1973) **73**, 441.
50. Pryor, W. A., "Free Radicals," Chap. 12–20, McGraw–Hill, New York, 1966.
51. Köhler, E., Brüser, W., Thiele, K. H., *J. Organomet. Chem.* (1974) **76**, 235.
52. Baker, E. C., Raymond, K. N., Marks, T. J., Wachter, W. A., *J. Am. Chem. Soc.* (1974) **96**, 7586.
53. Guggenberger, L. J., *Inorg. Chem.* (1973) **12**, 294.
54. Koenig, T., Fischer, H., in "Free Radicals," J. K. Kochi, Ed., Vol. I Chap. 4, Wiley, New York, 1973.
55. Kosower, E. M., "An Introduction to Physical Organic Chemistry," p. 352, Wiley, New York, 1968.
56. Greene, F. D., Berwick, M. A., Stowell, J. C., *J. Am. Chem. Soc.* (1970) **92**, 867.
57. Bartlett, P. D., McBride, J. M., *Pure Appl. Chem.* (1967) **15**, 89.
58. Marks, T. J., Kolb, J. R., *J. Am. Chem. Soc.* (1975) **97**, 3397.
59. Parshall, G. W., *Acc. Chem. Res.* (1970) **3**, 139.
60. Schunn, R. A., in Transition Metal Hydrides," E. L. Muetterties, Ed., Vol. I, p. 234, Marcel Dekker, New York, 1971.
61. Heck, R. F., "Organotransition Metal Chemistry," p. 38, Academic, New York, 1974.
62. Cundy, C. S., Lappert, M. F., Pearce, R., *J. Organomet. Chem.* (1973) **59**, 161.
63. Schwartz, J., Cannon, J. B., *J. Am. Chem. Soc.* (1972) **94**, 6226.
64. Bercaw, J. E., Marvich, R. H., Bell, L. G., Brintzinger, H. H., *J. Am. Chem. Soc.* (1972) **94**, 1219.
65. Teuben, J. H., *J. Organomet. Chem.* (1974) **69**, 241.
66. Keller, C., "The Chemistry of the Transuranium Elements," p. 212, Verlag Chemie, Weinheim/Bergstr., 1971.
67. Nujent, L. J., Baybarz, R. D., Burnett, J. L., Ryan, J. L., *J. Phys. Chem.* (1973) **77**, 1528.

68. Kanellakopulos, B., Dornberger, E., Baumgartner, F., *Inorg. Nucl. Chem. Lett.* (1974) **10**, 155.
69. Lucas, J., Rannon, J. P., *C. R. Acad. Sci. Ser. C* (1968) **266**, 1056.
70. Shiloh, M., *A. E. C. IA Rep.* (1966) **1128** (quoted in Ref. *66*).
71. Cotton, F. A., Lukehart, C. M., *Prog. Inorg. Chem.* (1972) **16**, 487.
72. Cardin, D. J., Cetinkaya, B., Lappert, M. F., *Chem. Rev.* (1972) **72**, 545.
73. Cardin, D. J., Cetinkaya, B., Doyle, M. J., Lappert, M. F., *Chem. Soc. Rev.* (1973) **2**, 99.
74. Casey, C. P., Burkhardt, T. J., *J. Am. Chem. Soc.* (1973) **95**, 5833.
75. Yamamoto, T., Garber, A. R., Wilkinson, J. R., Boss, C. B., Streib, W. E., Todd, L. J., *J. Chem. Soc. Chem. Commun.* (1974) **354**.
76. Sanders, A., Cohen, L., Giering, W. P., Kenedy, D., Magatti, C. V., *J. Am. Chem. Soc.* (1973) **95**, 5430.
77. Schrock, R. R., *J. Am. Chem. Soc.* (1974) **96**, 6796.
78. Day, V. W., Stults, B. R., Reimer, K. J., Shaver, A., *J. Am. Chem. Soc.* (1974) **96**, 1227.
79. *Ibid.* (1974) **96**, 4008.
80. Dürr, H., Werndorff, F., *Angew. Chem. Int. Ed. Engl.* (1974) **13**, 483.
81. Dürr, H., *Fortschr, Chem. Forsch.* (1973) **40**, 103.
82. Gleiter, R., Hoffman, R., *J. Am. Chem. Soc.* (1968) **90**, 5457.
83. Moss, R. A., Przybyla, J. R., *J. Org. Chem.* (1968) **33**, 3816.
84. Johnson, A. W., "Ylid Chemistry," p. 70, Academic, New York, 1966.
85. Ramirez, F., Levy, S., *J. Am. Chem Soc.* (1957) **79**, 67.
86. Marks, T. J., Kolb, J. R., *J. Am. Chem. Soc.* (1975) **97**, 27.
87. Tanny, S. R., Pickering, M., Springer, C. S., *J. Am. Chem. Soc.* (1973) **95**, 6227.
88. Gutowsky, H. S., Cheng, H. N., *J. Chem. Phys.* (1975) **63**, 2439.
89. Maddox, M. L., Stafford, S. L., Kaesz, H. D., *Adv. Organomet. Chem.* (1965) **3**, 71.
90. Zieger, H. E., Roberts, J. D., *J. Org. Chem.* (1969) **34**, 2826.
91. Lugli, G., Marconi, W., Mazzei, A., Paladino, N., Pedretto, U., *Inorg. Chim. Acta* (1969) **3**, 353.
92. Wilke, G., *et al.*, *Angew. Chem. Int. Ed. Engl.* (1966) **5**, 151.
93. Vrieze, K., van Leeuwen, P. W. N. M., *Prog. Inorg. Chem.* (1971) **14**, 1.
94. Krieger, J. K., Deutch, J. M., Whitesides, G. M., *Inorg. Chem.* (1973) **12**, 1535.
95. Zanella, P., Faleschini, S., Doretti, L., Faraglia, G., *J. Organomet. Chem.* (1971) **26**, 353.
96. Marks, T. J., Kennelly, W. J., "Abstracts of Papers," 168th National Meeting, ACS, Sept. 1974, INOR 9.
97. Kanellakopulos, B., Aderhold, C., Dornberger, E., *J. Organomet. Chem.* (1974) **66**, 447.
98. Doretti, L., Zanella, P., Faraglia, G., Faleschini, S., *J. Organomet. Chem.* (1972) **43**, 339.
99. Marks, T. J., Kennelly, W. J., Ernst, R. D., Day, V. W., unpublished data.
100. Jamerson, J. D., Takats, J., *J. Organomet. Chem.* (1974) **78**, C23.
101. Black, D. St. C., Markham, E., *Rev. Pure Appl. Chem.* (1965) **15**, 109.
102. Busch, D. H., *Helv. Chim. Acta, Fasc. Extraordinarius,* Alfred Werner Commemoration Volume (1967).
103. Curtis, N. F., *Coord. Chem. Rev.* (1968) **3**, 3.
104. Busch, D. H., Farmery, K., Katovic, V., Melnyk, A. C., Sperati, C. R., Tokel, N. E., *Adv. Chem. Ser.* (1971) **100**, 44.
105. Lindoy, L. F., Busch, D. H., *Prep. Inorg. React.* (1971) **6**, 1.
106. Lever, A. B. P., *Adv. Inorg. Chem. Radiochem.* (1965) **7**, 27.
107. Moser, F. A., Thomas, A. L., "Phthalocyanine Compounds," Reinhold, New York, 1963.
108. Frigerio, N. A., U.S. Patent **3,027,391** (1962).

109. Bloor, J. E., Walden, C. C., Demerdache, A., Schlabitz, J., *Can. J. Chem.* (1964) **42**, 2201.
110. Hagenberg, W., Gradl, R., Lux, F., GDCh-Hauptversammlung, GDCh-Fachgrouppe "Kern, Radio- und Strahlenchemie," Karlsruhe, September 1971 (quoted in Ref. *111*).
111. Bagnall, K. W., "The Actinide Elements," p. 225, Elsevier, Amsterdam, 1972.
112. Lux, F., *Proc. 10th Rare Earth Res. Conf., Carefree, Ariz., May 1973,* p. 871.
113. Day, V. W., Marks, T. J., Wachter, W. A., *J. Am. Chem. Soc.* (1975) **97**, 4519.
114. Marks, T. J., Stojakovic, D. R., *J. Chem. Soc. Chem. Commun.* (1975) 28.
115. Friedel, M. K., Hoskins, B. T., Martin, R. L., Mason, S. A., *Chem. Commun.* (1970) 400.
116. Allred, A. L., Busch, L. W., *J. Am. Chem. Soc.* (1968) **90**, 3352.
117. Chen, I., *J. Mol. Spectrosc.* (1967) **23**, 131.
118. Chen, I., *J. Chem. Phys.* (1969) **51**, 3241.
119. Schaffer, A. M., Gouterman, M., Davidson, E. R., *Theor. Chim. Acta* (1973) **30**, 9.

RECEIVED January 24, 1975. Work supported by generous grants from the National Science Foundation (GP-30623X and GP-43642X), the Research Corporation, and the Paint Research Institute.

21

Use of the Bulky Alkyl Ligand (Me₃Si)₂CH⁻ to Stabilize Unusual Low Valent Transition Metal Alkyls and Dialkylstannylene Derivatives

MICHAEL F. LAPPERT

School of Molecular Sciences, University of Sussex,
Brighton BN1 9QJ, England

New kinetically stable homoleptic metal alkyls MRₙ were prepared using principally the ligand $(Me_3Si)_2CH^-(R)^-$: (a) low coordination number complexes of Group III–VI transition metals such as $[YR_3]$ and $[Ti(\eta^5\text{-}C_5H_5)_2CHPh_2]$, and (b) stannylene complexes of transition metals such as trans-*$[Cr(CO)_4(SnR_2)_2]$, $[Fe_2(\eta^5\text{-}C_5H_5)_2(CO)_3SnR_2]$, and $[Fe(\eta^5\text{-}C_5H_5)(CO)_2SnR_2X]$ (X = Cl or Me). With type b, $\ddot{S}nR_2$ behaves as a good σ donor, and its coordination chemistry is much like that of tertiary phosphines with the added complication that, in some reactions, products of insertion may be obtained in which the Sn has increased its oxidation state but it is still attached to the transition metal. Monomeric (cyclo-C_6H_{12} solution), yellow $\ddot{G}eR_2$ was obtained from $Ge\{N(SiMe_3)_2\}_2$ and LiR. He(I) photoelectron data on bivalent Ge, Sn, and Pb alkyls and amides are presented and analyzed.*

R ecently there has been much interest in the synthesis, structure, characterization, and thermal decomposition of unusual stable homoleptic metal alkyls MRₙ (*1*). (The term homoleptic is used to describe a metal complex in which all the ligands are identical (*2*), *i.e.* TiR₄ and Sn(NR′₂)₂ are examples of homoleptic metal alkyls and dialkylamides whereas Ti(Cl)R₃ and Sn(NR′₂)R are heteroleptic compounds.) There are three classes (*2*) of these complexes: (a) transition metal compounds, (b) diamagnetic lower valent main group element derivatives such as the dialkyls of Sn and Pb (s^2 complexes), and (c) paramagnetic main group compounds, *i. e.* the metal-centered radicals such as the s^1-complexes MR_3 (M $=$ Si, Ge, or Sn). The key feature is that thermal stability is governed primarily by kinetic factors. Hence, by the choice of suitable ligands R^-, normally low activation energy decomposition pathways become energetically unfavorable. Such ligands are of four types: (a) bulky alkyls that are free from β-hydrogen atoms, including the neopentyl types $(R_3M')_m$ CH $_{3-m}^-$ ($m = 1, 2,$ or 3) (2) such as $Me_3SiCH_2^-$ (3, 4, 5) and Me_3PCH_2 (6), and $PhCH_2^-$ (7); (b) bulky alkyls that are free from α-H atoms such as 1-norbornyl (8) and *tert*-butyl (9); (c) simple alkyls if the metal is (nearly) coordinatively saturated, *e.g.* Me^- in WMe_6 (10); and (d) chelating alkyls such as *o*-MeO $\cdot$ C_6H_4 $\cdot$ Si(Me_2)CH_2^- (11, 12) and

$$Me_2P \begin{array}{c} \diagup CH_2 \\ \diagdown CH_2 \end{array} - (6).$$

Table I. Stable Homoleptic Metal Alkyls, MR_n of Neopentyl Type

$Type^b$	$R =$ Me_3SiCH_2	$R =$ $(Me_3Si)_2CH$		$R =$ Me_3CCH_2	$R =$ Me_3SnCH_2
$(MR)_x$	$(CuR)_4$ d^{10}				
$(MR_2)_x$		$\begin{matrix} SnR_2 \\ PbR_2 \end{matrix} \Big\}$	s^2		
$(MR_3)_x$		$\begin{matrix} SiR_3 \\ GeR_3 \\ SnR_3 \end{matrix} \Big\}$	s^1		
		YR_3	d^0		
		TiR_3	d^1		
		VR_3	d^2		
	$\begin{matrix} (MoR_3)_2 \\ (WR_3)_2 \end{matrix} \Big\} d_3{}^c$	CrR_3	d^3	$(MoR_3)_2{}^c$ d^3	
MR_4	$\begin{matrix} TiR_4 \\ ZrR_4 \\ HfR_4 \end{matrix} \Big\} d^0$			$\begin{matrix} TiR_4 \\ ZrR_4 \\ HfR_4 \end{matrix} \Big\} d^0$	$\begin{matrix} TiR_4 \\ ZrR_4 \\ HfR_4 \end{matrix} \Big\} d^0$
	$\begin{matrix} VR_4{}^c & d^1 \\ CrR_4{}^c & d^2 \end{matrix}$			$CrR_4{}^c$ d^2	

^{a} For details, *see* Ref. 2.
^{b} The degree of molecular aggregation (x) refers to the situation in C_6H_{12} solution.
^{c} These complexes were first prepared by G. Wilkinson *et al.,* the others by M. F. Lappert *et al.* (*see* Ref.2).

A list of stable (at 20°C in an inert atmosphere) neopentyl-type complexes is presented in Table I. The most bulky of these ligands, $(Me_3Si)_2CH^-$, clearly favors three-coordination for the metal. This may be illustrated by reference to $Y\{CH(SiMe_3)_2\}_3$ (11); with the less hindered $Me_3SiCH_2^-$, the metal alkyl can also be obtained in presence of a donor solvent as with $M(CH_2SiMe_3)_3 \cdot 2THF$ (M = Sc or Y) (12). An interesting structural feature is found in $[CuCH_2SiMe_3]_4$ (see Figure 1 for the crystal and molecular structure) which demonstrates the unusual single alkyl bridge between two Cu atoms (13). The conventional syn-

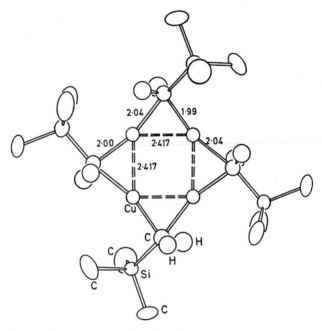

Figure 1. The crystal and molecular structure of (Cu-CH₂SiMe₃)₄ (13)—Cu-Cu, 2.417 Å; Cu-C (mean), 2.02 Å; ∠ C-Cu-C, 164°; Cu₄C₄Si₂ coplanar; centrosymmetric

thetic route to all alkyls, except for the s^1 derivatives, has been from corresponding chlorides, or chlorometal complexes, and the appropriate alkyl of Li or Mg (2).

The s^2 Sn and Pb complexes $M\{CH(SiMe_3)_2\}_2$ are unusual in being stable, monomeric in C_6H_6 solution (by cryoscopy), colored, and diamagnetic (14). Moreover, they behave as good donors and give rise to complexes such as $\{(Me_3Si)_2CH\}_2Sn-Cr(CO)_5$; the crystal and molecular structure of this stannylene complex has an unusual trigonal planar environment for the tin atom (Figure 2) (15).

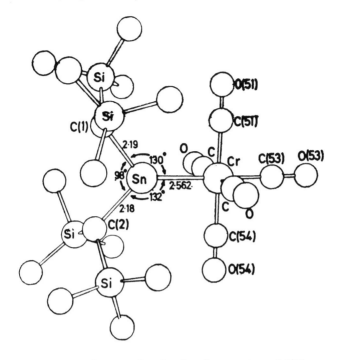

Figure 2.   The crystal and molecular structure of $\{(Me_3\text{-}Si)_2CH\}_2SnCr(CO)_5$ (15)—Sn–Cr, 2.562 A; Sn–C, 2.19 A; < C–Sn–C, 96°; < C–Sn–Cr (mean), 131°; C_2SnCr-$(CO)_3$ essentially coplanar with Cr in octahedral environment

Results

Synthesis. A new method has been devised (see Reaction 1). It

$$M(NR'_2)_n \xrightarrow{nLiR} MR_n + nLiNR'_2 \qquad (1)$$

was used effectively for preparing the formerly elusive (14)—i.e. not accessible from $GeCl_2/LiR$)—$Ge\{CH(SiMe_3)_2\}_2$ (16). The starting amide was $Ge\{N(SiMe_3)_2\}_2$ which is monomeric in cyclo-C_6H_{12} solution (by cryoscopy) (17). The products were separated by converting the Li amide to $Sn\{N(SiMe_3)_2\}_2$ (using $SnCl_2$) (17) which is extremely soluble in cyclo-C_6H_{12} and much more so than the Ge(II) alkyl. The same procedure was used successfully for the corresponding SnR_2 (16) and PbR_2 (also monomers in C_6H_{12}) (18). It is promising but as yet inconclusive for CrR_3 and FeR_3 [R' = Me_3Si, R = $(Me_3Si)_2CH$] (19).

THE BENZHYDRYL LIGAND (19). Reactions with $LiCHPh_2$ and metal chlorides such as $TiCl_4$, $HfCl_4$, VCl_4, $TaCl_5$, $NbCl_5$, $CrCl_3$, $MoCl_5$, $FeCl_3$, and $SnCl_2$ have not at this time afforded characterized crystalline products. However, it was noted that this particular lithium alkyl is a rather powerful reducing agent. The group IV metallocene dichlorides ($R = Ph_2CH$) may be summarized by Reaction 2.

$$M(\eta^5\text{-}C_5H_5)_2Cl_2 + 2RLi \rightarrow M(\eta^5\text{-}C_5H_5)_2R_2 \qquad (2)$$

$Ti(\eta^5\text{-}C_5H_5)_2R$
dark green; ESR doublet
$a(H) \sim 5$ G, $g = 1.98$ ($20°C$
in C_6D_6)

$M = Zr$: orange; dec. $> 130°C$;
τ; 3.4 (Ph), 5.2 (C_5H_5),
7.3 (CH)
$M = Hf$: yellow; m.p. 192–194°C
(dec.) ; τ; 3.4, 5.25, 7.4

THERMAL DECOMPOSITION. Various pathways are available for transition metal alkyl decomposition (1). Apparently small structural changes can change totally the nature of the products. This is illustrated by the tetra-alkyls of Ti, Zr, and Hf (see Reaction 3).

$$M(CH_2M'Me_3)_4 \xrightarrow[\substack{M = C \text{ or Si } (20)}]{\text{heat}} Me_3M'CH_3 \qquad (3)$$
(sole volatile product)

$\downarrow$ heat
$M = Sn$ (21)

$(Me_3M'CH_2)_2$
(principal volatile product)

A Monomeric Germanium(II) Alkyl (16). Di[bis(trimethylsilyl)-methyl]germanium, $Ge\{CH(SiMe_3)_2\}_2$, is a yellow solid, m.p. 179°–181°C (under argon), that sublimes at 110°C and 10^{-3} mm Hg. It is thermochromic, becoming red when molten. It is a monomer in C_6H_6 and cyclo-C_6H_{12} (by cryoscopy) (18). The monomeric parent ion is the highest peak in the mass spectrum. The Raman spectrum of the solid has a strong line at 300 cm^{-1} which may arise from a Ge–Ge stretch in the dimeric solid (by analogy with Figure 3) (22). The ^1H NMR spectrum shows a single sharp methyl resonance at 9.55 τ, whence it is inferred that the molecule in solution is the diamagnetic singlet (possibly of C_{2v} symmetry).

The Crystal and Molecular Structure of $[Sn\{CH(SiMe_3)_2\}_2]_2$ (23). This compound crystallizes as small red plates. The molecule is centrosymmetric and preliminary parameters are shown in Figure 3. The environment about each tin atom is perhaps best described as distorted tetrahedral, with the sum of bond angles at tin as 341°; for "pure" sp^3 Sn this would be ca. 329° and for "pure" sp^2, 360°. The two sets of alkyl

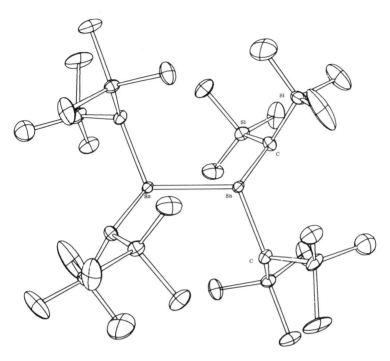

Figure 3. The crystal and molecular structure of $[Sn\{CH(Si-Me_3)_2\}_2]_2$ *(23)—Sn–Sn, 2.76 A; Sn–C, 2.17 A; < Sn–Sn–C (mean), 115°; < C–Sn–C, 112°*

ligands at each tin atom are directed away from one another as in Structure I. It was difficult to obtain good quality single crystals, but numerous samples were taken for indexing and each belonged to the same space

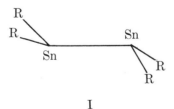

I

group, Pī⁻, as the eventual sample crystal. We hope to examine this particular crystal by Raman spectroscopy; we expect to locate a line at *ca.* 180–200 cm⁻¹ that is attributable to ν(Sn–Sn). At present, however, such attempts with the bulk sample have led to photochemical decomposition (22). (We demonstrated earlier that photolysis of SnR₂ in C₆H₆ yields the stable ṠnR₃ (24).) The Sn–Sn bond length is similar to that in Sn₂Ph₆ and somewhat shorter than that in tetrahedral tin. There may be a tenuous relation between the crystal structure of the dialkylstannylene and that of the green form of chlorobis[1,2-bis(diphenylphosphino)ethane]-cobalt(II) trichlorostannate(II), in which the $SnCl_3^-$ anion is located near a center of symmetry which results in a Sn–Sn distance of 3.597(4) A with the three chlorine atoms bonded to tin directed away from the Sn–Sn vector (25). [I thank a referee for drawing attention to this analogy.]

Our present tentative interpretation of the structural and analytical data—visible, IR, MS, and ^1H NMR; magnetism (diamagnetic solid by the Gouy method; no ESR signal in the solid or in solution in the absence of irradiation); ^{119}Sn Mössbauer; single crystal x-ray; and cryoscopic molecular weight—is that in solution and as a vapor the compound exists as the monomeric bent singlet with bonding formalized as in Structure II, whereas in the solid state bent bonds prevail as in Structure III. It is assumed that, although the Sn–Sn bond is a double bond, it is weak; this accounts for the solution molecular weight data and the ready formation of stannylene complexes such as that in Figure 2. The puzzling feature of the x-ray findings (Figure 3) concerns the angles at tin. The above interpretation would be more consistent with planar trigonal tin; d orbital participation in the bonding is likely.

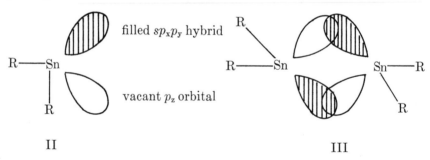

II   III

He(I) **Photoelectron Spectra of Monomeric Divalent Metal Alkyls and Dialkylamides (26).** In keeping with the good donor properties of the dialkylstannylene (14, 15), it and related group IV compounds have low energy orbitals that are attributable essentially to nonbonding metal lone-pair orbitals. Their position is clearly affected by the electro-negativity of the ligand (N > C), but it is broadly similar to that found for the gaseous elements. As expected, the appropriate bands in the photoelectron spectra are sharp. Salient results are summarized in Table II.

The Coordination Chemistry of a Dialkylstannylene. In terms of the solution behavior of $Sn\{CH(SiMe_3)_2\}_2$, formalized as Structure II, reactions may be classified as those in which it functions as a Lewis acid, as a Lewis base, or as a carbene-like reactive intermediate that gives rise to insertion reactions (14). Further illustrations of this behavior are given in Reactions 4 [R = $(Me_3Si)_2CH$] (27).

$$[Fe(\eta^5\text{-}C_5H_5)(CO)_2]_2 \longrightarrow \left[\begin{array}{c} R_2Sn \diagdown \ \overset{O}{\underset{\parallel}{C}} \ \diagup C_5H_5\text{-}\eta^5 \\ Fe \qquad Fe \\ \eta^5\text{-}C_5H_5 \diagup \ \overset{}{\underset{\parallel}{C}} \ \diagdown CO \\ O \end{array} \right] \begin{array}{l} \text{mp } 167°\text{--}170°C, \\ \text{brown} \end{array}$$

$$SnR_2 \xrightarrow{[Cr(CO)_4(\text{nor-}C_7H_8)]} \textit{trans-}[Cr(CO)_4(SnR_2)_2] \qquad (4)$$
$$\text{mp } 215°\text{--}216°C, \text{ orange}$$

$$\xrightarrow{[Fe(\eta^5\text{-}C_5H_5)(CO)_2X]} [Fe(\eta^5\text{-}C_5H_5)(CO)_2(SnR_2X)]$$
$$X = Cl, \text{ mp } 119°\text{--}120°C, \text{ orange}$$
$$X = Me, \text{ mp } 141°\text{--}142°C, \text{ yellow-brown}$$

Table II.   The Highest Occupied Electronic Energy Levels (in eV) of the Monomeric Divalent Metal Alkyls and Dialkylamides

Compound				
Ge{CH(SiMe₃)₂}₂	7.75	8.87		
Sn{CH(SiMe₃)₂}₂	7.42	8.33		
Pb{CH(SiMe₃)₂}₂	7.25	7.98		
Ge{N(SiMe₃)₂}₂	8.68	—	7.71	8.99
Sn{N(SiMe₃)₂}₂[a]	8.36	9.50	7.75	8.85
Pb{N(SiMe₃)₂}₂	8.16	9.39	7.92	8.81

[a] This compound was obtained by us (*17*) and also independently by Schaeffer and Zuckerman (*37*). We find it and the other amides in this Table to be monomeric in solution and vapor, but our colleagues formulate it as an *N*-bridged dimer.

The ¹¹⁹Sn Mössbauer Spectra of Some of the Complexes (*28*). The ¹¹⁹Sn Mössbauer spectra of a number of the complexes were examined. Data are summarized in Table III together with relevant reports from the literature. There are essentially two types: (a) the dialkylstannylene complexes such as $R_2SnCr(CO)_5$ in which divalent tin is three-coordinate, and (b) dialkylstannylene insertion products into C–Hal, M–H, M–Cl, M–Me, or M–M bonds (M = a transition metal) in which tetravalent tin is four-coordinate. ($SnR_2)_2$ and Class (a) complexes are characterized by

Table III. Data from ^{119}Sn Mössbauer Spectra[a]

Compound	Isomer Shift,[a] mm/sec	Quadrupole Splitting, mm/sec	Ref.
$(SnR_2)_2$[b]	2.16	2.31	28
$Sn(\eta^5\text{-}C_5H_5)_2$	3.74	0.86	29, 30
$SnCl(\eta^5\text{-}C_5H_5)$	3.70	1.05	31
$SnBr(\eta^5\text{-}C_5H_5)$	3.69	0.99	31
$Sn(Ph\text{-}CB_{10}H_{10}C)_2$	2.95	1.90	32
$3\text{-}Sn\text{-}1,2\text{-}C_2B_9H_{11}$	4.67	3.83	33
$[R_2SnCr(CO)_5]$[b]	2.21	4.43	28
$[R_2SnMo(CO)_5]$[b]	2.15	4.57	28
$trans\text{-}[(R_2Sn)_2Cr(CO)_4]$[b]	2.21	4.04	28
$trans\text{-}[(R_2Sn)_2Mo(CO)_4]$[b]	2.13	4.24	28
$[PtCl(PEt_3)(SnR_2Cl)SnR_2]$[b]	2.05, 1.73	4.23, 2.66	28
$Sn(\eta^5\text{-}C_5H_5)_2Br_3$	3.79	0.90	30
$[Bu^t_2(THF)SnCr(CO)_5]$	2.11	4.14	34
$[Bu^t_2(py)SnCr(CO)_5]$	2.01	3.44	34
$[SnR_2\{Fe(CO)_4\}_2]$[b]	1.73	1.53	28
SnR_3Cl[b]	1.27	2.18	28
SnR_3Br[b]	1.24	2.05	28
$SnR_2(Me)I$[b]	1.48	2.24	28
$[Fe(\eta^5\text{-}C_5H_5)(CO)_2(SnR_2Cl)]$[b]	1.54	2.37	28
$[Fe(\eta^5\text{-}C_5H_5)(CO)_2(SnR_2Me)]$[b, c]	1.48	—	28
$[Mo(\eta^5\text{-}C_5H_5)(CO)_3(SnR_2H)]$[b]	1.39	0.67	28
$[Mo(\eta^5\text{-}C_5H_5)(CO)_3(SnR_2Me)]$[b]	1.35	0.71	28
$[Fe(\eta^5\text{-}C_5H_5)(CO)_2SnCl_3]$	1.77	1.82	35
$[Fe(\eta^5\text{-}C_5H_5)(CO)_2SnMe_3]$[c]	1.35	> 0	36
$[Mo(\eta^5\text{-}C_5H_5)(CO)_3SnMe_3]$	1.36	1.25	36

[a] Relative to $BaSnO_3$ or SnO_2 at ca. 20°C; crystalline samples.
[b] $R = (Me_3Si)_2CH$.
[c] The full line width at half-height, Γ, was 1.34 for $Fe(\eta^5\text{-}C_5H_5)(CO)_2(SnR_2Me)$ and 1.16 for $Fe(\eta^5\text{-}C_5H_5)(CO)_2SnMe_3$.

isomer shifts at 2.15 ± 0.1 mm/sec relative to $BaSnO_3$ and large quadrupole splittings [2.31 for $(SnR_2)_2$ and 4.25 ± 0.2 mm/sec for Class (a) complexes]. Class (b) complexes have lower isomer shifts (1.49 ± 0.25 mm/sec) and quadrupole splittings (< 2.37 mm/sec). It is perhaps surprising that for the new tin(II) complexes, the isomer shifts are rather low and not by themselves diagnostic of the tin valency. Magnetic Mössbauer measurements of $(SnR_2)_2$ and $R_2SnCr(CO)_5$ demonstrate that for both complexes the sign of the quadrupole coupling constant eQV_{zz}, and hence V_{zz}, the principle component of the field gradient tensor, is positive.

Acknowledgment

The author thanks his collaborators, named in the bibliography; it has been a privilege to work with them.

Literature Cited

1. Davidson, J. J., Lappert, M. F., Pearce, R., *Chem. Rev.* (1976) **76.**
2. Davidson, P. J., Lappert, M. F., Pearce, R., *Acc. Chem. Res.* (1974) **7,** 209.
3. Collier, M. R., Kingston, B. M., Lappert, M. F., Truelock, M. M., British Patent **36021** (1969).
4. Collier, M. R., Lappert, M. F., Truelock, M. M., *J. Organomet. Chem.* (1970) **25,** C36.
5. Yagupsky, G., Mowat, W., Shortland, A. J., Wilkinson, G., *Chem. Commun.* (1970) 1369.
6. Schmidbaur, H., Franke, R., *Angew. Chem. Int. Ed. Engl.* (1973) **12,** 416.
7. Zucchini, U., Albizzatti, E., Giannini, U., *J. Organomet. Chem.* (1971) **26,** 357.
8. Bower, B. K., Tennent, H. G., *J. Am. Chem. Soc.* (1972) **94,** 2512.
9. Kruse, W., *J. Organomet. Chem.* (1972) **42,** C39.
10. Shortland, A. J., Wilkinson, G., *J. Chem. Soc. Dalton Trans.* (1973) 872.
11. Barker, G. K., Lappert, M. F., *J. Organomet. Chem.* (1973) **76,** C45.
12. Lappert, M. F., Pearce, R., *J. Chem. Soc. Chem. Commun.* (1973) 128.
13. Jarvis, J. A., Kilbourn, B. T., Pearce, R., Lappert, M. F., *J. Chem. Soc. Chem. Commun.* (1973) 475.
14. Davidson, P. J., Lappert, M. F., *J. Chem. Soc. Chem. Commun.* (1973) 317.
15. Cotton, J. D., Davidson, P. J., Goldberg, D. E., Lappert, M. F., Thomas, K. M., *J. Chem. Soc., Chem. Commun.* (1974) 893.
16. Harris, D. H., unpublished data.
17. Harris, D. H., Lappert, M. F., *J. Chem. Soc. Chem. Commun.* (1974) 895.
18. Power, P. P., Miles, S. J., unpublished data.
19. Holton, J., unpublished data.
20. Davidson, P. J., Lappert, M. F., Pearce, R., *J. Organomet. Chem.* (1973) **57,** 269.
21. Webb, M., unpublished data.
22. Gynane, M. J. S., unpublished data.
23. Golberg, D. E., Thomas, K. M., unpublished data.
24. Davidson, P. J., Hudson, A., Lappert, M. F., Lednor, P. L., *J. Chem. Soc. Chem. Commun.* (1973) 829.
25. Stalick, J. D., Corfield, P. W. R., Meek, D. W., *Inorg. Chem.* (1973) **12,** 1668.
26. Sharp, G. J., unpublished data.
27. Cotton, J. D., unpublished data.
28. Donaldson, J. D., Silver, J., unpublished data.
29. Harrison, P. G., Zuckerman, J. J., *J. Am. Chem. Soc.* (1969) **91,** 6885.
30. *Ibid.* (1970) **92,** 2577.
31. Bos, K. D., Bulten, E. J., Noltes, J. G., *J. Organomet. Chem.* (1972) **39,** C52.
32. Aleksandrov, A. Yu., Bregadse, V. I., Goldanskii, V. I., Zakharkin, L. I., Okhlobystin, O. Yu., Khrapov, V. I., *Dokl. Adak. Nauk SSSR* (1965) **165,** 593.
33. Rudulph, R. W., Chowdhry, V., *Inorg. Chem.* (1974) **13,** 248.
34. Grynkewich, G. W., Ho, B. Y. K., Marks, T. J., Tomaja, D. L., Zuckerman, J. J., *Inorg. Chem.* (1973) **12,** 2522.
35. Bird, S. R. A., Donaldson, J. D., Holding, A. F. LeC., Senior, B. J., Tricker, M. J., *J. Chem. Soc. A* (1971) 1616.
36. Bird, S. R. A., Donaldson, J. D., Keppie, S. A., Lappert, M. F., *J. Chem Soc. A* (1971) 1311.
37. Schaeffer, Jr., C. D., Zuckerman, J. J., *J. Am. Chem. Soc.* (1974) **96,** 7160.

RECEIVED January 24, 1975.

22

The Stabilization of Unusual Valence States and Coordination Numbers by Bulky Ligands

D. C. BRADLEY

Queen Mary College, Mile End Rd., London E1 4NS, Great Britain

The use of bulky organic ligands containing O- or N-donor atoms in stabilizing unusually low coordination numbers and oxidation states by metals is described. The work mainly concerns the effect of tertiary alkoxo groups and bulky dialkylamido groups such as NPr_2^i and $N(SiMe_3)_2$ on transition metals and lanthanides. Supplementary ligands such as NO, THF, and $P(C_6H_5)_3$ are sometimes involved. The systems characterized include four-coordinated Nb(IV), Ta(V), Cr(II), and Cr(IV); three-coordinated M(III) (M = Al, Ga, Sc, Ti, V, Cr, Fe, and lanthanides), M(II) (M = Mn, Co, and Ni), and M(I) (M = Co, Ni, and Cu); and two-coordinated M(II) (M = Mn, Co, Ni, Zn, Cd, and Hg) and M(I) (M = Au).

By the use of organic ligands containing suitable donor atoms—*viz.*: XR_x with X = O or S and x = 1: X = N or P and x = 2; and X = C and x = 3—and by suitable choice of the alkyl or aryl R groups, it is possible to vary the bulkiness of the ligand and thereby control the coordination number of the metal to which these ligands are bonded. In general, this steric control leads to the imposition of an unusually low coordination number that often produces a discrete monomeric compound that is appreciably volatile and soluble in organic solvents. An early example of a deliberate attempt to restrict the coordination in metal alkoxides, which are normally oligomeric $[M(OR)_x]_n$ as a result of the presence of alkoxo bridges, was the synthesis of the volatile monomeric zirconium tetra-*tert*-butoxide $Zr(OCMe_3)_4$ (*1*). (For reviews on the metal alkoxides, *see* Refs. *2, 3, 4,* and *5*.) In certain cases, the consequences of ligand steric hindrance are severe enough to cause either a breakdown of the ligand or an electronic rearrangement involving the central metal that leads to the stabilization of an unusual valency state. The steric effect of the dialkylamido group ($X = N$; $x = 2$) is much greater than that of the alkoxo group ($X = O$; $x = 1$) containing the same alkyl group, and recent work has concentrated in the field of the metal dialkylamides (*4, 5*).

Besides binary compounds (ML_x) containing the uninegative ligand $L = XR_x$, a number of ternary ($ML_xL'_y$) and quaternary ($ML_xL'_yL''_z$) complexes were also isolated by use of supplementary ligands L' and L'', and there is scope for further exploitation of the field. In this account, we restrict the discussion to metal alkoxides and metal dialkylamides since the metal alkyls are the subject of papers by Lappert *et al.* (*6, 7*).

For convenience of presentation, the available material, which relates predominantly to the first-row transition metals, is organized according to the particular uninegative ligand type.

Tertiary Alkoxo Ligands

Monomeric species $M(OR\text{-}tert)_x$ have been characterized for titanium, vanadium, chromium, zirconium, and hafnium ($x = 4$) and for niobium and tantalum ($x = 5$). With chromium it was found that limiting Cr(III) to coordination number 4 in the dimeric $Cr_2(OBu^t)_6$ caused instability and a remarkable facility toward valency disproportionation or oxidation to the stable quadricovalent $Cr(OBu^t)_4$ (*8, 9*). In contrast, molybdenum formed a stable dimeric tri-*tert*-butoxide $(Bu^tO)_3Mo\equiv Mo(OBu^t)_3$ which is diamagnetic and presumably bound by a metal–metal triple bond (*10, 11*). Yet another interesting feature of chromium is the synthesis of a stable diamagnetic nitrosyl $Cr(NO)(OBu^t)_3$ in which the nitric oxide is believed to act as a three-electron donor with formation of a four-coordinated low spin chromium(II) compound (*12*). The instability of $Cr_2(OBu^t)_6$ and the stability of both $Cr(NO)(OBu^t)_3$ and $Cr(OBu^t)_4$ must result from the steric effects of the tertiary butoxo groups since the less bulky normal alkoxo groups form very stable polymeric $[Cr(OR)_3]_x$ compounds in which the Cr(III) has its usual coordination number of 6 (octahedral).

Dialkylamido Ligands

With the sterically less demanding dimethylamido group, a range of binary compounds $MNMe_2$, $M(NMe_2)_2$, $M(NMe_2)_3$, $M(NMe_2)_4$, $M(NMe_2)_5$, and $M(NMe_2)_6$ has been characterized (*4, 5*). However, with the higher homologues NR_2 ($R = Et$, Pr^n, Bu^n, etc.), complications arise. With $Nb(V)$ and $Ta(V)$ the predominant products are $Nb(NR_2)_4$ and $Ta(NR)(NR_2)_3$ respectively (13, 14). This implies a greater stability of covalent $Nb(IV)$ compared with $Ta(IV)$ and a greater tendency of the $5d$ element to form a stable imido double bond $Ta{=}NR$. With $Cr(III)$, the unstable four-coordinated dimer $Cr_2(NR_2)_6$ was readily converted to the stable $Cr(IV)$ compound $Cr(NR_2)_4$ (8, 9, 15) thus paralleling the reaction with *tert*-butoxo ligands. Stable $Cr(IV)$ compounds CrR_4 with bulky alkyl groups were recently prepared (16, 17, 18), and there is no doubt that four-coordinated $Cr(IV)$ is quite stable in non-aqueous systems. Molybdenum and tungsten behave differently from chromium in that stable trisdialkylamide dimers $(R_2N)_3M{\equiv}M$-$(NR_2)_3$ are the predominant products starting from $Mo(V)$ and $W(VI)$ chlorides (10, 11). Moreover there are significant differences between Mo and W with the $5d$ element preferring to form imido compounds $W(NR)_2(NR_2)_2$.

With the $3d$ elements Mn, Fe, Co, and Ni, reactions of MCl_2 or MCl_3 with $LiNEt_2$ caused drastic modification of the NEt_2 ligands with resultant formation of organo-nitrogen ligands typified by the cobalt bis-chelated complex involving NN'-diethylbutane-1,3-diimine (19). The mechanisms of these unusual reactions are now unknown, but it seems likely that steric hindrance plays a subsidiary role with electronic factors predominating since the NEt_2 group may function as a σ-N donor and a π-N donor ligand which leads to a buildup of negative charge on the central metal.

The diisopropylamido group NPr_2^i is extremely bulky. This enabled us to isolate a discrete three-coordinated $Cr(III)$ compound $Cr(NPr_2^i)_3$ which, since it is a coordinatively unsaturated molecule, was exceedingly reactive chemically but stable thermodynamically with no tendency to disproportionate to $Cr(IV)$ and $Cr(II)$ (20, 21, 22). With nitric oxide, a stable diamagnetic mononitrosyl $Cr(NO)(NPr_2^i)_3$ formed readily which tolerated exchange of NPr_2^i groups by *tert*-butoxo groups without loss of NO (Reactions 1 and 2) (12). It was demonstrated by variable tempera-

$$Cr(NO)(NPr_3^i)_3 + 2Bu^tOH \rightarrow Cr(NO)(OBu^t)_2(NPr_2^i) + 2Pr_2^iNH \tag{1}$$

$$Cr(NO)(NPr_2^i)_3 + 3Bu^tOH \rightarrow Cr(NO)(OBu^t)_3 + 3Pr_2^iNH \tag{2}$$

ture ^1H and ^{13}C NMR studies that the interesting quaternary compound $Cr(NO)(OBu^t)_2(NPr_2^i)$ contains the diisopropylamido ligand locked into a conformation in which the two isopropyl groups are non-equivalent (23).

Bistrimethylsilylamido Compounds

Bürger and co-workers (*24, 25, 26*) prepared a number of binary metal compounds with the very bulky silazane ligand $N(SiMe_3)_2$, *e.g.* ML_3 (M = Al, Ga, Cr, and Fe), ML_2 (M = Be, Mn, Co, Ni, Zn, Cd, and Hg), ML (M = Li, Na, and Cu). It is reasonably certain that the ML_2 compounds contain two-coordinated metals and that the metals are three-coordinated in ML_3 compounds. Recently Bradley and co-workers isolated some additional three-coordinated species of the transition metals ML_3 (M = Sc, Ti, and V) (*27, 28, 29*) and the lanthanides (*30, 31*). X-ray single-crystal diffraction analysis revealed that the three-coordinated compounds exhibit either trigonal planar MN_3 units (Ti, V, Cr, Fe, Al, and Ga) or pyramidal MN_3 (Sc, Eu, and Yb) (*32, 33, 34, 35*). The reason for this structural difference is not yet clear, but it does appear that the more covalently bonded compounds have the trigonal planar configuration.

The isomorphous series ML_3 (M = Ti, V, Cr, and Fe) is of interest for ligand field studies (d^1, d^2, d^3, and d^5 configurations), and detailed spectroscopic and magnetic studies demonstrated that a substantial crystal field stabilization of the d electrons occurs (*36*). However, the silylamide ligand is not high in the nephelauxetic series and the Fe(III) compound is high spin. Thus the formation of diamagnetic Cr(NO)-$[N(SiMe_3)_2]_3$ from the paramagnetic CrL_3 compound must result from the very strong π-acceptor properties of NO^+ in the formally Cr(II) compound (*12*).

Attempts to synthesize ML_4 compounds have not yet succeeeded, and $MClL_3$ compounds (M = Ti, Th) were obtained instead. The NMR spectra reveal that in the titanium compound the silylamide ligands are locked into a conformation in which the two $SiMe_3$ groups on each ligand are non-equivalent because of restricted rotation whereas in the thorium compound there is free rotation (*37, 38*).

Additional unusual compounds were obtained with the silylamide groups augmented by other neutral donor ligands, *e.g.* tetrahydrofuran (THF). For example, the three-coordinated Mn(II) compound Mn-$[N(SiMe_3)_2]_2(THF)$ was characterized by x-ray single-crystal analysis and electron spin resonance studies (*39*). By contrast, the Cr(II) bis-silylamide took up two molecules of THF giving $Cr[N(SiMe_3)_2]_2(THF)_2$ with a trans-square planar configuration (*40*). Not surprisingly, both the chromium(II) and the manganese(II) were in the high spin state.

Some interesting compounds were also obtained by using triphenylphosphine as the additional ligand. From the reaction of $CoCl_2[P-(C_6H_5)_3]_2$ with $LiN(SiMe_3)_2$, the very reactive, green, crystalline three-coordinated Co(II) complex $Co[N(SiMe_3)_2]_2[P(C_6H_5)_3]$ was isolated while under comparable conditions the analogous nickel(II) salt gave the reactive, yellow, crystalline three-coordinated Ni(I) complex Ni-[N(SiMe₃)₂][P(C₆H₅)₃]₂ (*41*). With the latter, there must be a fine balance between electronic and steric factors that results in this unique type of compound. No doubt the phosphine is a better π-acceptor ligand than the silylamide, and it helps to stabilize the lower valency nickel. A number of other alkylarylphosphines also promote the formation of the three-coordinated nickel silylamide bisphosphine. Steric hindrance of the bulky ligands presumably prevents the formation of square planar metal bissilylamide bisphosphine complexes. However, the relatively stable, orange, crystalline Co(I) complex Co[N(SiMe₃)₂][P(C₆H₅)₃]₂ was recently synthesized starting from CoCl[P(C₆H₅)₃]₃ (*42*). Another example of a phosphine-stabilized, low coordinated metal silylamide is the gold(I) compound Au[N(SiMe₃)₂](PMe₃) (*43*). We also prepared the diamagnetic Cu(I) complex Cu[N(SiMe₃)₂][P(C₆H₅)₃] (*42*) giving a series (d^8, d^9, and d^{10}).

While attempting to prepare complexes of the lanthanide trissilyl-amides, in the course of reactions involving triphenylphosphine oxide and La[N(SiMe₃)₂]₃, we isolated a remarkable new μ-peroxo complex (O₂)-La₂[N(SiMe₃)₂]₄[PO(C₆H₅)₃]₂ (*44*). The peroxo group acts as a doubly bidentate bridge between the two lanthanum atoms which are each bonded to two silylamide and one phosphineoxide ligand, thereby giving a coordination number of 5. A four-coordinated complex La-[N(SiMe₃)₂]₃[PO(C₆H₅)₃] was also isolated.

Finally we mention one other unusual reaction involving the silyl-amide ligand, this time with bis-π-cyclopentadienyl titanium dichloride. All of the chlorine was replaced, but, presumably because of steric reasons, the compound (π-C₅H₅)₂Ti[N(SiMe₃)₂]₂ could not be formed and one of the methyl groups was deprotonated instead forming a novel azasilatitanacyclobutane derivative, a heterocyclic four-membered ring that contains four different elements (*45*). It is clear that the scope for interesting chemistry of the bistrimethylsilylamido ligand is by no means exhausted, and further work is in progress.

2,5-Dimethylpyrrole Compounds

It seemed desirable to investigate other bulky uninegative N-donor ligands besides NPr₂ⁱ and N(SiMe₃)₂, particularly in consideration of the greater versatility of the silylamido ligand. Although both these ligands are bulky, they should differ considerably in electronic behavior. Thus the diisopropylamido group can be both a σ-donor and a π-donor ligand whereas the bistrimethylsilylamido group is (a) potentially a σ donor, (b) a weaker π donor as the result of N$_π$–Si$_π$ interactions, and (c) possibly a weak π acceptor in using the vacant π^*-antibonding NSi_2 orbital. It occurred to us that the 2,5-dimethylpyrrolyl group might mimic to a certain degree the silylamido ligand rather than the diisopropylamido group, and accordingly the lithio derivative was used in reactions with nickel(I) and nickel(II) chloride triphenylphosphine complexes.

The Ni(I) compound $Ni(NC_6H_8[P(C_6H_5)_3]_2$ gave lime-green, re-active crystals, and the compound was paramagnetic with an ESR signal that corresponds to an axially symmetric d^9 species. The Ni(II) complex $Ni(NC_6H_8)_2[P(C_6H_5)_3]$ was a purple, diamagnetic, crystalline com-pound which was relatively stable and which gave a parent molecular ion in the mass spectrum.

It therefore seems likely that the bulky 2,5-dimethylpyrrole ligand will have considerable scope in stabilizing unusual coordination numbers and unusual valency states for the transition metals. It points the way to the development of additional ligands of this kind.

Literature Cited

1. Bradley, D. C., Wardlaw, W., *J. Chem. Soc.* (1951) 280.
2. Bradley, D. C., "Progress in Inorganic Chemistry," Vol. II, pp. 303–361, Interscience, New York 1960.
3. Mehrotra, R. C., *Inorg. Chim. Acta* (1967) 1, 99.
4. Bradley, D. C., Fisher, K. J., in "M.T.P. International Review of Science," Vol. 5, Part I, pp. 65–91, Butterworths, London, 1972.
5. Bradley, D. C., "Advances in Inorganic and Radiochemistry," Vol. 15, pp. 259–322, Academic, London and New York, 1972.
6. Barker, G. K., Cotton, J. D., Davidson, P. J., Lappert, M. F., ADV. CHEM. SER. (1976) 150, 256.
7. Barker, G. K., Lappert, M. F., *J. Organomet. Chem.* (1974) 76, C45.
8. Basi, J. S., Bradley, D. C., *Proc. Chem. Soc. London* (1963) 305.
9. Alyea, E. C., Basi, J. S., Bradley, D. C., Chisholm, M. H., *J. Chem. Soc. A* (1971) 772.
10. Bradley, D. C., Chisholm, M. H., unpublished data.
11. Chisholm, M. H., Extine, M., Reichett, W., ADV. CHEM. SER. (1976) 150, 273.
12. Bradley, D. C., Newing, C. W., *Chem. Commun.* (1970) 219.
13. Bradley, D. C., Thomas, I. M., *Can. J. Chem.* (1962) 40, 449.
14. *Ibid.* (1962) 40, 1355.
15. Basi, J. S., Bradley, D. C., Chisholm, M. H., *J. Chem. Soc. A* (1971) 1433.
16. Bower, B. K., Tennent, H. G., *J. Am. Chem. Soc.* (1972) 94, 2512.
17. Mowat, W., Shortland, G., Hill, N. J., Yagupsky, M., Wilkinson, G., *J. Chem. Soc. Dalton Trans.* (1972) 533.
18. Mowat, W., Shortland, A. J., Hillfi N. J., Wilkinson, G., *J. Chem. Soc. Dalton Trans.* (1973) 770.
19. Bonnett, R., Bradley, D. C., Fisher, K. J., Rendall, J. F., *J. Chem. Soc. A* (1971) 1622.
20. Alyea, E. C., Basi, J. S., Bradley, D. C., Chisholm, M. H., *Chem. Commun.* (1968) 495.
21. Bradley, D. C., Newing, C. W., Chien, J. C. W., Kruse, W., *Chem. Commun.* (1970) 1177.

22. Bradley, D. C., Hursthouse, M. B., Newing, C. W., *Chem. Commun.* (1971) 411.
23. Airoldi, C., Bradley, D. C., Newing, C. W., unpublished data.
24. Bürger, H., Wannagat, U., *Monatsh. Chem.* (1963) 94, 1007.
25. *Ibid.* (1964) 95, 1099.
26. Bürger, H., Forker, C., Goubeau, J., *Monatsh. Chem.* (1965) 96, 597.
27. Bradley, D. C., Copperthwaite, R. G., *Chem. Commun.* (1971) 765.
28. Alyea, E. C., Bradley, D. C., Copperthwaite, R. G., *J. Chem. Soc. Dalton Trans.* (1972) 1580.
29. Bradley, D. C., Copperthwaite, R. G., Cotton, S. A., Gibson, J., Sales, K. D., *J. Chem. Soc. Dalton Trans.* (1973) 191.
30. Bradley, D. C., Ghotra, J. S., Hart, F. A., *Chem. Commun.* (1972) 349.
31. Bradley, D. C., Ghotra, J. S., Hart, F. A., *J Chem. Soc. Dalton Trans.* (1973) 1021.
32. Bradley, D. C., Hursthouse, M. B., Rodesiler, P. F., *Chem. Commun.* (1969) 14.
33. Heath, C. E., Hursthouse, M. B., unpublished data.
34. Sheldrick, G. M., Sheldrick, W. S., *J. Chem. Soc. A* (1969) 2279.
35. Ghotra, J. S., Hursthouse, M. B., Welch, A. J., *Chem. Commun.* (1973) 669.
36. Alyea, E. C., Bradley, D. C., Copperthwaite, R. G., Sales, K. D., *J. Chem. Soc. Dalton Trans.* (1973) 185.
37. Bradley, D. C., Ghotra, J. S., Hart, F. A., *Inorg. Nucl. Chem. Lett.* (1974) 10, 209.
38. Airoldi, C., Bradley, D. C., *Inorg. Nucl. Chem. Lett.* (1975) 11, 155.
39. Bradley, D. C., Copperthwaite, R. G., Hursthouse, M. B., Welch, A. J., unpublished data.
40. Bradley, D. C., Hursthouse, M. B., Newing, C. W., Welch, A. J., *Chem. Commun.* (1972) 567.
41. Bradley, D. C., Hursthouse, M. B., Smallwood, R. J., Welch, A. J., *Chem. Commun.* (1972) 872.
42. Bradley, D. C., Smallwood, R. J., unpublished data.
43. Shiotani, A., Schmidbaur, H., *J. Am. Chem. Soc.* (1970) 92, 7003.
44. Bradley, D. C., Ghotra, J. S., Hart, F. A., Hursthouse, M. B., Raithby, P. R., *Chem. Commun.* (1974) 40.
45. Bradley, D. C., Bennett, C. R., *Chem. Commun.* (1974) 29.

RECEIVED January 24, 1975.

Preparation, Characterization, and Reactions of Dialkylamides and Alkoxides of Molybdenum and Tungsten

M. H. CHISHOLM, M. EXTINE, and W. REICHERT

Frick Inorganic Chemical Laboratories, Princeton University, Princeton, N. J. 08540

The preparation and characterization of Mo_2L_6 where L = NMe_2, NMeEt, NEt_2, OBu^t, $OSiMe_3$, $OCMe_2Ph$, and $W_2(NMe_2)_6$ are reported. Physical measurements including single-crystal x-ray studies indicate that in this series dimerization occurs by metal–metal triple bond formation in the absence of bridging ligands. ^1H NMR studies on $M_2(NR_2)_6$ reveal the large diamagnetic anisotropy induced by the metal–metal triple bond. Reactions of $W(NMe_2)_6$ are presented and compared with those of $W(Me)_6$. The reaction of $W(NMe_2)_6$ with carbon dioxide leads to $W(NMe_2)_3(O_2CNMe_2)_3$ which contains a fac-WN_3O_3 moiety and very short W–N bonds. A general mechanism for the insertion of carbon dioxide into the covalent metal–nitrogen bond is proposed.

We set out to synthesize a series compounds of the general formula ML_3, ML_4, and ML_6 where M = molybdenum and tungsten and L = alkyl, (R), dialkylamide, (NR_2), and alkoxide (OR) ligands. We then wanted to compare in detail the chemistry associated with covalent metal–carbon, –nitrogen, and –oxygen bonds based on (a) the reactivity of the organic ligand and (b) the coordination properties of the transition metal. This summary covers our first two years' work which focused on the synthesis and coordinate properties of the MoL_3 series and the reactions of $W(NMe_2)_6$.

The MoL$_3$ Series

Wilkinson and co-workers (1, 2) recently described the preparation and characterization of a number of stable alkyl complexes of the early transition elements. The reaction of MoCl$_5$ with LiR or RMgX where R = a β-hydride elimination stabilized alkyl led to the isolation of dimeric diamagnetic orange crystalline compounds Mo$_2$R$_6$ (1). For the trimethylsilylmethyl derivative, crystal structure (3) revealed the presence of a very short Mo–Mo bond (2.167 A) and the staggered arrangement of the Mo$_2$C$_6$ moiety. It occurred to us that it might be possible to make structurally related dialkylamido and alkoxy compounds, thus providing a unique opportunity for the study of strong homonuclear interactions in the closely related series L$_3$Mo≡MoL$_3$ where L = R, NR$_2$, and OR.

However there are certain pertinent factors that would seem to discount this view. First, dimerization, or more generally polymerization, by metal–metal bond formation in the absence of bridging ligands was unprecedented in the chemistry of metal alkoxides and dialkylamides (4, 5). Second, it had already been shown (6) that the reaction of MoCl$_5$ with LiNMe$_2$ and LiNEt$_2$ led to the isolation of monomeric tetrakis-(dialkylamido)molybdenum(IV) compounds (cf. Wilkinson's preparation of Mo$_2$R$_6$ from MoCl$_5$). Third, in the chemistry of chromium, vanadium, and titanium, it is known (4, 7) that the dimeric dimethyl- and diethylamides undergo disproportionation under relatively mild conditions (high vacuum and warming to ca. 60°C) via Reaction 1 with resultant isolation of the monomeric metal(IV) dialkylamides.

$$\begin{array}{c} R_2N \diagdown \quad \overset{R_2}{N} \diagup NR_2 \\ M \quad M \\ R_2N \diagup \quad \underset{R_2}{N} \diagdown NR_2 \end{array} \longrightarrow M(NR_2)_4 + [M(NR_2)_2]_n \quad (1)$$

where M = Ti, V, Cr

A consideration of the second and third of these factors led us to use MoCl$_3$ and to attempt low temperature crystallizations in the preparation of Mo(NMe$_2$)$_3$. A typical synthesis is outlined below.

The addition of MoCl$_3$ to an ice-cooled, magnetically stirred solution of LiNMe$_2$ (three mole equivalents in 50:50 THF:hexane) gave a dark brown solution. The reaction mixture was stirred 12 hr at 0°C under an atmosphere of dry, oxygen-free nitrogen. Solvent was then removed by vacuum distillation, and the residue was dried under high vacuum at 60°C. Subsequent extraction of this dried residue with pentane yielded a dark, red-brown solution which, when concentrated and cooled to

$-78°C$, gave a very dark, powdery crystalline product—crude $Mo(NMe_2)_3$. This was collected by filtration (Schlenk technique) and dried *in vacuo*. Analytically pure $Mo(NMe_2)_3$ is pale yellow; it was obtained from the crude product by vacuum sublimation ($100°C$, 10^{-4} cm Hg).

During the vacuum sublimation of $Mo(NMe_2)_3$, small quantities of the purple monomeric $Mo(NMe_2)_4$ were initially evolved (50–$60°C$, 10^{-4} cm Hg). Because of its greater volatility, $Mo(NMe_2)_4$ could be easily separated from $Mo(NMe_2)_3$. It is not clear why $Mo(NMe_2)_4$ forms. Its formation could result from impurities in our $MoCl_3$ or from some redox reaction which competes with the formation of $Mo(NMe_2)_3$. It is, however, quite certain that $Mo(NMe_2)_4$ is not formed from $Mo(NMe_2)_3$ by a disproportionation reaction such as Reaction 1.

The addition of alcohols to hydrocarbon solutions of $Mo(NMe_2)_3$ causes quantitative formation of alkoxides *via* Reaction 2. The nature of the alkoxide depends on the nature of R (*see* below). The *tert*-butoxide

$$Mo(NMe_2)_3 + 3\ ROH \rightarrow M(OR)_3 + 3\ HNMe_2 \qquad (2)$$

$$\text{where } R = Me, Et, Pr^n, Pr^i, Bu^t, CMe_2Ph, SiMe_3$$

and trimethylsiloxide are volatile orange crystalline solids that are readily purified by vacuum sublimation ($100°C$, 10^{-4} cm Hg).

The molybdenum(III) derivatives $Mo(NMe_2)_3$ and $Mo(OR)_3$, where $R = Bu^t$ and $SiMe_3$, are moisture- and oxygen-sensitive diamagnetic compounds. Cryoscopic molecular weight determinations in benzene and mass spectral data demonstrate that these compounds are dimeric (*i.e.* Mo_2L_6) in solution and in the vapor state. The mass spectra reveal strong parent ions, $Mo_2L_6^+$, and several other Mo_2-containing species. Indeed the virtual absence of ions containing only one molybdenum is very striking. In this connection, Mo- and Mo_2-containing species are readily distinguishable because of the several isotopes of molybdenum. The mass distribution of the ion $Mo_2(OSiMe_3)_6^+$ is presented in Figure 1. The basic Mo_2 pattern is further complicated by the presence of silicon (^{28}Si, ^{29}Si, and ^{30}Si) which causes the complex envelope.

The diamagnetic nature of these molybdenum(III) compounds and the presence of only Mo_2-containing ions in the mass spectrum constitute good evidence for strong metal–metal bonding. Further evidence for metal–metal bonding was sought by Raman spectroscopy. Remarkable resonance Raman enhancements of metal–metal stretching vibrations were noted recently ($8, 9$). For the dimeric compounds Mo_2L_6, where $L = NMe_2$, OBu^t, and $OSiMe_3$, the Raman spectra do not indicate this type of resonance enhancement; indeed, assignment of $\nu_{str}(Mo–Mo)$ is not immediately obvious. The color of $Mo_2(NMe_2)_6$ and $Mo_2(OR)_6$,

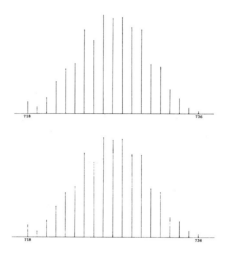

Figure 1.   Parent ion of $Mo_2(OSiMe_3)_6$ as observed (top) and calculated (bottom) for naturally occurring isotopes of Mo, Si, O, and C

where R = Bu^t and $SiMe_3$, is attributable to a tailing of a UV absorption into the visible [*c.f.* in Refs. *10* and *11* $Mo_2(O_2CR)_4$ and related compounds which show $\delta \rightarrow \delta^*$ transitions in the visible]. Consequently laser light in the visible region of the spectrum does not induce a resonance enhancement of $\nu_{str}(Mo-Mo)$. In the Raman spectrum of a crystalline sample of $Mo_2(NMe_2)_6$ (Figure 2), the strong bands at 550, 320, and 230 cm^{-1} are all polarizable, and qualitatively we assign these to ν_{str}-

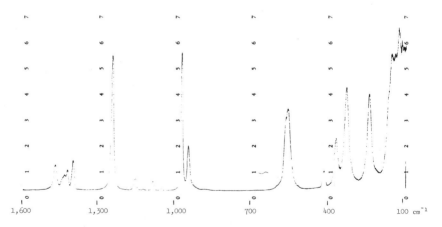

Figure 2.   Raman spectrum (100–1600 cm^{-1}) of a polycrystalline sample of $Mo_2(NMe_2)_6$

(Mo–N), ν_{str}(Mo–Mo), and ρ(NC$_2$) respectively. However, a considerable degree of mixing is possible, and a quantitative interpretation of this spectrum awaits future calculations.

The variable temperature ^{1}H NMR spectra of Mo$_2$(OR)$_6$, where R = Bu^t, CMe$_2$Ph, and SiMe$_3$, in toluene-d_8 reveal single methyl resonances at temperatures of $+90°$ to $-90°$C. These spectra are thus consistent with the presence of only terminally bonded alkoxy ligands L$_3$Mo≡MoL$_3$. However, the observation of a single methyl resonance

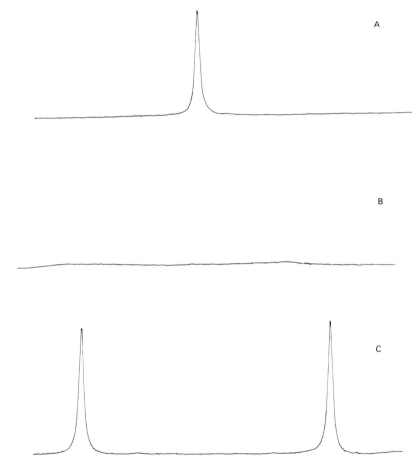

Figure 3. Variable temperature ^{1}H NMR spectrum of Mo$_2$(NMe$_2$)$_6$ recorded in toluene-d_8 at 100 MHz and A at 30°, B at −20°, and C at −60°C. For A, δ(Me) = 3.3 ppm; for C, δ(Me) = 2.4 and 4.2 ppm relative to hexamethyldisiloxane.

does not exclude the possibility of rapid (NMR time scale) bridge–terminal ligand exchange, nor, of course, can we exclude the possibility of accidental magnetic degeneracy of bridging and terminal ligands for dimeric molecules $(RO)_2Mo(\mu\text{-}OR)_2Mo(OR)_2$. The variable temperature ^1H NMR spectrum of $Mo_2(NMe_2)_6$ is however, definitive (Figure 3). At 20°C and above, a single sharp resonance is observed at 3.3 δ from hexamethyldisiloxane. On cooling, this resonance broadens until at $\sim -20°C$ it virtually disappears into the base line. On further cooling to $-40°C$, two resonances appear at 4.2 and 2.4 δ in the integral ratio of 1:1. No further change is observed when the sample is cooled to $-80°C$. The NMR spectrum is independent of $Mo_2(NMe_2)_6$ concentration and is unaffected by added $HNMe_2$. These observations are incompatible with a fluxional dimeric molecule that has a ground state geometry involving bridging dimethylamido ligands (cf. $[Ti(NMe_2)_3]_2$ and $[Al(NMe_2)_3]_2$ in Refs. 4 and 12). The variable temperature ^1H NMR spectrum of $Mo_2(NMe_2)_6$ is consistent with the presence of only terminal dimethylamido ligands. The low temperature limiting spectrum, which consists of two resonances in the integral ratio of 1:1, indicates freezing out of proximal and distal methyl groups with respect to the metal–metal triple bond. The large chemical shift difference (ca. 2.0 ppm) between proximal and distal methyl groups results from the large diamagnetic

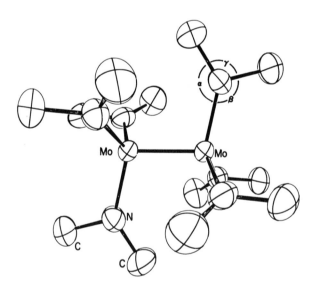

Figure 4. Molecular structure of $Mo_2(NMe_2)_6$. Bond lengths: Mo–Mo 2.142(2) Å, Mo–N 1.98(1) Å, and N–C 1.48(2) Å; angles: Mo–Mo–N 103.7(3)°, α 133(1)°, β 116(1)° and γ 110(1)°

anisotropy induced by the metal–metal triple bond (*13*). At this point, a crystal structure was obviously necessary to verify these predictions.

Figure 4 is an ORTEP drawing of the structure of $Mo_2(NMe_2)_6$ as determined by Cotton *et al.* (*14*). Certain aspects of the structure are evident from this view of the molecule: (a) the absence of bridging dimethylamido ligands: the Mo–Mo–N angles are 103.7°; (b) the short molybdenum–molybdenum bond length [2.214(2) A] and the staggered arrangement of the Mo_2N_6 moiety; and (c) $MoNC_2$ units are planar and lead to six proximal and six distal methyl groups with respect to the metal–metal triple bond. Both (a) and (b) confirm the presence of a molybdenum–molybdenum triple bond.

The arrangement of the Mo–NC_2 planes is revealed more clearly in Figure 5 in which $Mo_2(NMe_2)_6$ is viewed down the metal–metal axis.

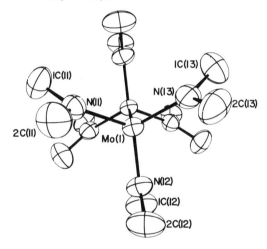

Figure 5.   Molecular structure of $Mo_2(NMe_2)_6$ viewed along the Mo–Mo axis showing the Mo–NC_2 planes; $Mo_2(NC_2)_6$ has S_6 symmetry that differs little from D_{3d}

This view of the molecule clearly demonstrates the staggered ethane-like conformation of the Mo_2N_6 moiety, and it also shows that the Mo–NC_2 planes are almost colinear with the metal–metal bond. In fact, they are tilted 7° out of colinearity which leads to absolute S_6 symmetry for the $Mo_2(NC_2)_6$ moiety. Finally, it should be noted that the proximal and distal Mo–N–C angles differ: $\alpha = 133°$ and $\beta = 116°$. The larger proximal angle probably reflects the greater internal steric congestion in this molecule.

The bonding in $Mo_2(NMe_2)_6$ can be qualitatively described as follows. Each molybdenum forms four σ bonds involving approximate sp^3

hybridization. Then, if we define the Mo–Mo axis as the z axis, molybdenum d_{xz} and d_{yz} atomic orbitals form two metal–metal π bonds. The molybdenum–molybdenum triple bond so formed parallels the situation with Wilkinson's $Mo_2(CH_2SiMe_3)_6$ (*1, 3*). With $Mo_2(NMe_2)_6$, however, the lone pairs on nitrogen may also enter into bonding. In the ground state configuration (*i.e.* Figure 4) molybdenum d_{xy} and $d_{x^2-y^2}$ atomic orbitals may accommodate four electrons from lone pairs on nitrogen. This leads to 16 valence shell electrons per molybdenum. The planarity of the Mo–NC$_2$ units and the somewhat short Mo–N bond lengths [1.98(1) A] suggest that nitrogen-to-metal π bonding is significant. In this situation, ligand π electrons are donated to metal orbitals which are δ and δ^* orbitals with respect to metal–metal bonding. It is perhaps this fact that accounts for the significantly longer Mo–Mo bond in $Mo_2(NMe_2)_6$ [2.214(2) A] than in Wilkinson's $Mo_2(CH_2SiMe_3)_6$ [2.167(?) A]. It is, of course, difficult to estimate the difference in metal–metal bond length which might arise solely from differences in the steric requirements of the ligands. With this in mind, we extended our synthesis of $Mo(NR_2)_3$ to bulkier NR_2 groups—NR_2 = NMeEt, NEt$_2$, and NPr$_2^i$.

The physical properties of $Mo(NMeEt)_3$ and $Mo(NEt_2)_3$ closely parallel those of $Mo_2(NMe_2)_6$. They are dimeric diamagnetic compounds. The low-temperature-limiting ^1H NMR spectrum of $Mo_2(NMeEt)_6$ is consistent with a ground state geometry that has six proximal methyl groups and six distal ethyl groups. The high-temperature-limiting spectrum of $Mo_2(NEt_2)_6$ shows the time-averaged proximal and distal resonances but also complex proton–proton couplings which suggest that motions about the N–CH$_2$ bond are severely hindered. A molecular model of $Mo_2(NEt_2)_6$ indicates that there is indeed severe congestion. Our characterization of $Mo(NPr_2^i)_3$ is incomplete, but it does appear that this compound is quite different from the dimethylamide.

Finally we should comment on the mechanism of proximal and distal alkyl interconversions in $Mo_2(NR_2)_6$ compounds. At the most elementary level, this could be described by Mo–N bond rotation. A sophisticated treatment of this type of intramolecular rearrangement requires the approach adopted by Mislow and co-workers in their graphical analyses of the dynamic stereochemistry of aryl compounds of the form Ar$_3$Z (*15, 16*) and Ar$_4$C (*17*). In this context, we note the stereochemical correspondence between $Mo_2(NMe_2)_6$ and Si_2Ph_6 (*18*). However, in addition to flip rearrangements (Mo–N bond rotation), a fluxional process in which dimethylamido groups are transferred from one molybdenum to the other could be operative. The most attractive fluxional mechanism involves the concerted pairwise interchange of ligands, *i.e.* a transition state that has the bridged $Al_2(NMe_2)_6$ structure. For

$Mo_2(NMe_2)_6$, these mechanisms are permutationally indistinguishable. However, we hope to be able to synthesize $Mo_2(NMe_2)_5(NR_2)$ which, barring accidental anisochrony, should enable us to distinguish between the flipping and the fluxional mechanisms.

The importance of steric factors in determining the coordinate properties of the metal in the ML_3 series cannot be overemphasized. For bulky ligands $L = CH_2SiMe_3$, $CH_2C(CH_3)_3$, NMe_2, $NMeEt$, NEt_2, OBu^t, $OSiMe_3$, and $OCMe_2Ph$, molybdenum adopts the $L_3Mo{\equiv}MoL_3$ structure. Increasing the steric demand of the ligands even further should greatly weaken the metal–metal interaction and could even lead to the isolation of monomeric compounds MoL_3—*cf.* CrL_3 where $L = CH(SiMe_3)_2$ (*19*), NPr_2^i (*20, 21*), and $N(SiMe_3)_2$ (*22*). On the other hand, less sterically demanding ligands should allow the metal to increase its coordination number by the formation of metal–ligand–metal bridges. This indeed appears to be the case for $Mo(OR)_3$ where $R = Me$, Et, and Pr^i. These are black, polymeric, nonvolatile, paramagnetic compounds for which polymerization probably occurs by the formation of MoO_6 units [*cf.* $Cr(OR)_3$ (*4, 5*) where $R = Me$ and Et]. A notable difference between the coordination chemistry of chromium(III) and molybdenum(III) is the following: chromium(III) will, though somewhat reluctantly, form tetrahedral complexes as in the salts $LiCrL_4$, where $L = CH_2SiMe_3$ (*1, 2*) and OBu^t (*20, 23*), whereas molybdenum-(III) prefers to adopt the $L_3Mo{\equiv}MoL_3$ structure at the expense of forming a fourth Mo–L bond. This presumably reflects the greater ability of molybdenum to form strong metal–metal multiple bonds in its tervalent state. Indeed to our knowledge there are no known chromium–chromium multiple bonded compounds of chromium(III).

$W_2(NMe_2)_6$

The great propensity of molybdenum to form metal–metal multiple bonded compounds might lead one to expect the same of tungsten. This, however, does not appear to be the case (*24*). Indeed, at the start of this work only $W_2(CH_2SiMe_3)_6$ had been prepared, and it was reported as forming crystals isomorphous to those of $Mo_2(CH_2SiMe_3)_6$ (*3*). The virtual absence of W–W multiple bonded compounds leads to intriguing questions. Are W–W multiple bonds inherently weaker than those of molybdenum and, if so, for what reasons, or, have certain subtle factors of the coordination chemistry of tungsten thus far precluded their preparation?

In 1969 we reported (*25*) the preparation and characterization of $W(NMe_2)_6$ from the reaction of WCl_6 with $LiNMe_2$. We noted that this reaction was always accompanied by some reduction of tungsten and, on the basis of analytical data, we formulated that the reduced tungsten was of the form $W(NMe_2)_3$. Our characterization of $Mo_2(NMe_2)_6$ encouraged us to pursue the synthesis of $W(NMe_2)_3$ since this compound could clearly reveal invaluable information about tungsten–tungsten bonding and allow a direct comparison with molybdenum. Since reactions involving WCl_6 and $LiNMe_2$ proceeded with some reduction of tungsten, it seemed that a reaction using a reduced tungsten halide might well lead to only, or at least to the predominance of, $W(NMe_2)_3$. Such was not the case. When they were reacted with four equivalents of $LiNMe_2$, $WCl_4(THF)_2$ (26) and $WCl_4(OEt_2)_2$ (26) yielded pure $W(NMe_2)_6$ as the only isolatable dimethylamide of tungsten; no $W(NMe_2)_3$ was obtained. The reaction of the cluster compound WCl_2 with two equivalents of $LiNMe_2$ gave a mixture of $W(NMe_2)_6$ and $W(NMe_2)_3$. This mixture was much richer in $W(NMe_2)_6$ than many samples obtained from reactions that involved WCl_6. However, we found that if $WCl_4(OEt_2)_2$ is allowed to decompose in ether under a nitrogen atmosphere at room temperature and the resultant black sludge is then reacted with three equivalents of $LiNMe_2$, a mixture of $W(NMe_2)_3$ and $W(NMe_2)_6$ is obtained which contains a high percentage of $W(NMe_2)_3$ with $W(III):W(VI) = 2:1$ based on tungsten. All attempts to isolate pure $W(NMe_2)_3$ from this mixture have failed so far; $W(NMe_2)_3$ and $W(NMe_2)_6$ cosublime and cocrystallize. Attempts to isolate $W(NMe_2)_3$ by chromatography have failed; $W(NMe_2)_3$ is the more reactive compound and it decomposes on the column (dehydrated florisil) which allows the isolation of $W(NMe_2)_6$. The spectroscopic properties of the mixture were, however, informative with regard to the nature of $W(NMe_2)_3$. In the mass spectrum a strong parent ion is observed for $W_2(NMe_2)_6^+$, and variable temperature ^1H NMR studies reveal properties analogous to those observed for $Mo_2(NMe_2)_6$, namely a single methyl resonance at room temperature which broadens when the sample is cooled and finally gives two peaks in the integral ratio of 1:1 separated by ca. 2.0 ppm. At this point we resorted to crystallographic techniques. Recent findings of Cotton and co-workers (27) reveal that $W(NMe_2)_6$ and $W_2(NMe_2)_6$ do indeed cocrystallize. The molecular structure of $W_2(NMe_2)_6$ parallels that of $Mo_2(NMe_2)_6$ with a W–W bond length of 2.30 A. Although this bond length is significantly longer than that of Mo–Mo in $Mo_2(NMe_2)_6$, it is now clear that tungsten–tungsten multiple bonding is not inherently weak. Furthermore, the abundance of Mo_2- and the paucity of W_2-multiple bonded complexes must reflect subtle differences in the coordination chemistry of these metals.

Reactions of W (NMe₂)₆

The chemistry of $W(NMe_2)_6$ (*25*) and $W(Me)_6$ (*28*) provides a great opportunity for comparing covalent transition metal–nitrogen and –carbon σ bonds. In some ways their reactions are similar; for example, both compounds react with alcohols to form alkoxides $W(OR)_6$. However, in most instances their reactions differ; for example, $W(Me)_6$ reacts (*28*) with NO to give the eight-coordinate complex $W(Me_4$-$(O_2N_2Me)_2$ whereas $W(NMe_2)_6$ does not react with NO under comparable conditions.

Superficially $W(Me)_6$ is the more reactive compound. It is pyrophoric in air, thermally unstable at room temperature, and even prone to detonation (*29*). $W(NMe_2)_6$ on the other hand is thermally stable up to 200°C, and it is appreciably less moisture-sensitive than other metal dimethylamides. However, $W(NMe_2)_6$ is the more reactive compound toward some substrates. For example, whilst $W(Me)_6$ is soluble in CS_2 and acetone (*28*), $W(NMe_2)_6$ readily reacts with both of these organic substrates. $W(NMe_2)_6$ reacts with CS_2 to give $W(S_2CNMe_2)_4$. Reduction of tungsten is accompanied by oxidation of the dithiocarbamato ligand to $Me_2NC(S)S–S(S)CNMe_2$.

A rather remarkable reaction occurs with CO_2. Even in the presence of six or more equivalents of CO_2, only three equivalents of CO_2 are consumed which yields the novel compound $W(NMe_2)_3(O_2CNMe_2)_3$. Tungsten remains in its hexavalent state. This reaction is remarkable in several ways. First, all other dimethylamides of the early transition elements react according to the general reaction, Reaction 3. Complete

$$M(NMe_2)_n + nCO_2 \rightarrow M(O_2CNMe_2)_n \qquad (3)$$

$M = Ti, Zr, Hf, V, Mo$	where $n = 4$
$M = Nb$ and Ta	where $n = 5$
$M = Mo$ and W	where $n = 3$

insertion occurs which yields the fully substituted carbamato compounds. Second, $W(NMe_2)_6$ is an extremely sterically congested molecule yet it reacts with CO_2 very rapidly, much more rapidly than it reacts with alcohols which give $W(OR)_6$. Third, although $W(NMe_2)_6$ consumes only three mole equivalents of CO_2, the product, $W(NMe_2)_3(O_2CNMe_2)_3$, exchanges CO_2 (Reaction 4). This exchange is readily followed by [1]H

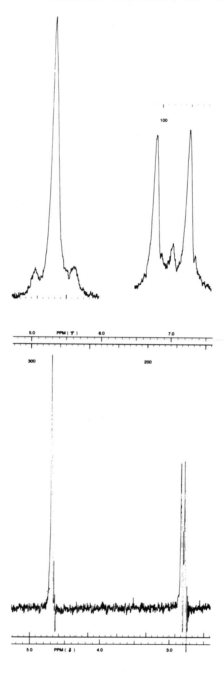

Figure 6. ^{1}H NMR spectrum of $W(NMe_2)_3(O_2{}^{13}CNMe_2)_3$ in the region of 2.5–5.5 ppm TMS (bottom) and scale expanded (top) to show ^{183}W coupling to the W–NMe$_2$ protons and ^{13}C coupling to the $O_2{}^{13}CNMe_2$ protons

$$W(NMe_2)_3(O_2CNMe_2)_3 + 3*CO_2 \rightleftharpoons \tag{4}$$

$$W(NMe_2)_3(O_2*CNMe_2)_n(O_2CNMe_2)_{3-n} + nCO_2 + (3-n)*CO_2$$

$$\text{where } n = 0\text{--}3$$

NMR spectroscopy since the ^{13}C-labelled O_2*CNMe_2 ligand shows $^3J_{13C-H} = 3.0$ Hz. The ^{1}H NMR spectrum of $W(NMe_2)_3(O_2^{13}CNMe_2)_3$ recorded in toluene-d_8 at 30°C and 60 MHz is presented in Figure 6. Since the $^{13}CO_2$ was only 92% pure, a small peak attributable to the presence of $O_2^{12}CNMe$ ligands is also observed. It is interesting to note that the W–NMe$_2$ protons show coupling to ^{183}W; $I = \frac{1}{2}$, natural abundance $= 14\%$, $^3J_{183W-H} = 3.0$ Hz. Coupling to ^{183}W was not observed for $W(NMe_2)_6$ but was noted (28) for $W(Me)_6$, $^2J_{183W-H} = 6$ Hz.

We believe that $W(NMe_2)_3(O_2CNMe_2)_3$ provides a key to the understanding of insertion reactions involving CO$_2$ and covalent metal–nitrogen bonds. The mechanism of these insertion reactions has received relatively little attention (30) although it is reasonable that it should involve a four-center transition state as indicated by Reaction 5. Such a mechanism is similar to that generally accepted (31) for the insertion of olefins and acetylenes into metal–hydrogen and metal–carbon σ bonds, but it differs with regard to the role of the lone pair on nitrogen. Clearly, both steric and electronic factors should be important in Reaction 5.

$$M-NMe_2 + CO_2 \rightleftharpoons M-\overset{O=C=O}{\underset{Me}{N}} \rightleftharpoons M-O-\overset{O}{\underset{}{C}}-NMe_2 \tag{5}$$

We obtained numerous spectroscopic data on $W(NMe_2)_3(O_2CNMe_2)_3$, all of which indicated the equivalence of the three NMe$_2$ ligands and the equivalence of the three O_2CNMe_2 ligands. Classical techniques could not, however, readily distinguish between six-coordinate tungsten, *i.e.* a fac-WN_3O_3 octahedron, and a nine-coordinate geometry involving bidentate carbamato ligands. We decided to resolve this interesting and important structural question by using commercial crystallographic services (Molecular Structure Corp., College Station, Texas).

An ORTEP view of $W(NMe_2)_3(O_2CNMe_2)_3$ is presented in Figure 7. The molecule has C_3 symmetry. Tungsten is six-coordinate and the local geometry about tungsten is suitably described as a fac-WN_3O_3 octahedron. Of particular note are the following observations: the N(2)–W–N(2) angles (94.8(3)°) are greater than the O(1)–W–O(1) angles (82.1(2)°), (b) the W–N(2) bond distances (1.922(7) A) are consid-

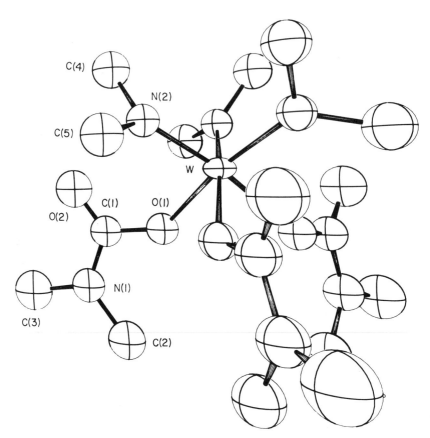

Figure 7. Molecular structure of W(NMe₂)₃(O₂CNMe₂)₃ *showing fac-WN₃O₃ moiety; the molecule has C₃ symmetry. Bond lengths: W–N(2) 1.922(7) A and W–O(1) 2.041(6) A; angles: N–W–N 94.8°, O–W–O 82.1°, N–W–O 170.3°, N–W–O 93.2°, and N–W–O 89.3°.*

erably shorter than the W–O(1) bond distances (2.041(6) A), and (c) the O₂CNC₂ and W–NC₂ moieties are planar.

The structures of W(NMe₂)₆ (25) and W(NMe₂)₃(O₂CNMe₂)₃ suggest that steric factors are not the sole controlling factors in limiting insertion. We believe that insertion is limited by the nucleophilicity of the NMe₂ ligands. Six dimethylamido ligands, ⁻NMe₂, offer tungsten a total of 24 electrons although available metal valence orbitals can accommodate only 18 electrons and 6 electrons occupy a triply degenerate nonbonding molecular orbital. Thus for W(NMe₂)₆, ligand-to-metal π bonding may lead to a maximum W–N bond order of 1.5. However, with W(NMe₂)₃(O₂CNMe₂)₃, replacement of three NMe₂ ligands by weaker π-donating oxygen ligands leads to greater N-to-W π bonding as is evidenced by the very short W–N bond length of 1.922(7) A, *cf.* W–N bond length of 2.032(25) A in $W(NMe_2)_6$ (25). It should also be noted that the fac-WN_3O_3 geometry allows for maximum W–N bonding. This enhanced nitrogen-to-tungsten π bonding reduces the nucleophilicity of the dimethylamido lone pairs, and further insertion of CO_2 is not favored.

In view of the previous observation that $W(NMe_2)_6$ readily reacts with alcohols to give $W(OR)_6$, it would seem that the WO_6 moiety is more thermodynamically stable than either the WN_6 or the fac-WN_3O_3 moieties of $W(NMe_2)_6$ and $W(NMe_2)_3(O_2CNMe_2)_3$ respectively. Insertion of CO_2 into W–N bonds in $W(NMe_2)_3(O_2CNMe_2)_3$ must present a relatively high energy of activation. It is therefore particularly pertinent to ask by what mechanism does $W(NMe_2)_3(O_2CNMe_2)_3$ exchange with CO_2 in Reaction 4. Kinetic data suggest that this exchange reaction involves an inner sphere process with a rate-determining step that involves expulsion of CO_2 which may be described by the deinsertion reaction (Reaction 6).

$$W(NMe_2)_3(O_2CNMe_2)_3 \rightleftharpoons W(NMe_2)_4(O_2CNMe_2)_2 + CO_2 \quad (6)$$

We conclude that $W(Me)_6$ and $W(NMe_2)_6$ generally react *via* different mechanisms. $W(Me)_6$ reacts (28) *via* initial nucleophilic attack at tungsten whereas $W(NMe_2)_6$ reacts *via* electrophilic attack on nitrogen nonbonding lone pairs.

Literature Cited

1. Mowat, W., Shortland, A. J., Yagupsky, G., Hill, N. J., Yagupsky, M., Wilkinson, G., *J. Chem. Soc. Dalton Trans.* (1972) 533.
2. Mowat, W., Shortland, A. J., Hill, N. J., Wilkinson, G., *J. Chem. Soc. Dalton Trans.* (1973) 770.
3. Huq, F., Mowat, W., Shortland, A. J., Skapski, A. C., Wilkinson, G., *Chem. Commun.* (1971) 1079.
4. Bradley, D. C., *Adv. Inorg. Chem. Radiochem.* (1972) **15**, 259.
5. Bradley, D. C., Fisher, K., in "MTP International Review of Science," Vol. 5, p. 65, Butterworths, London, 1972.
6. Bradey, D. C., Chisholm, M. H., *J. Chem. Soc. A* (1971) 2741.
7. Basi, J. S., Bradley, D. C., Chisholm, M. H., *J. Chem. Soc. A* (1971) 1433.
8. Angell, C. A., Cotton, F. A., Frenz, B. A., Webb, T. R., *J. Chem. Soc. Chem. Commun.* (1973) 399
9. Clark, R. J. H., Franks, M. L., *J. Chem. Soc. Chem. Commun.* (1974) 316.
10. Cowman, C. D., Gray, H. B., *J. Am. Chem. Soc.* (1973) **95**, 8177.
11. Norman, Jr., J. G., Kolari, H. J., *J. Chem. Soc. Chem. Commun.* (1974) 303.
12. Lappert, M. F., Sanger, A. R., *J. Chem. Soc. A* (1971) 874.
13. Filippo, J. S., *Inorg. Chem.* (1972) **11**, 3140.
14. Chisholm, M. H., Reichert, W., Cotton, F. A., Frenz, B. A., Shive, L., *J. Chem. Soc. Chem. Commun.* (1974) 480.
15. Gust, D., Mislow, K., *J. Am. Chem. Soc.* (1973) **95**, 1535.

16. Mislow, K., Gust, D., Finocchiaro, P., Boetcher, R. J., Top. Current Chem. (1974) **47**, 1.
17. Hutchings, M. G., Mislow, K., Nourse, J. G., Tetrahedron (1974) **30**, 1535.
18. George, M. V., Peterson, D. J., Gilman, H., J. Am. Chem. Soc. (1960) **82**, 403.
19. Barker, G. K., Lappert, M. F., J. Organomet. Chem. (1974) **76**, C45.
20. Alyea, E. C., Basi, J. S., Bradley, D. C., Chisholm, M. H., Chem. Commun. (1968) 495.
21. Bradley, D. C., Hursthouse, M. B., Newing, C. W., Chem. Commun. (1971) 411.
22. Bradley, D. C., Hursthouse, M. B., Rodesiler, P. F., Chem. Commun. (1969) 14.
23. Chisholm, M. H., Ph.D. Thesis, London, 1969.
24. Cotton, F. A., Chemical Society Centenary Lecture, 1974; Chem. Soc. Rev. (1975) **4**, 27.
25. Bradley, D. C., Chisholm, M. H., Heath, C. E., Hursthouse, M. B., Chem. Commun. (1969) 1261.
26. Grahlert, W., Thiele, K. H., Z. Anorg. Allg. Chem. (1971) **383**, 144.
27. Cotton, F. A., personal communication; J. Amer. Chem. Soc. (1975) **97**, 1242.
28. Shortland, A. J., Wilkinson, G., J. Chem. Soc. Dalton Trans. (1973) 872.
29. Wilkinson, G., private communication.
30. Lappert, M. F., Prokai, B., Adv. Organomet. Chem. (1967) **5**, 225.
31. Cotton, F. A., Wilkinson, G., in "Advanced Inorganic Chemistry," 3rd ed., Chap. 24, Interscience, New York, 1972.

RECEIVED January 24, 1975. Work supported by the Research Corp., the Petroleum Research Fund administered by the American Chemical Society (grant PRF-7722 AC3), and the National Science Foundation (grant GP-42691X). M. Extine was the recipient of a graduate student fellowship from the American Can Co.

24

Some Recent Advances in Polypyrazolylborate Chemistry

SWIATOSLAW TROFIMENKO

Plastics Dept., Experimental Station, E. I. du Pont de Nemours and Co., Wilmington, Del. 19898

Selected studies of polypyrazolylborate chemistry are reviewed. X-ray and spectral studies have clarified the nature of transannular interaction in bidentate chelates containing the puckered $R_2B(pz)_2M$ ring where the pseudoaxial R group (H, ethyl) approaches the metal close enough for three-center bonding ($B–H\cdots Mo$; $B–C–H\cdots Mo$) to occur. The puckered nature of the metallocyclic ring in bidentate systems $R_2B(pz)_2CoR_fC_5H_5$ leads to non-interconvertible conformational isomer pairs. Some were separated by chromatography; in others interconversion via ring flipping was relatively facile, and the isomers were detected only spectroscopically. The stabilizing effect of an $RB(pz)_3$ ligand was exploited in making a stable copper carbonyl derivative $HB(pz)_3CuCO$ and a variety of five-coordinate platinum compounds of type $RB(pz)_3PtMeL$. Hybrid sandwiches containing an $RB(pz)_3$ ligand and a carbocyclic moiety (C_5H_5, C_6H_6, $C_4\phi_4$) and new heavy metal (Ta, U) complexes were made recently.

The polypyrazolylborates are relatively new, uninegative chelating agents of general structure $[R_nB(pz)_{4-n}]^-$ where R is a noncoordinating substituent, pz is a 1-pyrazolyl group, and n may be 0, 1, or 2 (1). They bond to metals and metalloids through the terminal nitrogens. Despite their seemingly exotic nature, these ligands were studied extensively since 1967 because of their versatility as chelating agents and because of the following attractive features which distinguish them from most other ligands.

(a) Alkali metal salts of the parent ions and of their substituted analogs are easily synthesized (from boranes and pyrazolide ion), and they are stable to storage.

(b) Stable free acids derived from the $R_nB(pz)_{4-n}$ anions may be prepared and used for the synthesis of other salts, e.g. R_4N^+, that are unavailable by the direct route.

(c) The pyrazole hydrogens can be effectively used for proton count, and as a symmetry probe in diamagnetic and in some paramagnetic compounds.

(d) It is possible to introduce as many as ten substituents into the parent $HB(pz)_3$ ligand, and thus to alter the steric and electronic effects around the metal without destroying the original C_{3v} symmetry of the ligand.

(e) For each polypyrazolylborate ligand, there exists an isosteric and isoelectronic $R_nC(pz)_{4-n}$ counterpart which has the same coordinative behavior, but it is neutral and gives complexes with a charge greater by +1 per ligand. This is a truly unique feature.

Since the earlier work (up to 1972) in this area was reviewed (2, 3), numerous new studies have appeared. Before these more recent developments are discussed, some background information on the characteristic features of the polypyrazolylborate ligands and their chelates is presented.

The bonding of polypyrazolylborates to metal is determined primarily by the number of pyrazolyl groups attached to boron. Dipyrazolylborates are necessarily bidentate, and they form with divalent transition metal ions complexes, 1, that are similar to β-diketonates but that are always monomeric for steric reasons. The major difference between dipyrazolylborates and β-diketonates is that the $R_2B(pz)_2M$ ring is not planar but is puckered in the boat form as in Structure 2. This results in asym-

1   2

metry and non-identity of the R groups, one of which (the pseudoaxial) is close to and sometimes interacts with the metal, even when R is H or an alkyl group.

If one of the R groups in $R_2B(pz)_2$ is another pyrazolyl ring, then it can also bond to the metal. With divalent transition metals, compounds with Structure 3 are formed whereas with trivalent metals the analogous cations are formed. The ligand $RB(pz)_3$ ($R \neq pz$) is the first known example of a uninegative tridentate ligand with C_{3v} symmetry, and it is analogous to a cyclopentadienide ion in being uninegative, in supplying six electrons, and in occupying three coordination sites. In this context, the octahedral compounds (3), which are of D_{3d} symmetry, are analogs of metallocenes. The rigid central cage structure persists in all their substituted derivatives. Accordingly, their properties change less dramatically with substitution than do those of the bidentate chelates, 1.

$$R = H, \text{ alkyl, aryl, pz}$$

$$Y = Z = Pd-\pi-CH_2CRCH_2$$
$$Y = Z = BEt_2$$
$$Y = BEt_2;\ Z = Pd-\pi-CH_2CRCH_2$$

3   4

When the fourth boron substituent is also a pyrazolyl group, the resulting $B(pz)_4$ ligand is still basically tridentate of local C_{3v} symmetry, and it forms chelates (Structure 3) with R = pz. In most such compounds, the three coordinated and one uncoordinated pz groups maintain their separate identity, as was demonstrated by NMR spectroscopy of the Co(II) derivative (4). By contrast, in the isomorphous zinc analog, all four pz groups are spectroscopically equivalent and thus exchange rapidly (5). The $B(pz)_4$ ligand may also act in tetradentate (bis-bidentate) fashion, forming spiro cations of the 4 type (6) which may contain identical or dissimilar bridging units.

Polypyrazolylborates have been widely used in organometallic chemistry. A large number of compounds containing the bi- or tridentate ligand bonded to a metal already containing various organic moieties were synthesized and studied. The tridentate $RB(pz)_3$ ligand forms organometallic compounds analogous to the half-sandwiches based on C_5H_5 (*e.g.* 5) which, as a rule, are more stable than their cyclopentadienyl counterparts, and sometimes these half-sandwiches can exist only in the polypyrazolylborate series. These were discussed in the earlier reviews (2, 3). A general idea of the way an $RB(pz)_3$ group functions in organometallic chemistry may be obtained from Schemes I and II.

Scheme I

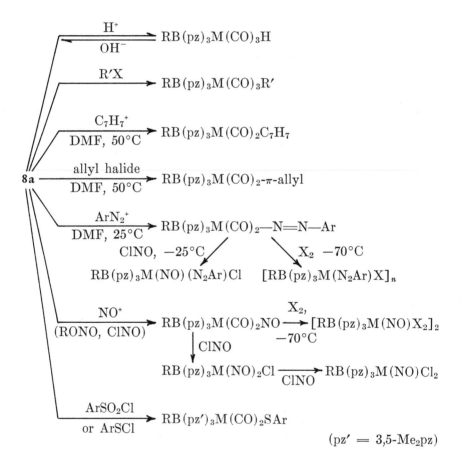

$$RB(pz)_3^- + M(CO)_6 \xrightarrow[100°C]{DMF} RB(pz)_3M(CO)_3^- + 3CO$$

(M = Cr, Mo, W)   **8a**

In addition to the previously cited examples, several recent findings illustrate the principle of stabilization through coordination with a poly-pyrazolylborate ligand. An interesting case is that of five-coordinate tri-gonal bipyramidal complexes of Pt(II). Few such complexes were known previously, and they were unstable. Clark and Manzer demonstrated

Scheme II

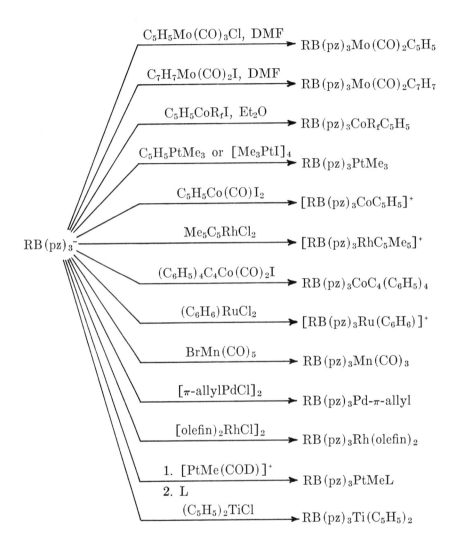

that quite stable five-coordinate Pt(II) complexes of the type RB(pz)₃-PtMeL (L = acetylene, olefin, allene, CO) (Structures **6, 7,** and **8**) can be prepared. In a typical reaction, the polymeric RB(pz)₃PtMe is treated in methylene chloride with the appropriate ligand, whereupon the polymer dissolves with formation of the stable, crystalline five-coordinate products (**7, 8**). Their structure was supported by NMR data, and by an x-ray crystallographic structure determination (**9**) which reveals the RB(pz)₃ ligand as symmetrically tridentate (mean angle

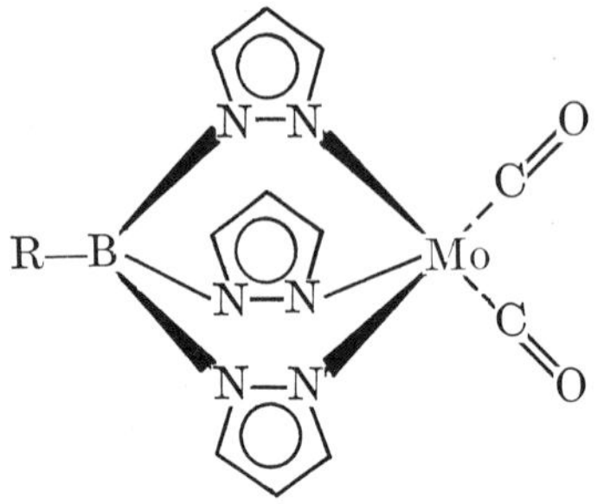

5

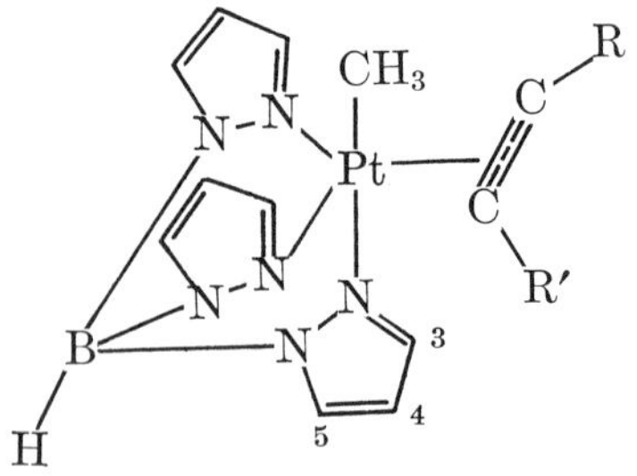

6

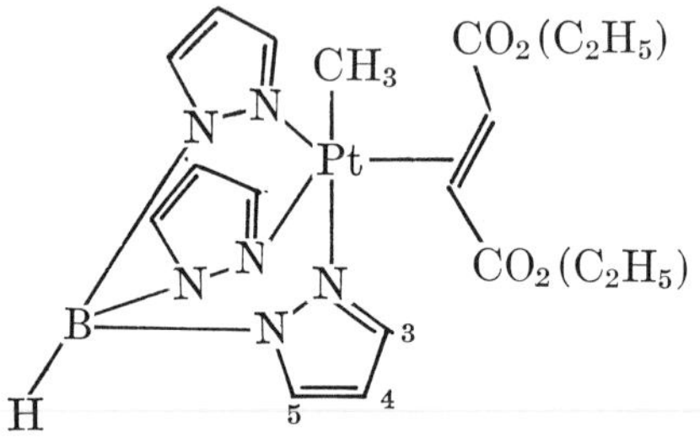

7

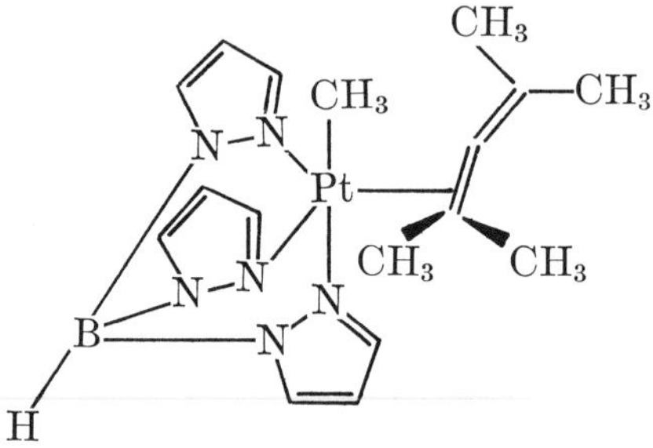

8

between pz groups, 120.1°), with two pz rings coordinating in the equatorial plane and the third apically. The methyl group is at the other apex, and the acetylene is in the equatorial plane with the CF_3 groups bent back by 34.4°.

The analogous five-coordinate olefin complexes show the presence of isomers because of an asymmetric environment above and below the equatorial plane. The isomer distribution depends on the steric requirement of the olefin substituent. In a number of fluoroolefins, a direct through-space H–F coupling was observed (10). With allenes, Pt is coordinated to one of the double bonds, usually the less substituted one, and the uncoordinated double bond is probably bent backward by some 38°, by analogy with the bending found in $PtP(C_6H_5)_3(CH_2=C=CH_2)$ (11).

$RB(pz)_3PtMeCO$ is interesting since its x-ray structure (12) reveals that the $HB(pz)_3$ ligand is bidentate, with the nonbonded Pt–B distance of 3.36 A substantially longer than the 3.20 A found in the five-coordinate complex $HB(pz)_3PtMe$(hexafluoro-2-butyne), and with the uncoordinated pz group turned so that its 2-N points away from Pt. Nevertheless, the evidence for a five-coordinate structure in solution is convincing: (a) these stereochemically nonrigid compounds have a limiting NMR spectrum at $-120°C$, at which temperature the three pz groups fall into a 2:1 pattern (a bidentate structure would have a 1:1:1 pattern); and (b) the limiting spectrum reveals [195]Pt coupling to ^{1}H and ^{13}C of all three pz groups, but not to the fourth one in the B(pz)$_4$ analog which at no time exchanges with the three coordinated pz groups. This is in contrast to B(pz)$_4$Pd-π-allyl, where all four pz groups exchange rapidly at room temperature (*13*). The most likely mechanism by which the three coordinated pz groups become equivalent in the five-coordinate HB(pz)$_3$PtMeCO is a rotation around the C_{3v} axis, a mode of fluxional behavior previously observed in other RB(pz)$_3$ complexes (*5, 14*). The presence of a bidentate HB(pz)$_3$ ligand in the crystal must be a packing effect.

A similar family of compounds based on dipyrazolylborate ligands— R$_2$B(pz)$_2$PtMeL (R = ethyl, phenyl; L = acetylene, phosphine, CO, isonitrile)—was prepared with good yields. In contrast to the RB(pz)$_3$ analogs, methyl migration occurs in this series with negatively substituted acetylenes, *e.g.* with dimethyl acetylenedicarboxylate to give **9**,

9   10

R$_2$B(pz)$_2$PtC(COOMe)CMe(COOMe). Isomerization occurs in this compound so that the methyl group ends up trans to the Pt, the fourth coordination site of which is occupied by the carbomethoxy oxygen. In the related complex R$_2$B(pz)$_2$PtMe(hexafluoro-2-butyne)P(C$_6$H$_5$)$_3$, the CF$_3$ groups are trans (*15*). X-ray of the stable Et$_2$B(pz)$_2$PtMe(C$_6$H$_5$-CCMe) reveals a structure which could be five-coordinate if one of the B–Et methylene hydrogens were to form a C–H–Pt bond (*16*).

Another case of stabilization *via* RB(pz)$_3$ ligand is the synthesis of the first stable CuCO complexes by Bruce and Ostazewski (*17*). The monomeric HB(pz)$_3$CuCO has a sharp CO band at 2,083 cm^{-1} and it melts at 165–167°C; it has structure **10** as was confirmed by x-ray (*18*). On heating, it decarbonylates to the dimer [HB(pz)$_3$Cu]$_2$ which may also be obtained from Cu(I) and HB(pz)$_3$$^-$. The structure of this dimer, as established by x-ray (19), has a centrosymmetric molecule with each HB(pz)$_3$ unit contributing two pz ligands (one to each copper) and the third pz group bridging crosswise two coppers. Each copper is in a highly distorted tetrahedral environment, with N–Cu–N angles ranging from 93.74° to 144.75°. The NMR spectrum of this dimer indicates equivalence of all pz groups down to −130°C which implies rapid rotation of the ligand around the B–B axis. The CuCO complex reacts readily with a variety of ligands forming complexes of the type HB(pz)$_3$-CuL(L = phosphines, phosphites, isonitriles, arsines, stibenes). Analogous reactions were also reported for HB(pz)$_3$AgL (20).

Several recent studies have helped to clarify some unresolved questions relating to the R$_2$B(pz)$_2$ complexes of Mo. When Mo(CO)$_6$ is heated in polar solvents with various dipyrazolylborates, unstable anions R$_2$B(pz)$_2$Mo(CO)$_4^-$ are formed which react rapidly with allylic halides to form π-allyl derivatives 11a and 11b. Both of these are formally 16-e compounds (21). The 11b compounds were less stable than the 11a, and they reacted readily with nucleophiles to give stable 18-e derivatives. The thermal stability of 11a and its unreactivity toward nucleophiles

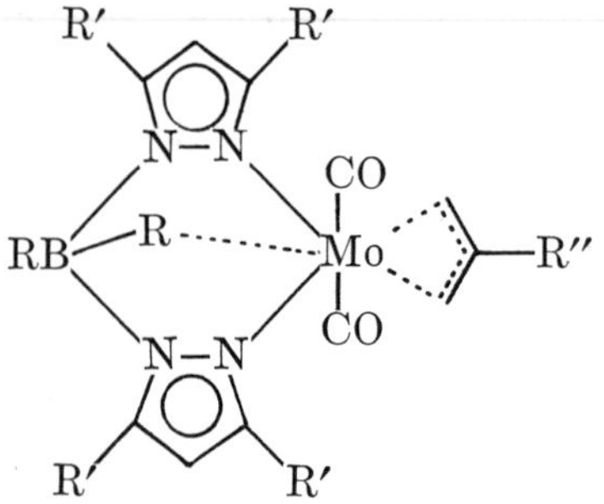

11a R′ = CH$_3$; R = H

11b R′ = H; R = —C—H with H above and CH$_3$ below

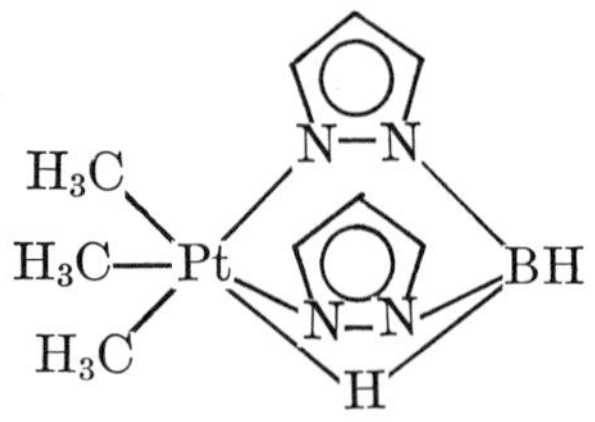

12

were surprising, as was its anomalous B–H stretch in the IR spectrum: one sharp spike that is normally characteristic of the HB(pz)$_3$ ligand. X-ray crystallographic studies of this compound established the presence of a three-center B–H–Mo bond (22).

A similar situation prevails with H$_2$B(3,5-Me$_2$pz)$_2$Mo(CO)$_2$C$_7$H$_7$ (23) where the presence of a B–H–Mo three-center bond makes Mo effectively 18-e, with bond distances of 1.05, 1.26, and 2.14 A for H–B (terminal), B–H (bridge), and H–Mo respectively. The C_7H_7 ligand was trihapto, as was also indicated by spectral studies (*24*).

A three-center B–H–Pt bond was also invoked to account for the stability of $H_2B(pz)_2PtMe_3$ (*12*) and its IR band at 2039 cm^{-1} (*25*). The IR spectra of **11b** had an anomalously low C–H stretch (2704, 2664 cm^{-1}), and in the NMR spectra the pseudoaxial methylene hydrogens were at 12.4τ. This was interpreted as indicating that these methylene hydrogens were sufficiently close to the Mo orbitals for interaction to occur (*21*).

The x-ray structure of $Et_2B(pz)_2Mo(CO)_2$-π-$CH_2C_\phi CH_2$ as determined by Cotton and co-workers indicates that such interaction does indeed occur but that only one of the pseudoaxial methylene hydrogens forms a C–H–Mo three-center bond with a H–Mo distance of about 2.15 A (*26*). NMR studies of this and the related π-allyl compound revealed that the structure found in the crystal also exists in solution at low temperature as the 12.4 τ peak splits into two: at 14.3 τ (bridging) and at 10.8 τ (nonbridging). One type of fluxionality observed in this molecule is an oscillatory exchange of methylene hydrogens bonded to Mo. This process has an activation energy of about 14 kcal/mole. At higher temperatures, a ring inversion occurs with an activation energy of 17–20 kcal/mole. This approximates the strength of the C–H–Mo interaction (*27*).

In the 18-*e* pyrazole adduct of this compound, the base attacked trans to the C–H–Mo bond; the flip-back of the B–Et group resulted in a nonbonded B–Mo distance of 3.806 A, the longest found in any compound of this type, and the $B(pz)_2Mo$ ring assumed a distorted chair conformation (*28*).

The formation of a three-center C–H–Mo bond competes successfully with olefinic coordination in the compound $Et_2B(pz)_2Mo(CO)_2C_7H_7$ (*29*). With a choice of attaining 18-*e* configuration through pentahapto bonding to C_7H_7 or through C–H–Mo bonding, the molecule opted for the latter. It is not clear, however, whether the pentahapto structure is sterically possible.

The complex $(C_6H_5)_2B(pz)_2Mo(CO)_2$-$\pi$-$CH_2CMeCH_2$ possesses a true 16-*e* configuration in the crystal (*30*). There is no interaction with an ortho C–H as might be expected because the necessary orientation of the pseudoaxial phenyl group would impart too much steric strain on the rest of the molecule.

The value of the dihedral angle between pz groups in various bidentate polypyrazolylborate compounds varies greatly (Table I). It reflects the depth of the boat conformation in the $B(pz)_2M$ ring and depends on the nature of M, on the nature of the pseudoaxial R group,

Table I. Dihedral Angles between Pyrazolyl Rings in Bidentate Complexes

Compound	Dihedral Angle	Ref.
$Et_2B(4\text{-}Brpz)_2BEt_2$	180°	*31*
$RB(pz)_3MXYZ$	120°	*9*
$[H_2B(pz)_2]_2Co$	121°	*33*
	114°	
$[Et_2B(pz)_2]_2Ni$	113°	*32*
$Et_2B(pz)_2Mo(CO)_2\text{-}\pi\text{-}CH_2C\phi CH_2$	111°	*27*
$H_2B(3,5\text{-}Me_2pz)_2Mo(CO)_2C_7H_7$	104°	*23*
	102°	

and on the mode of interaction of the pseudoaxial group with M. R groups, which are not normally considered prone to bonding, are brought close enough to the metal for interaction/bonding to occur because of the very flexible nature of polypyrazolylborate ligands and their adjustable bite, which permit the adoption of a wide range of conformations. At the one extreme, where M is BR_2 as in **13**, the dihedral angle is almost

13

180° (*31*). The tridentate $HB(pz)_3$ ligand should and does have an angle of 120°. Presence of a B–H–Mo bond gives rise to the smallest angle (102° and 104°). $Et_2B(pz)_2$ ligands, with less strain required to bring the C–H close to M, have a more relaxed angle (111° and 113°) even though the latter angle is found in a compound that is devoid of C–H–M bonds, $[Et_2B(pz)_2]_2Ni$ (*32*). It is noteworthy that in the tetrahedral $[H_2B(pz)_2]_2Co$, two distinctly different conformations of the ligands are encountered (*33*); one has a dihedral angle of 121°, the other 114°.

A good illustration of geometric and conformational stereoisomerism in bidentate polypyrazolylborate complexes is provided by the stable, red perfluoroalkyl $Co(C_5H_5)$ derivatives prepared by King and Bond (*34*). These compounds have the general structure $H_nB(pz)_{4-n}CoR_f$-(C_5H_5). The polypyrazolylborate ligand is always bidentate, and there is no exchange between the coordinated and uncoordinated pz groups. With $R_2B(pz)_2$ ligands where both R groups are identical, two conformational isomers are possible (**14** and **15**) because of the puckered nature of the ring. Although these isomers should be capable of interconversion through inversion of the $Co(pz)_2B$ ring, this was not observed. In fact, these isomers could be separated in pure state by column chromatography. With $HB(pz)_3$, four isomers are theoretically possible (the two geometric isomers, **16** and **17**, plus their respective conformational isomers in which C_5H_5 and R_f are reversed). When R_f was CF_3 or n-C_3F_7, however, only a single isomer was obtained; it probably has Structure **16** with CF_3 and Structure **17** with n-C_3F_7. On the other hand, C_2F_5 gave an inseparable mixture of isomers. It is remarkable that conformational isomers were isolated, but geometric isomers were not.

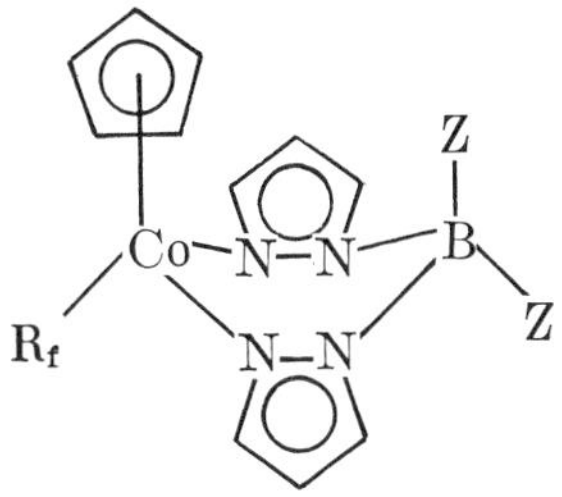

14 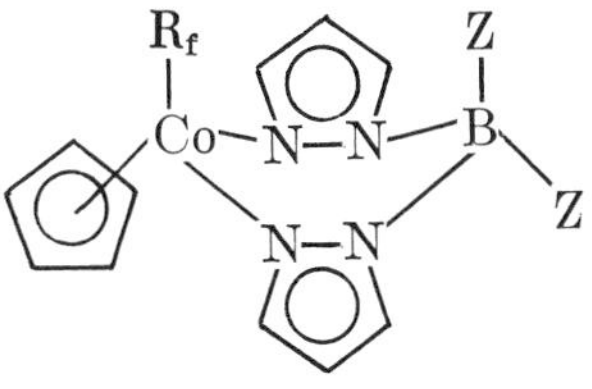   **15**

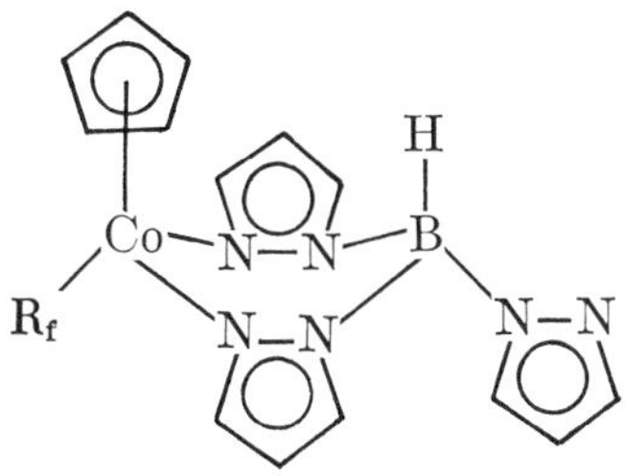

16 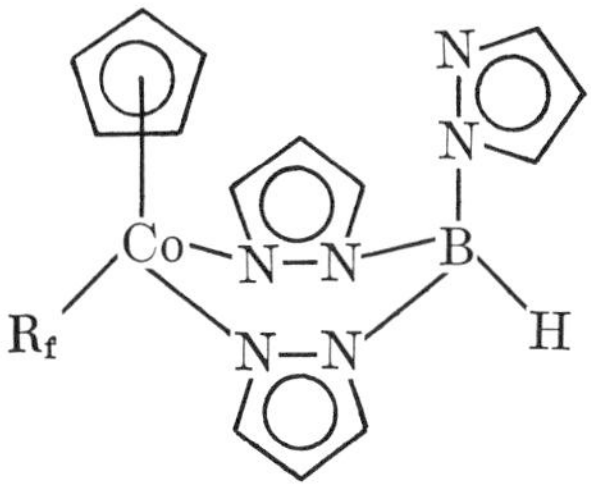   **17**

Another interesting recent development is the successful synthesis of hybrid sandwiches that contain one $RB(pz)_3$ ligand and one carbocyclic ligand—*e.g.* $[RB(pz)_3CoC_5H_5]^+$, $(C_6H_5)_4C_4Co(pz)_3BR$, $[C_6H_6Ru(pz)_3$-$Bpz]PF_6$, and $[C_5Me_5Rh(pz)_3BH]PF_6$. They were prepared by displacement reactions on the appropriate halide precursors (35). X-ray crystallographic studies confirmed the structural assignments for the last two compounds (36, 37).

Finally, two communciations reporting polypyrazolylborates of heavier metals have appeared: Wilkinson and co-workers (38) described the synthesis of the seven-coordinate $[H_2B(pz)_2]_2TaMe_3$ while Bagnall and Edwards (39) prepared the sublimable, green, air-sensitive $HB(pz)_3$-$UCl_2C_5H_5$ as well as $[H_2B(pz)_2]_4U$, $[HB(pz)_3]_4U$, and $[HB(pz)_3]_2$-UCl_2 (40). The above findings bear witness to a sustained and rapid growth of polypyrazolylborate chemistry.

Literature Cited

1. Trofimenko, S., *J. Am. Chem. Soc.* (1966) **88**, 1842.
2. Trofimenko, S., *Acc. Chem. Res.* (1971) **4**, 17.
3. Trofimenko, S., *Chem. Rev.* (1972) **72**, 497.
4. Jesson, J. P., Trofimenko, S., Eaton, D. R., *J. Am. Chem. Soc.* (1967) **89**, 3148.
5. Trofimenko, S., *J. Am. Chem. Soc.* (1969) **91**, 3183.
6. Trofimenko, S., *J. Coord. Chem.* (1972) **2**, 75.
7. Clark, H. C., Manzer, L. E., *Inorg. Chem.* (1974) **13**, 1291.
8. *Ibid.* (1974) **13**, 1996.
9. Davies, B. W., Payne, N. C., *Inorg. Chem.* (1974) **13**, 1843.
10. Clark, H. C., Manzer, L. E., *J. Chem. Soc. Chem. Commun.* (1973) 870.
11. Kadonaga, M., Yasuoka, N., Kasai, N., *J. Chem. Soc. Chem. Commun.* (1971) 1597.
12. Rush, P. E., Oliver, J. D., *J. Chem. Soc. Chem. Commun.* (1974) 996.
13. Trofimenko, S., *J. Am. Chem. Soc.* (1969) **91**, 588.
14. Meakin, P., Trofimenko, S., Jesson, J. P., *J. Am. Chem. Soc.* (1972) **94**, 5677.
15. Clark, H. C., personal communication.
16. Davies, B. W., Payne, N. C., personal communication.
17. Bruce, M. I., Ostazewski, A. P. P., *J. Chem. Soc. Dalton Trans.* (1973) 2433.
18. Bruce, M. I., personal communication.
19. Arcus, C. S., Wilkinson, J. L., Mealli, C., Marks, T. J., Ibers, J. A., *J. Am. Chem. Soc.* (1974) **96**, 7564.
20. Abu Salah, O. M., Bruce, M. I., *J. Organomet. Chem.* (1975) **87**, C 15.
21. Trofimenko, S., *Inorg. Chem.* (1970) **9**, 2493.
22. Kosky, C. A., Ganis, P., Avitabile, G., *Acta Cryst. Sect. B* (1971) **27**, 1859.
23. Cotton, F. A., Jeremic, M., Shaver, A., *Inorg. Chem. Acta* (1972) **6**, 543.
24. Calderon, J. L., Cotton, F. A., Shaver, A., *J. Organomet. Chem.* (1972) **42**, 419.
25. King, R. B., Bond, A., *J. Am. Chem. Soc.* (1974) **96**, 1338.
26. Cotton, F. A., LaCour, T., Stanislowski, A. G., *J. Am. Chem. Soc.* (1974) **96**, 754.
27. Cotton, F. A., Stanislowski, A. G., *J. Am. Chem. Soc.* (1974) **96**, 5074.
28. Cotton, F. A., Frenz, B. A., Stanislowski, A. G., *Inorg. Chim. Acta* (1973) **7**, 503.
29. Cotton, F. A., Day, V. W., *J. Chem. Soc. Chem. Commun.* (1974) 415.
30. Cotton, F. A., personal communication
31. Holt, E. M., personal communication.
32. Echols, H. M., Dennis, D., *Acta Cryst. Sect. B* (1974) **30**, 2173.

33. Guggenberger, L. J., Prewitt, C. T., Meakin, P., Trofimenko, S., Jesson, J. P., *Inorg. Chem.* (1973) **12**, 508.
34. King, R. B., Bond, A., *J. Am. Chem. Soc.* (1974) **96**, 1334.
35. O'Sullivan, D. J., Lalor, F. J., *J. Organomet. Chem.* (1973) **57**, C58.
36. Ferguson, G., Restivo, R. J., *J. Chem. Soc. Chem. Commun.* (1973) 847.
37. Ferguson, G., private communication.
38. Williamson, D. H., Santini-Scampucci, C., Wilkinson, G., *J. Organomet. Chem.* (1974) **77**, C25.
39. Bagnall, K. W., Edwards, J., *J. Organomet. Chem.* (1974) **80**, C14.
40. Bagnall, K. W., Edwards, J., du Preez, J. G. H., Warren, R. F., *J. Chem. Soc. Dalton Trans.* (1975) 140.

RECEIVED February 6, 1975.

25

Metalloborane Derivatives with Ligand– Metal Single Bonds

LEE J. TODD

Indiana University, Bloomington, Ind. 47401

This paper reviews the chemistry of metalloborane deriva-
tives that contain borane or heteroatom borane groups which
function as monohapto ligands. These compounds can be
divided into three classes according to the number of elec-
trons formally donated by the borane ligand to the metal.
The electron pair acceptor class, represented by the com-
pound $Na[(OC)_5Mn \rightarrow BH_3]$, has received little attention
thus far. The one-electron donor class, exemplified by the
complex, $1,2-(CH_3)_2-3-[(C_5H_5)Fe(CO)_2]-B_{10}C_2H_9$, has a
rich chemistry of metal–carbon and metal–boron derivatives.
The third class includes two-electron donor derivatives that
are represented by the compound $(CH_3)_4N[7,8-B_9H_{10}CHP$
$\rightarrow Cr(CO)_5]$.

Metalloborane chemistry has developed rapidly over the past few years. It is now apparent that there are as many diverse structural types in this area as there are in the more established field of organo-transition metal chemistry. There are now many known examples where the borane or hetroatom borane moiety functions as a mono-, di-, tri-, tetra- or penta-hapto ligand. This paper is limited to borane and heteroatom borane groups which function as monohapto ligands.

As a matter of classification, these single bond complexes are divided into three groups according to the number of electrons formally donated by the borane ligand to the metal. The zero-electron donor (or electron pair acceptor) complex is exemplified by the compound $Na[(OC)_5Mn \rightarrow BH_3]$ (1), the one-electron donor by the compound $1,2-(CH_3)_2-3-[(C_5H_5)Fe(CO)_2]-B_{10}C_2H_9$ (2), and the two-electron donor by the compound $(CH_3)_4N[7,8-B_9H_{10}CHP \rightarrow Cr(CO)_5]$ (3).

Electron Pair Acceptor Derivatives

A derivative from this class of complexes was the first compound reported, but overall they have been the least studied group. Reaction of diborane with salts of $Re(CO)_5^-$, $Mn(CO)_5^-$, and $Ph_3PMn(CO)_4^-$ gave monoborane complexes in which BH_3 is coordinated to the metal *(1)*. The complex $BH_3 \cdot Mn(CO)_5^-$ was stable at $-78°C$ but unstable at room temperature. The borane complexes of $Re(CO)_5^-$ and $(Ph_3P)Mn-(CO)_4^-$ are both stable at room temperature. It is suggested that the greater base strength of the latter two metal carbonyl anions enhances the thermal stability of these donor–acceptor complexes. A stable bis-(borane) complex, $(n\text{-}C_4H_9)_4P[(H_3B)_2Re(CO)_5]$, was isolated by using excess diborane in the reaction described above, but its structure is not known. Potassium germyltrihydroborate, $K[GeH_3BH_3]$, is formed by the reaction of diborane with potassium germyl *(4)*. This compound is thermally stable to 200°C and is fairly stable in alkaline aqueous solutions.

One-Electron Donor Derivatives

Treatment of 2-chloropentaborane *(9)* with $Na[M(CO)_5]$ (M = Mn or Re) in ether solution generated 2-$[M(CO)_5]B_5H_8$ *(5)*. The rhenium derivative is stable at room temperature for several hours, but the manganese compound develops a yellow color upon melting at $-10°C$.

There have been several reports of transition metal–carborane complexes with covalent two-electron bonds. An early report indicates that carbon–mercury bonds were formed by the reaction of C-lithiocarboranes with mercuric halides *(6, 7)*.

$$2 \text{ R}\underset{\underset{B_{10}H_{10}}{\diagdown O \diagup}}{\text{C—CLi}} + \text{HgCl}_2 \rightarrow [\text{R}\underset{\underset{B_{10}H_{10}}{\diagdown O \diagup}}{\text{C—C}}]_2\text{Hg}$$

This synthetic route has also been applied to many other transition metal derivatives. Reaction of $(C_5H_5)Fe(CO)_2I$ with 1-lithio-2-methyl-1,2-$C_2B_{10}H_{10}$ or 1-lithio-10-methyl-1,10-$C_2B_8H_{10}$ afforded 1-$[(C_5H_5)Fe-(CO)_2]$-2-methyl-1,2-$C_2B_{10}H_{10}$ and 1-$[(C_5H_5)Fe(CO)_2]$-10-methyl-1,10-$C_2B_8H_8$ respectively *(8)*. The derivative $(C_5H_5)Fe(CO)_2C_2B_4H_7$ was also formed by this type of reaction *(9)*. However based on ^1H and ^{11}B NMR data, this compound appears to have the carborane ligand attached to the iron atom by a B–Fe–B three-center two-electron bridge bond.

A series of chelated bis(carborane) metal complexes have been made by the following type of reaction *(10)*.

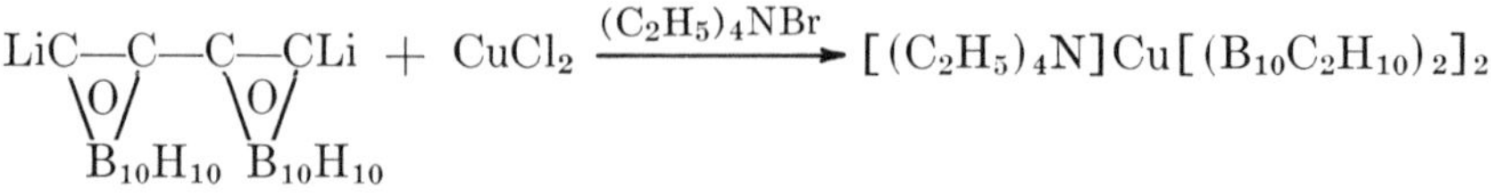

$$\text{LiC—C—CLi} + \text{CuCl}_2 \xrightarrow{(\text{C}_2\text{H}_5)_4\text{NBr}} [(\text{C}_2\text{H}_5)_4\text{N}]\text{Cu}[(\text{B}_{10}\text{C}_2\text{H}_{10})_2]_2$$

with the two $\text{B}_{10}\text{H}_{10}$ bridges shown below the carbon chain.

Derivatives of copper, nickel, and cobalt in both the (2+) and (3+) formal oxidation states and a Zn (2+) complex have been isolated and characterized. In all these complexes it is proposed that the chelate is bonded to the metal by single bonds to the cage carbons. A single-crystal x-ray study of $[\text{N}(\text{C}_2\text{H}_5)_4]_2\text{Co}[(\text{B}_{10}\text{C}_2\text{H}_{10})_2]_2$ has confirmed this bonding scheme (11). However it was also observed that one hydrogen atom attached to a boron adjacent to both carbon atoms in a chelate ligand appeared to be hydrogen bonded to the cobalt center. The ^{11}B NMR spectrum of this complex consisted of a 2 : 2 : 1 : 2 : 1 : 2 pattern of doublets reading from low to high field. The lowest field double is at −103.8 ppm [relative to $\text{BF}_3 \cdot \text{O}(\text{C}_2\text{H}_5)_2$]. This is an unusually low field position for such a boron resonance. This doublet corresponds to eight boron atoms in the entire bis(chelate)–metallocarborane molecule. It is proposed that the one B–H–Co interaction observed in the solid state is a time-averaged (fluxional) process in solution such that the eight potentially equivalent boron atoms feel the presence of the metal atom equally.

Various nickel (12), palladium (13), platinum (13), cobalt (14), and rhodium (15) derivatives which contain 1,2-or 1,7-$\text{B}_{10}\text{H}_{10}\text{C}_2\text{R}$ groups bonded by means of a carbon–metal single bond have been prepared by Bresadola and co-workers in recent years. The steric effect of the metal-bonded carboranyl group is one of the more interesting aspects of these studies. Reaction of cis-$(\text{Ph}_3\text{P})_2\text{PtCl}_2$ with 1-lithio-2-methyl-1,2-$\text{C}_2\text{B}_{10}\text{H}_{10}$ formed 1-$[cis$-$(\text{Ph}_3\text{P})_2\text{PtCl}]$-2-methyl-1,2-$\text{C}_2\text{B}_{10}\text{H}_{10}$ (12). It was not possible to attach two carboranyl groups to the platinum atom by this reaction procedure. In contrast, lithiocarboranes react with $trans$-$[(\text{C}_2\text{H}_5)_3\text{P}]_2\text{PtCl}_2$ to form a halogen-free metallocarborane for which the following structure has been proposed based on available spectroscopic and chemical data (13).

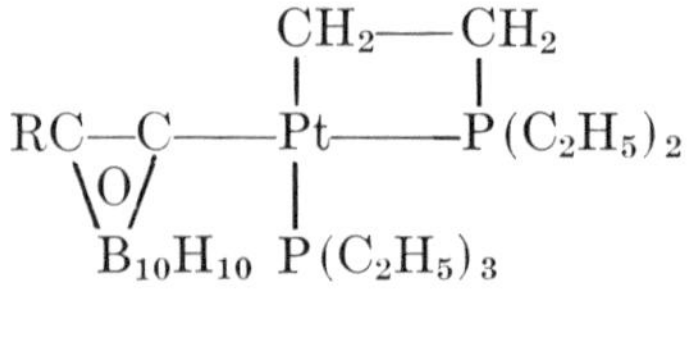

$$\begin{array}{ccccc}
 & & \text{CH}_2\!\!-\!\!-\!\!\text{CH}_2 & & \\
 & & | \qquad\quad | & & \\
\text{RC—C} & \!\!-\!\!-\!\!-\!\! & \text{Pt} & \!\!-\!\!-\!\!-\!\! & \text{P}(\text{C}_2\text{H}_5)_2 \\
\diagdown\text{O}\diagup & & | & & \\
\text{B}_{10}\text{H}_{10} & & \text{P}(\text{C}_2\text{H}_5)_3 & &
\end{array}$$

Platinum hydride complexes were prepared by the following reaction (16):

$$trans\text{-}[(C_2H_5)_3P]_2PtHCl + \text{Li-carb} \rightarrow cis\text{-} \text{ or}$$
$$trans\text{-}[C_2H_5)_3P]_2PtH \ (\sigma\text{-carb})$$

The general formula "Li-carb" represents C-lithio derivatives of 1,2- and 1,7-$B_{10}H_{10}CHCR$ (R = H, CH_3, C_6H_5). Cis or trans metallocarborane products were obtained as indicated by the multiplicity of the metal hydride signal in the ^1H NMR spectrum. The reaction of $(Ph_3P)_3RhCl$ with C-lithio derivatives of 2-R-1,2- and 7-R-1, 7-$B_{10}C_2H_{11}$ (R = CH_3, C_6H_5) generate the unusual, formally three-coordinate compounds (Ph_3-$P)_2Rh$(carborane) (*15*). An X-ray structure of the derivative 1-[($Ph_3P)_2$-Rh]-2-phenyl-1,2-$C_2H_{10}H_{10}$ (*17*) revealed a B–H–Rh interaction similar to that observed (*vide supra*) in the bischelate compound, [N($C_2H_5)_4]_2$-Co[($C_2B_{10}H_{10})_2]_2$.

Zakharkin and co-workers have developed a method to attach a variety of substituents at B(3) of ortho-carborane. Using the acid chloride derivative, they were able to prepare new sigma-bonded derivatives as indicated in the following sequence (*18*).

$$\underset{B_{10}H_9CCl}{\overset{RC-CR}{\underset{\overset{\|}{O}}{\diagdown O \diagup}}} + Na[Fe(CO)_2C_5H_5] \rightarrow \underset{B_{10}H_9CFe(CO)_2C_5H_5}{\overset{RC-CR}{\underset{\overset{\|}{O}}{\diagdown O \diagup}}}$$

$$\xrightarrow[-CO]{170°C} \underset{B_{10}H_9Fe(CO)_2C_5H_5}{\overset{RC-CR}{\diagdown O \diagup}}$$

Treatment of 3-$[C_5H_5Fe(CO)_2]$-1,2-$B_{10}H_9C_2H_2$ with mercuric chloride at 110°C formed the stable mercury derivative, 3-$[ClHg]$-1,2-$B_{10}H_9C_2H_2$. The boron–mercury single bond was cleaved with bromine to generate 3-Br-1,2-$B_{10}H_9C_2H_2$, which is a known derivative that has been prepared by an alternate route. These chemical transformations indicated that the metals are attached to boron at position B(3). The rhenium carbonyl complex, 3-$[Re(CO)_5]$-1,2-$B_{10}H_9C_2H_2$ can be prepared in a similar manner to the iron derivative described above (*19*).

Another route to metalloborane derivatives with metal–boron single bonds is the reaction of certain cationic transition metal compounds with borane anions (*20*). Addition of solid $[(C_5H_5)Fe(CO)_2(cyclohexene)]$-$PF_6$ to an ether solution of $NaB_{10}H_{13}$ at room temperature formed yellow 6-$[(C_5H_5)Fe(CO)_2]B_{10}H_{13}$ in 50% yield. The pattern of the ^{11}B NMR spectrum of this derivative was very similar to other 6-substituted decaborane derivatives, and on this basis it was proposed that an iron–boron single bond had been formed at B(6) of the decaborane cage as illus-

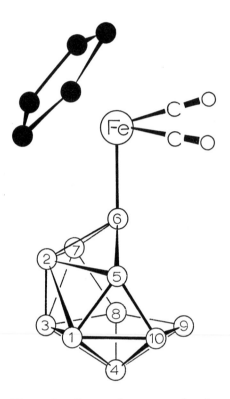

Figure 1. Proposed structure for 6-
[(C$_5$H$_5$)Fe(CO)$_2$]-B$_{10}$H$_{13}$ (terminal and
bridge hydrogen atoms have been omit-
ted for clarity)

trated in Figure 1. Treatment of this iron–borane derivative with bromine at 20°C resulted in $(C_5H_4B_{10}H_{13})Fe(CO)_2Br$ (20). In this interesting reaction the borane group has migrated from the iron atom to the cyclopentadienyl ring. Based on the ^{11}B NMR spectrum it was proposed that the $B_{10}H_{13}$ unit is attached at B(6) by means of a carbon–boron single bond to the cyclopentadienyl ring. There is one previously reported example of this type of rearrangement which is illustrated in the following equation (21):

$$HC—CH \atop \diagdown O \diagup \atop B_{10}H_9Fe(CO)_2(C_5H_5) \qquad + Br_2 \rightarrow \qquad HC—CH \atop \diagdown O \diagup \atop B_{10}H_9— \qquad Fe(CO)_2Br$$

Treatment of 6-[(C$_5$H$_5$)Fe(CO)$_2$]B$_{10}$H$_{13}$ with triethylamine in benzene rapidly formed [(C$_5$H$_5$)Fe(CO)$_2$]$_2$ in high yield (20). The only boron product isolated was $[(C_2H_5)_3NH]_2B_{10}H_{10}$ which was obtained in 10% yield.

Reaction of $(CH_3)_3NH[7,8-B_9H_{10}C_2R_2]$ (R = H, CH_3) at reflux in acetone solution with $[(C_5H_5)Fe(CO)_2(cyclohexene)]PF_6$ formed red $9-[(C_5H_5)Fe(CO)_2]-7,8-B_9H_{10}C_2R_2$ in 85% yield (*20*). The parent carborane derivative (R = H) was deprotonated by trimethylamine to form the substituted $B_9C_2H_{12}^-$ salt, $[(CH_3)_3NH]\{9-[C_5H_5)Fe(CO)_2]7,8-B_9C_2H_{11}\}$. In this case also ^{11}B NMR provided the major structural information for the proposed position of the boron-iron single bond, and this structure is indicated in Figure 2. One other metal derivative of this type

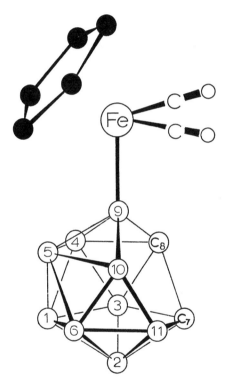

Figure 2.   Proposed structures for 9-
[(C₅H₅)Fe(CO)₂]-7,8-B₉C₂H₁₂ and 9-
[(C₅H₅)Fe(CO)₂]-7,8-B₉C₂H₁₁⁻ (termi-
nal and bridge hydrogen atoms have
been omitted for clarity)

has been reported. Mikhailov and Potapova (*22*) reported the synthesis of $(CH_3)_4N[(7,8-B_9C_2H_{11})HgC_6H_5]$ by reaction of C_6H_5HgCl with $Na_2[7,8-B_9C_2H_{11}]$. It was proposed that a mercury–carbon single bond was formed in this reaction, but more structural data are needed to substantiate this suggestion. Certain Lewis bases are capable of displacing the $7,8\text{-}B_9C_2H_{12}^-$ ion from the coordination sphere of the iron atom as illustrated in the following reactions (20):

$$(C_5H_5)Fe(CO)_2(7,8\text{-}B_9C_2H_{12}) \rightarrow$$

$$\xrightarrow{\text{CH}_3\text{CN}} [(C_5H_5)Fe(CO)_2CH_3CN](7,8\text{-}B_9C_2H_{12})$$

$$\xrightarrow{\text{PPh}_3} [(C_5H_5)Fe(CO)_2PPh_3](7,8\text{-}B_9C_2H_{12})$$

$$\xrightarrow{\text{C}_6\text{H}_{11}\text{NC}} [(C_5H_5)Fe(CO)(C_6H_{11}NC)_2](7,8\text{-}B_9C_2H_{12})$$

Another recently exploited route for the formation of boron–metal single bonds is the oxidative addition reaction. When $[Ir(C_8H_{14})_2Cl]_2$ was allowed to react at room temperature with six equivalents of 1-$[(CH_3)_2P]\text{-}1,2\text{-}C_2B_{10}H_{11}$, a yellow compound, assumed to be (carboranylphosphine)$_3$IrCl, was rapidly formed (23). Reflux of this product in cyclohexane for 2 hr generated white $IrHCl[C_2B_{10}H_{10}P(CH_3)_2]$-$[C_2B_{10}H_{11}P(CH_3)_2]_2$ in high yield. Further chemical studies suggest that this is the first example of a intramolecular oxidative addition reaction in which a boron–metal single bond is formed. Transition metal-catalyzed exchange of deuterium gas with terminal boron–hydrogen bonds in boranes, carboranes, and metallocarboranes has been recently observed (24).

$$1,2\text{-}C_2H_2B_{10}H_{10} + (PPh_3)_3RuHCl + D_2 \xrightarrow[\text{3 days}]{\substack{100° \\ \text{toluene}}} 1,2\text{-}C_2H_2B_{10}D_8H_2$$

Presumably these processes also involve transient formation of boron–metal single bonds. Recently it was reported that oxidative addition of (1- or 2-)BrB$_5$H$_8$ to IrCl(CO)(PMe$_3$)$_2$ formed *cis*-dibromo-*trans*-bis-(trimethylphosphine)(2-pentaboranyl)carbonyliridium(III) (29).

Two-Electron Donor Derivatives

The first derivatives of this type were formed by the photochemical reaction of $(CH_3)_4N[B_9H_{10}CHE]$ (E = P or As) with chromium, molybdenum, and tungsten hexacarbonyls that generated complexes with the general formula, $(CH_3)_4N[B_9H_{10}CHE \cdot M(CO)_5]$ (3). Nuclear magnetic resonance data suggest that the phosphorus or arsenic atom of the heteroatomborane ligand is sigma-bonded to the metal. Similarly photochemical irradiation of a tetrahydrofuran solution of the icosahedral anion,

$(CH_3)_4N[1,2-B_{10}CHGe]$ and $Cr(CO)_6$ in a 1:1 mole ratio formed $(CH_3)_4$-$N[1,2-B_{10}H_{10}CHGe \cdot Cr(CO)_5]$ *(25)*. The corresponding molybdenum and tungsten complexes were formed by the same type of reaction. It is proposed that a germanium–metal sigma bond exists in these derivatives and that this may be similar to compounds of the type Cl_3Ge-Metal$(L)_n$.

Attempts to remove a proton from $[7,9-B_9H_{10}CHP \cdot Cr(CO)_5]^-$ and to π-bond the resulting dianion with iron(II) chloride were not successful. However photochemical reaction of $(1,7-B_9H_9CHE)_2Fe^{2-}$ (E = P or As) with Group VI metal carbonyls has produced complexes with the general formula, $[1,7-B_9H_9CHE \cdot M(CO)_5]_2Fe^{2-}$ *(26)*. It is proposed that the $1,7-B_9H_9CHE$ carborane simultaneously functions as a pentahapto ligand for the iron atom and a monohapto ligand for the $M(CO)_5$ group in these compounds. In all the phosphorus and arsenic complexes described above we had assumed that the carborane ligands were functioning like a R_3E (E = P or As) group with the added feature of a delocalized negative charge. However, it could be argued that the negative charge was localized on the Group V atom and that this was necessary for metal complex formation. Reaction of $Cr(CO)_6$ with $7,8-B_9H_9(3$-bromopyridine$)CHP$ with UV irradiation formed neutral $7,8-B_9H_9(3$-bromopyridine$)CHP \cdot Cr(CO)_5$ in which the mode of ligand coordination appears to be through the phosphorus atom *(26)*.

This type of reaction is not limited only to Group VI carbonyls. Photolysis of $Fe(CO)_5$ and $K[7,8-B_9H_{10}CHP]$ in THF formed yellow $K[7,8-B_9H_{10}CHP \cdot Fe(CO)_4]$ in 70% yield *(27)*. In addition the following two photolytic reactions were observed with $Mn_2(CO)_{10}$.

$$Mn_2(CO)_{10} \begin{cases} \xrightarrow{7,8-B_9H_{10}CHP^-/UV} [7,9-B_9H_{10}CHP \cdot Mn(CO)_4]_2^{2-} \\ \\ \xrightarrow{7,9-B_9H_{10}CHP^-/UV} [7,8-B_9H_{10}CHP \cdot Mn_2(CO)_9]^- \end{cases}$$

Our attempts to displace iodide ion from $(C_5H_5)Fe(CO)_2I$ with heteroatom borane anions have not been successful. We therefore have sought a complex with a better leaving group. Reaction of the cationic complexes $(C_5H_5)Fe(CO)_2(cyclohexene)^+$ and $(C_7H_7)Mo(CO)_3^+$ with $[1,2-GeCHB_{10}H_{10}]^-$ formed the stable neutral derivatives $(C_5H_5)Fe(CO)_2GeCHB_{10}H_{10}$ and $(C_7H_7Mo(CO)_2GeCHB_{10}H_{10}$ respectively *(28)*. Similar transition metal complexes were obtained with the heteroatom borane anions, $[7,8-B_9H_{10}CHP]^-$, $[7,8-B_9H_{10}As_2]^-$, $[B_{10}H_{12}P]^-$, and $[B_{10}H_{12}As]^-$. Available NMR evidence suggests that each heteroatom borane is sigma-bonded to either the iron or molybdenum atom by a germanium, phosphorus, or arsenic atom.

Literature Cited

1. Parshall, G. W., *J. Amer. Chem. Soc.* (1964) **86**, 361.
2. Zakharkin, L. I., Orlova, L. V., *Izv. Akad. Nauk SSSR Ser. Khim.* (1970) 2417.
3. Silverstein, H. T., Beer, D. C., Todd, L. J., *J. Organometal. Chem.* (1970) **21**, 139.
4. Rustad, D. S., Jolly, W. L., *Inorg. Chem.* (1968) **7**, 213.
5. Gaines, D. F., Iorns, T. V., *Inorg. Chem.* (1968) **7**, 1041.
6. Zakharkin, L. I., Bregadze, V. I., Okhlobystin, O. Yu., *J. Organometal. Chem* (1965) **4**, 211.
7. Bregadze, V. I., Okhlobystin, O. Yu., *Organometal. Chem. Rev.*, A (1969) **4**, 368.
8. Owen, D. A., Smart, J. C., Garrett, P. M., Hawthorne, M. F., *J. Amer. Chem. Soc.* (1971) **93**, 1362.
9. Sneddon, L. G., Grimes, R. N., *J. Amer. Chem. Soc.* (1972) **94**, 7161.
10. Owen, D. A., Hawthorne, M. F., *J. Amer. Chem. Soc.* (1970) **92**, 3194; (1971) **93**, 873.
11. Love, R. A., Bau, R., *J. Amer. Chem. Soc.* (1972) **94**, 8274.
12. Bresadola, S., Cecchin, G., Turco, A., *Gazz. Chim. Ital.*, (1970) **100**, 682.
13. Bresadola, S., Frigo, A., Longato, B., Rigatti, G., *Inorg. Chem.* (1973) **12**, 2788.
14. Bresadola, S., Cecchin, G., Turco, A., *Proc. Int. Conf. Coord. Chem. 13th* (1970) 160.
15. Bresadola, S., Longato, B., *Inorg. Chem.* (1974) **13**, 539.
16. Bresadola, S., Longato, B., Morandini, F., *J. Chem. Soc., Chem. Commun.* (1974) 510.
17. Allegra G., Calligaris, M., Furlanetto, R., Nardin, G., Randaccio, L., *Cryst. Struct. Commun.* (1974) **3**, 69.
18. Zakharkin, L. I., Orlova, L. V., *Izv. Akad. Nauk, SSSR Ser. Khim.* (1970) 2417.
19. Zakharkin, L. I., Orlova, L. V., *Abstracts, 6th Int. Conf. Organometal. Chem., Amherst, Mass., Aug.*, 1973 (No. 101).
20. Sato, F., Yamamoto, T., Wilkinson, J. R., Todd, L. J., *J. Organometal. Chem.* (1975) **86**, 243.
21. Zakharkin, L. I., Orlova, L. V., Lokshin, B. V., Fedorov, L. A., *J. Organometal. Chem.* (1972) **40**, 15.
22. Mikhailov, B. M., Potapova, T. V., *Izv. Akad. Nauk, SSSR, Ser. Khim.* (1970) 2634.
23. Hoel, E. L., Hawthorne, M. F., *J. Amer. Chem. Soc.* (1973) **95**, 2712.
24. Hoel, E. L., Hawthorne, M. F., *J. Amer. Chem. Soc.* (1974) **96**, 4676.
25. Wikholm, G. S., Todd, L. J., *J. Organometal. Chem.* (1974) **71**, 219.
26. Beer, D. C., Todd, L. J., *J. Organometal. Chem.* (1972) **36**, 17.
27. Beer, D. C., Todd, L. J., *J. Organometal. Chem.* (1973) **55**, 363.
28. Yamamoto, T., Todd, L. J., *J. Organometal. Chem.* (1974) **67**, 75.
29. Churchill, M. R., Hackbarth, J. J., Davison, A., Traficante, D. D., Wreford, S. J. *Amer. Chem. Soc.* (1974) **96**, 4041.

RECEIVED January 24, 1975. Work supported in part by National Science Foundation grant GP42757.

Borane Anion Ligands—New Bonding Combinations with Metals

DONALD F. GAINES, MARK B. FISCHER,
STEVEN J. HILDEBRANDT, JEFFREY A. ULMAN,
and JOHN W. LOTT

University of Wisconsin, Madison Wis. 53706

This paper discusses new types of borane–transition metal complexes (metalloboranes). The borane anions investigated include $B_3H_8^-$, $B_5H_8^-$, and $B_9H_{14}^-$; typical metalloborane complexes are $(CO)_3MnB_3H_8$, $(CO)_4ReB_3H_8$, $(\pi\text{-}C_5H_5)(CO)_2\text{-}FeB_5H_8$, $(CO)_3MnB_8H_{13}$, and salts of $(CO)_3MnB_9H_{13}^-$. Significant variations in the bonding between the borane ligands and the transition metal in these complexes appear to depend largely on the size of the borane ligand. Low metal oxidation states are favored, and the most stable complexes result when the borane functions as a tridentate ligand. Chemical studies have revealed, however, that reversible bidentate–tridentate borane ligand functionality is possible in some cases. X-ray and spectroscopic studies indicate that most of these metalloboranes obey conventional polyhedral skeletal electron-counting schemes although there are several interesting variations.

M etal complexes containing borane or borane anion ligands have often been referred to as metalloboranes, especially when the metal can be considered as occupying a boron position in a borane polyhedral fragment. Outlined here are some of our recent experimental findings in the area of metalloborane chemistry. Much of the earlier work was reviewed elsewhere (1, 2, 3).

$B_3H_8^-$ Complexes

The bidentate functionality of the $B_3H_8^-$ anion has been well established for the chromium group complexes $[(CO)_4MB_3H_8]^-$ (M = Cr,

Mo, W) (4), and for the copper complex $(Ph_3P)_2CuB_3H_8$ (5). We extended the chromium work by preparing the isoelectronic neutral manganese group complexes $(CO)_4MB_3H_8$ (M = Mn, Re) as well as an iron derivative h^5-Cp(CO)FeB$_3$H$_8$. These new species are prepared by the general substitution reaction which is illustrated by Reaction 1 (6) for the manganese complex.

$$(CO)_5MnBr + Me_4NB_3H_8 \rightarrow (CO)_4MnB_3H_8 + CO + Me_4NBr \quad (1)$$

The ^{11}B and ^{1}H NMR spectra of these complexes indicate that the M–B$_3$H$_8$ bonding and geometry are analogous to those established by X-ray studies of the chromium and copper complexes. The proposed structure of $(CO)_4MnB_3H_8$, **1**, illustrates the general structural features

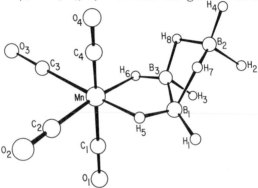

Structure **1**

that appear to characterize all the known bidentate B$_3$H$_8^-$ complexes.

A very interesting property of $(CO)_4MnB_3H_8$ is its reversible loss of a carbonyl group to form the neutral $(CO)_3MnB_3H_8$ (Reaction 2).

$$(CO)_4MnB_3H_8 \underset{CO}{\overset{UV \text{ or } \Delta}{\rightleftarrows}} (CO)_3MnB_3H_8 \quad (2)$$

This reaction illustrates what appears to be the first example of reversible bidentate–tridentate borane ligand functionality. Spectroscopic studies (^{1}H and ^{11}B NMR, and IR) of $(CO)_3MnB_3H_8$ indicate that the B$_3$H$_8^-$ anion functions as a tridentate ligand bound to the metal by three Mn–H–B bridge hydrogen bonds (one from each boron). The static structure proposed for this complex is Structure **2**. The ^{1}H NMR of $(CO)_3MnB_3H_8$ spectrum consists of two resonances, a high field group of area three that indicates three Mn–H–B protons, and a broad low

Structure 2

field group of area five that indicates an intramolecular exchange between BH and BHB protons. Upon decouplng ^{11}B, these two resonances become sharp singlets. The ^{11}B NMR spectrum consists of a single resonance which has temperature-dependent fine structure. Upon decoupling all ^1H, a very narrow resonance is observed which is indicative of the magnetic equivalence of all three boron atoms. Selective decoupling (*e.g.* Mn–*H*–B alone decoupled) combined with Fourier transform line narrowing techniques (7) gives further credence to our postulate that the hydrogens that are not involved in Mn–H–B bonding are involved in rapid tautomerism around the periphery of the B_3 triangle.

$B_5H_8^-$ *Ligands*

It has been found that the square pyramidal $B_5H_8^-$ ligand acts as a monodentate ligand. In all cases reported to date, the ligand is bound to the metal by a sigma bond involving a boron atom in the base of the B_5 pyramid [labeled $B(2)$]. The first examples were prepared by nucleophilic displacement of a halogen from a halopentaborane(9) by a pentacarbonyl metallate anion (Reaction 3) (8). The isolated products always

$$1\text{- or } 2\text{-ClB}_5H_8 + \text{NaM(CO)}_5 \rightarrow 2\text{-}[\text{M(CO)}_5]\text{B}_5H_8 + \text{NaCl} \quad (3)$$
$$(\text{M} = \text{Mn or Re})$$

have the B–M bond at the 2-position on the base of the B_5 pyramid regardless of the position of the B–Cl bond in the starting material. More recently it was discovered that B_5H_9 and BrB_5H_8 add oxidatively to $IrCl(CO)(PMe_3)_2$ to produce $2\text{-}[\text{IrClH(CO)(PMe}_3)_2]\text{B}_5H_8$ and $2\text{-}[\text{IrBr}_2(CO)(PMe_3)_2]\text{B}_5H_8$, respectively (9). We have discovered a third reaction variation, namely nucleophilic substitution of the $B_5H_8^-$ anion for a halogen in h^5-CpFe$(CO)_2X$ that produces 2-$[h^5$-CpFe-$(CO)_2]B_5H_8$.

Reaction of the $B_5H_8^-$ anion with $Fe(CO)_4X_2$ (X = Br, I) produces a mixture of products that probably contains $(CO)_3FeB_4H_8$ (perhaps two isomers) and a complex that is tentatively identified as $(CO)_3FeB_5H_9$. It was demonstrated previously that reaction of B_5H_9 with $Fe(CO)_5$ produces the apical isomer of $(CO)_3FeB_4H_8$ (10) and that reaction of the $B_5H_8^-$ anion with $CoCl_2$ and NaC_5H_5 produces two isomers of h^5-CpCoB$_4$H$_8$, 1,2-$(h^5$-CpCo$)_2B_4H_6$ (and derivatives thereof) and 5-$(h^5$-Cp)-5-CoB$_9$H$_{13}$ (11).

$B_8H_{13}^-$ Ligand

The only known complex of the tridentate $B_8H_{13}^-$ ligand is the recently described $(CO)_3MnB_8H_{13}$ whose structure, as determined by X-ray, is depicted in Structure 3 (12). The ^{11}B and ^1H NMR spectra

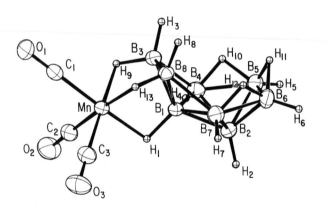

Structure 3

of this complex indicate the presence of three Mn–H–B bridge hydrogen atoms, but the data do not distinguish between a structure that incorporates the Mn atom in the polyhedral cage and the alternate correct structure, shown in 3, in which the metal is not incorporated into the borane polyhedral framework but rather is bonded to a triangular boron face *via* three Mn–H–B bridge hydrogen bonds. A $B_8H_{12}^{-2}$ complex, $(Et_3P)_2$-PtB$_8$H$_{12}$, has been reported, and although its structure is unknown, the Pt is probably incorporated into the polyhedral framework of the bi- or tridentate B_8 ligand (13).

$B_9H_{13}^{-2}$ Ligands

These complexes are prepared by substitution reactions between the $B_9H_{14}^-$ anion and metal carbonyl halides (Reaction 4) (*14*). Oxidation

$$B_9H_{14}^- + BrMn(CO)_5 \xrightarrow{R_2O} 6\text{-}(CO)_3\text{-}6\text{-}MnB_9H_{13}^- + 2CO + HBr \quad (4)$$

of these complexes in the presence of appropriate Lewis bases produces complexes of the ligand $B_9H_{12}L^-$ (Reaction 5).

$$6\text{-}(CO)_3\text{-}6\text{-}MnB_9H_{13}^- + HgCl_2 + THF \rightarrow$$

$$2\text{-}THF\text{-}6\text{-}(CO)_3\text{-}6\text{-}MnB_9H_{12} + Hg + HCl + Cl^- \quad (5)$$

The bonding of the tridentate $B_9H_{13}^{-2}$ ligand and its derivatives, $B_9H_{12}L^{-1}$, to manganese group metals is illustrated by the x-ray determined structure of 2-THF-6-(CO)$_3$-6-MnB$_9$H$_{12}$, Structure 4 (*15*). In all

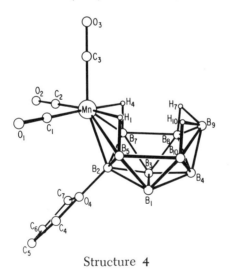

Structure 4

of these complexes, the $B_9H_{13}^{-2}$ or $B_9H_{12}L^{-1}$ ligands are tridentate and are bound to the metal *via* two Mn–H–B bridge hydrogen bonds and a B–Mn sigma-type bond. The metal can be considered as substituting for a boron position in a borane framework, which differs from the case of the tridentate $B_8H_{13}^{-1}$ ligand. Attempts to displace the THF from the complex produced in Reaction 5 by using Et$_3$N result in cleavage of the THF at an alpha carbon atom followed by a complex internal rearrangement to a zwitterion, 10-[Et$_3$N(CH$_2$)$_4$O]-6-(CO)$_3$-6-MnB$_9$H$_{12}$, whose x-ray-determined structure is depicted in Structure 5 (*16*).

Structure 5

Skeletal Electron Counting in Metalloboranes

The general polyhedral electron-counting schemes (17, 18, 19, 20, 21, 22), that were developed largely by Wade, allow one to correlate general structure types for metalloborane polyhedra and their fragments and to predict probable formulations and structure types for other metalloborane systems that are as yet unknown. The skeletal electron counts for a number of characterized metalloboranes and corresponding boranes are tabulated in Table I. The starred examples in Table I warrant special comment. The structure of $Mn_3(CO)_{10}(BH_3)_2H$ resembles that of B_4H_{10} in which B_2 and B_4 (and their terminal hydrogens) have been replaced

Table I.   Skeletal Electron Counting in Boranes and Metalloboranes

Structure Type	Number of Vertices, n	Compounds[a]	Skeletal Electron Pairs
Close	6	$B_6H_6^{-2}$, $(\pi\text{-}CpCo)_2B_4H_6$	7
Nido	4	$(CO)_3MnB_3H_8$, $B_4H_7^-$	6
	5	$(CO)_3FeB_4H_8$, $\pi\text{-}CpCoB_4H_8$, B_5H_9	7
	6	$(CO)_3FeB_5H_9$, B_6H_{10}	8
	9	$(Et_3P)_2PtB_8H_{12}$, $B_9H_{12}^-$	11
	10	$5\text{-}(\pi\text{-}Cp)CoB_9H_{13}$, $6\text{-}(CO)_3MnB_9H_{12}L$ $(L = Et_2O, THF, O(CH_2)_4NEt_3, or H^-)$	12
Arachno	4	$(CO)_4MnB_3H_8$, $(CO)_4CrB_3H_8^{-1}$, $\pi\text{-}Cp(CO)FeB_3H_8$, $(\phi_3P)_2CuB_3H_8$, $Mn_3(CO)_{10}(BH_3)_2H^*$	7
	8	$(CO)_3MnB_8H_{13}^*$, $B_8H_{13}^-$, B_8H_{14}	11

[a] See text for explanation of starred compounds.

by $(CO)_3Mn$ groups and a $(CO)_4Mn$ group bridges B_1-B_3, being bound to these boron atoms *via* two Mn–H–B bridge hydrogen atoms *(23)*. The $(CO)_4Mn$ group is not considered part of the polyhedral cage but does contribute three electrons to it. The Mn atoms of the two $(CO)_3Mn$ moieties that replace B_2 and B_4 are bound together by a single bridging hydrogen atom. In $(CO)_3MnB_8H_{13}$, the $(CO)_3Mn$ group bridges the $B_1-B_3-B_8$ face of the $B_8H_{13}^{-1}$ ligand and is bound to it by three Mn–H–B bridge hydrogen bonds. In this case, the $(CO)_3Mn$ moiety is not part of the polyhedral cage, but it does contribute one electron to it. Another example of a metal carbonyl group that contributes electrons but not vertices to a polyhedral cage is $Os_6(CO)_{18}$ *(24)*.

The broad scope of metalloborane structure types and the diversity of the few chemical studies that have been undertaken suggest that the future holds many surprises in terms of potential chemical applications.

Literature Cited

1. James, B. D., Wallbridge, M. G. H., *Prog. Inorg. Chem.* (1970) **11**, 99.
2. Muetterties, E. L., *Pure Appl. Chem.* (1972) **29**(4), 585.
3. Greenwood, N. N., Ward, I. M., *Chem. Soc. Rev.* (1974) 231.
4. Guggenberger, L. J., *Inorg. Chem.* (1970) **9**, 367.
5. Lippard, S. J., Melmed, K. M., *Inorg. Chem.* (1969) **8**, 2755.
6. Gaines, D. F., Hildebrandt, S. J., *J. Am. Chem. Soc.* (1974) **96**, 5574.
7. Clouse, A. O., Moody, D. C., Rietz, R. R., Roseberry, T., Schaeffer, R., *J. Am. Chem. Soc.* (1973) **95**, 2496.
8. Gaines, D. F., Iorns, T. V., *Inorg. Chem.* (1968) **7**, 1041.
9. Churchill, M. R., Hackbarth, J. J., Davison, A., Traficante, D. D., Wreford, S. S., *J. Am. Chem. oSc.* (1974) **96**, 4041.
10. Greenwood, N. N., Savory, C. G., Grimes, R. N., Sneddon, L. G., Davison, A., Wreford, S. S., *J. Chem. Soc. Chem. Commun.* (1974) 718.
11. Miller, V. R., Grimes, R. N., *J. Am. Chem. Soc.* (1973) **95**, 5078.
12. Calabrese, J. C., Fischer, M. B., Gaines, D. F., Lott, J. W., *J. Am. Chem. Soc.* (1974) **96**, 6318.
13. Kane, A. R., Guggenberger, L. J., Muetterties, E. L., *J. Am. Chem. Soc.* (1970) **92**, 2571.
14. Lott, J. W., Gaines, D.F., *Inorg. Chem.* (1974) **13**, 2261.
15. Schaeffer, R., Shenhav, H., Lott, J. W., Gaines, D. F., *J. Am. Chem. Soc.* (1973) **95**, 3042.
16. Gaines, D. F., Lott, J. W., Calabrese, J. C., *Inorg. Chem.* (1974) **13**, 2419.
17. Wade, K., *Chem. Commun.* (1971) 792.
18. Rudolph, R. W., Pretzer, W. R., *Inorg. Chem.* (1972) **11**, 1974.
19. Mingos, D. M. P., *Nature (London) Phys. Sci.* (1972) **236**, 99.
20. Wade, K., *Nature (London) Phys. Sci.* (1972) **240**, 71.
21. Wade, K., *Inorg. Nucl. Chem. Lett.* (1972) **8**, 559.
22. Jones, C. J., Evans, W. J., Hawthorne, M. F., *Chem. Commun.* (1973) 543.
23. Kaesz, H. D., Fellman, W., Wilkes, G. R., Dahl, L. F., *J. Am. Chem. Soc.* (1965) **87**, 2753.
24. Mason, R., Thomas, K. M., Mingos, D. M. P., *J. Am. Chem. Soc.* (1973) **95**, 3802.

RECEIVED January 24, 1975. Work supported in part by grants from the Office of Naval Research and from the National Science Foundation.

27

Some Unusal Dimeric and Cluster Species of the Group VI Transition Metals

R. E. McCARLEY, J. L. TEMPLETON, T. J. COLBURN, V. KATOVIC, and R. J. HOXMEIER

Ames Laboratory–USAEC and Department of Chemistry, Iowa State University, Ames, Iowa 50010

Several new compounds of molybdenum and tungsten that have multiple metal–metal bonds are described. It is concluded that the compounds $[Mo_2(O_2CR)_4]I_3$ ($R = C_2H_5$, $C(CH_3)_3$, or C_6H_5), formed by reaction of $Mo_2(O_2CR)_4$ with iodine in noncoordinating solvents, are I_3^- salts of the cations $[Mo_2(O_2CR)_4]^+$ with Mo–Mo bond order of 3.5. The first species containing quadruply bonded tungsten, $MoW(O_2C-(CH_3)_3)_4$, was prepared in mixtures with $Mo_2(O_2CR)_4$ with which it is isostructural and isomorphous. The mixtures were separated by iodination in benzene, which provided $[MoW(O_2CC(CH_3)_3)_4]I$ containing the one-electron oxidized cation $[MoW(O_2CC(CH_3)_3)_4]^+$ with Mo–W bond order of 3.5. The compound $[(C_3H_7)_4N]_2W_2Br_9$, prepared by reaction between $[(C_3H_9)_4N]W(CO)_5Br$ and dibromoethane, contains the new confacial bioctahedral anion $W_2Br_9^{2-}$, $d(W–W) = 2.58(1)$ Å, and $\mu(exp.) = 1.72$ BM. An unusual mixed halide complex $[(C_4H_9)_4N]_2Mo_4I_{10}Cl$ is also reported.

O ne objective of our work is the development of synthetic methods for forming strong metal–metal bonds, with bond order $n > 1$. Although many published reports of the synthesis of dimeric transition metal species having metal–metal bonds of this kind have appeared (*see* References 1 and 2 and the references cited therein), nonetheless the repertoire of reactions which may be called upon to weld such metal–metal bonds is quite limited. It is a lament that reactions which do lead to formation of metal–metal bonds with $n > 1$ are poorly understood, highly specific for a particular transition metal, and of little use when applied to construction of new molecules with specified structural features. Outstanding examples of these highly specific reactions are those that lead respectively to the quadruple bonds of the well known Mo_2 and Re_2 derivatives (*see* Reference *1* and the references cited therein). In both cases, an impressive list of compounds is now known, yet the number of reactions actually resulting in formation of the metal–metal bond is unfortunately restricted.

In the case of the $Mo \equiv Mo$ linkage, all derivatives are derived ultimately from the carboxylate dimers (*3, 4, 5*) $Mo_2(O_2CR)_4$ which, in turn, are obtained by variants of essentially a single route, *viz.* by Reaction 1 (*3, 4, 5, 6, 7*). Although it was recently observed that $Cr_2(O_2CR)_4$

$$2Mo(CO)_6 + 4RCOOH \rightarrow Mo_2(O_2CR)_4 + 12CO + 2H_2 \qquad (1)$$

derivatives can be prepared from $Cr(CO)_6$ under similar conditions (*8*), there has been a notable lack of success in forming the structurally analogous $W_2(O_2CR)_4$ using similar reactions (*9, 10*). Since there is no reason *a priori* for supposing that alternate methods do not exist, nor that the $Mo \equiv Mo$ linkage has unique stability, a search for alternative approaches to forming such quadruple metal–metal bonds was initiated. The initial fruits of this work are reported herein.

Experimental

Materials. The reagents $Mo(CO)_6$, $W(CO)_6$, iodine, tetrapropylammonium bromide, tetrabutylammonium iodide, propionic acid, benzoic acid, pivalic acid, and 1,2-dibromoethane were obtained from commercial sources and used without purification. The solvents chlorobenzene, 1,2-dichloroethane, *o*-dichlorobenzene, toluene, decahydronaphthalene (decalin), and cyclohexane were purged 10–30 min with a stream of dry nitrogen prior to use. Acetonitrile was dried over molecular sieves (4A) and also purged with nitrogen prior to use. Benzene used in the preparation of $MoW(O_2CC(CH_3)_3)_4I$ was carefully dried and stored over calcium hydride, then vacuum distilled into the reaction vessel when needed.

Analyses. For single metal analyses, samples were decomposed with nitric acid and ignited to the metal trioxide in tared porcelain crucibles. Halogen analyses were performed by dissolving the sample in dilute $KOH–H_2O_2$ solution, boiling to decompose peroxides, acidifying with acetic acid, and then titrating potentiometrically with standard silver nitrate solution. For molybdenum analyses of samples containing iodine, it was necessary to fume the samples extensively with hot sulfuric acid to eliminate iodine; molybdenum was then determined by reduction to $Mo(III)$ in a zinc amalgam column followed by reaction of $Mo(III)$ with excess $Fe(III)$ and titration of the resulting $Fe(II)$ with standard $Ce(IV)$ solution. In the mixed-metal compounds, molybdenum was separated from tungsten and determined by the Yagoda and Fales procedure (*11*).

Carbon and hydrogen analyses were performed by the Ames Laboratory analytical services and by Galbraith Laboratories, Knoxville, Tenn. **Physical Measurements.** IR spectra were recorded on nujol mulls, carefully sealed from air with NaCl and CsI windows, on Beckman IR11, 12, and 4250 instruments. Frequencies are accurate to ± 2 cm^{-1}. Electron spin resonance (ESR) spectra were obtained on powdered samples sealed in quartz ampoules using a locally constructed cavity, Strand 601 spectrometer, and Magnion magnet system. During the linear magnetic field sweep, the spectrum was calibrated with proton frequency marks at regular intervals. For magnetic susceptibility measurements, a Faraday balance was used as described previously (12). X-ray powder diffraction patterns were determined with samples sealed in Lindemann glass capillaries using a 114.6-mm DeBye–Scherrer camera, CuKα radiation, and a 12-hr exposure time. Mass spectra were obtained with the AEI MS902 high resolution spectrometer.

General Procedures. Unless otherwise noted, all reactions and sample manipulations were performed under nitrogen using standard Schlenk

Table I.   IR Absorption Frequencies (cm^{-1}) for

$Mo_2(O_2CC_2H_5)_4$	$[Mo_2(O_2CC_2H_5)_4]I_3$	$Mo_2(O_2CC_6H_5)_4$
604 s	601 s	665 w
675 s	695 s	680 s
808 s	810 s	705 s
888 s	890 m	805 w
1008 m	1010 m	—
1075 m	1080 m	840 m
1085 m	1085 w	860 w
1260 w	1248 w	
1295 s	1295 s	930 m
1301 s		
1380 m	1380 m	975 w
1385 m		
1410 sh	1410 m	1000 w
1430 s		1030 m
1450 s	1440 s	1070 m
1512 s	1520 w	1100 w
		1140 m
		1160 w
		1290 w
		1310 w
		1320 w
		1405 s
		1440 s
		1450 m
		1490 s
		1505 sh
		1590 m
		1600 m

^{a} Relative intensities are indicated as s, strong; m, medium; w, weak; and sh,

and dry box techniques or on the vacuum manifold capable of a limiting pressure of *ca.* 10^{-6} torr. The samples most sensitive to oxidation, such as $MoW(O_2CC(CH_3)_3)_4$, were stored in evacuated and ougassed ampoules.

TETRACARBOXYLATODIMOLYBDENUM(II) COMPLEXES. These compounds were prepared by modification of earlier procedures (3, 4, 5, 6, 7) in order to provide the most convenient route for specific derivatives. $Mo_2(O_2CC_2H_5)_4$ was prepared most conveniently by refluxing $Mo(CO)_4$-(tmed) (where tmed = tetramethylethylenediamine) with neat propionic acid until CO evolution was complete. The product $Mo_2(O_2CC_2H_5)_4$ crystallized as large yellow needles when cooled to $0°C$; it was then filtered, washed with cyclohexane, and dried *in vacuo.* IR absortpion frequencies are listed in Table I. Lines were observed at the following d values (in A) in the X-ray powder pattern: 9.80 vw, 8.72 s, 6.21 s, 4.99 s, 4.42 m, 4.10 s, 3.64 w, 2.93_1 w, and 2.87_3 m.

$Mo_2(O_2CC_6H_5)_4$ was prepared by reacting $Mo(CO)_6$ with a slight excess (5–10%) of benzoic acid in refluxing decalin. The mixture was

Dimolybdenum Tetracarboxylate Derivatives[a]

$[Mo_2(O_2CC_6H_5)_4]I_3$	$Mo_2(O_2CC(CH_3)_3)_4$	$[Mo_2(O_2CC(CH_3)_3)_4]I_3$
665 w	—	230 w
680 s	—	290 w
710 m	318 w	—
805 w	340 m	335 m
810 w	420 m	445 sh
840 w	450 s	465 s
845 w	615 s	625 s
935 w	775 m	778 m
975 w	795 m	788 m
985 w	895 m	888 m
1000 w	935 w	938 w
1025 m	975 w	975 w
1070 m	1025 w	1022 w
1095 w	1218 s	1215 s
1140 m	1300 w	1300 sh
1160 w	1362 s	1365 s
1290 w	1420 s	1420 s
1310 w	1485 s	1485 s
1320 w	1550 w	1570 w
1405 s		
1440 s		
1490 s		
1505 sh		
1590 sh		
1595 s		

[a] shoulder.

brought rapidly to reflux and allowed to reflux until gas evolution ceased. Upon cooling to room temperature, the orange crystals of $Mo_2(O_2CC_6H_5)_4$ were collected and washed with benzene and then cyclohexane and then dried *in vacuo*. IR data are given in Table I. The following X-ray powder pattern was observed: 11.14 s, 10.22 s, 7.53 m, 5.64 m, 5.09 m, 4.92 m, 4.63 m, 4.39 w, 4.01 w, and 3.49 s.

The pivalate derivative $Mo_2(O_2CC(CH_3)_3)_4$ was prepared most conveniently by reaction of $Mo(CO)_6$ with the calculated amount of pivalic acid in refluxing *o*-dichlorobenzene. The compound crystallized as slender yellow needles when the reaction mixture was cooled to room temperature; it was filtered, washed with benzene and cyclohexane, and dried *in vacuo*. IR data for $Mo_2(O_2CC(CH_3)_3)_4$ are included in Table I. The following lines were observed in the X-ray powder pattern: 11.49 w, 10.46 s, 9.16 s, 6.36 s, 5.69 m, 5.51 m, 5.01 s, 4.59 s, 4.26 w, and 4.13 m.

TETRAPROPIONATODIMOLYBDENUM(2.5) TRIIODIDE, $[Mo_2(O_2CC_2-H_5)_4]^+I_3^-$. In a typical preparation, 1.5 g (3.1 mmoles) $Mo_2(O_2CC_2H_5)_4$, 1.3 g (5.2 mmoles) I_2, and 50 ml 1,2-dichloroethane were placed in a Schlenk flask. The mixture was stirred 1 hr at room temperature. The brown crystals (85% yield) were collected on a frit in air, washed twice with chlorobenzene, twice with cyclohexane, and then dried *in vacuo*. Analysis for $Mo_2C_{12}H_{20}O_8I_3$: calcd: Mo 22.19, I 44.02; found: Mo 21.54, I 43.83, I/Mo 1.54.

This compound was moderately air-sensitive; decomposition was noticeable after 30 min in air. It was slightly soluble in benzene and 1,2-dichloroethane and insoluble in cyclohexane. In acetonitrile it formed a dark brown solution from which no solid could be recovered. Removal of solvent *in vacuo* yielded a tar which was not characterized. IR absorption frequencies are listed in Table I. The following lines were observed in the X-ray powder pattern: 10.21 vw, 7.18 s, 6.76 m, 4.62 m, 4.06 w, 3.90 w, 3.66 m, and 3.59 w.

TETRABENZOATODIMOLYBDENUM(2.5) TRIIODIDE, $[Mo_2(O_2CC_6H_5)_4]^+$-I_3^-. For this preparation, 2.10 g (3.1 mmoles) $Mo_2(O_2CC_6H_5)_4$, and 1.3 g (5.2 mmoles) iodine were placed in a Schlenk flask, and 50 ml 1,2-dichloroethane was added. The mixture was stirred at 25°C 1 hr, then filtered. The brown crystals (96% yield) were washed with chlorobenbene and cyclohexane, and dried *in vacuo*. Analysis for $Mo_2C_{28}H_{20}O_8I_3$: calcd: Mo 18.15, I 36.02; found: Mo 17.99, I 36.36, I/Mo 1.53.

This compound was stable in air for several days; it was sparingly soluble in dichloromethane and 1,2-dichloroethane. When $[Mo_2(O_2CC_6-H_5)_4]I_3$ was stirred in acetonitrile, a dark solution formed and a light colored solid material appeared almost immediately. Both elemental analysis and X-ray powder pattern confirmed that this light material was pure $Mo_2(O_2CC_6H_5)_4$. This procedure was repeated several times on different samples; the yield of precipitated $Mo_2(O_2CC_6H_5)_4$ was about 33% and was independent of reaction time. No solid product could be isolated from the mother liquor; removal of solvent *in vacuo* yielded only a dark tar. IR absorption frequencies for $[Mo_2(O_2CC_6H_5)_4]I_3$ are given in Table I. The following lines were observed in the X-ray powder pattern: 12.66 w, 10.87 s, 9.90 m, 9.35 w, 7.25 s, 6.36 m, 5.40 w, 5.04 m, 4.70 m, 4.49 m, and 4.25 s.

TETRAPIVALATIDIMOLYBDENUM(2.5) TRIIODIDE, $[Mo_2(O_2CC(CH_3)_3)_4]^+$-$I_3^-$. In a typical preparation, 1.0 g (1.7 mmoles) $Mo_2(O_2CC(CH_3)_3)_4$ and 0.72 g (2.83 mmoles) iodine were placed in a Schlenk flask together with 40 ml of either benzene or 1,2-dichloroethane (both solvents work equally well). The mixture was stirred 12 hr at 25°C, then cooled 2 hr at 0°C. Red-brown crystals of the product (80% yield) were collected on a frit in air, washed with chlorbenzene and cyclohexane, and then dried *in vacuo*. Analysis for $Mo_2C_{20}H_{36}O_8I_3$: calcd: Mo 19.64, I 38.96, C 24.58, H 3.71; found: Mo 19.34, I 38.85, C 24.45, H 3.89, I/Mo 1.52.

This compound was air-sensitive, slightly soluble in benzene, chlorobenzene, and 1,2-dichloroethane, and insoluble in cyclohexane. In acetonitrile it gave a dark solution from which solid could not be recovered. IR absorption frequencies for $[Mo_2(O_2CC(CH_3)_3)_4]I_3$ are reported in Table I. In the X-ray powder pattern the following lines were observed: 8.33 s, 7.55 s, 6.64 m, 4.78 w, 4.55 m, 4.13 m, and 2.89 m. Magnetic susceptibility data are given in Table II.

Table II. Magnetic Susceptibility Data for $[Mo_2(O_2CC(CH_3)_3)_4]I_3{}^a$

$T°, K$	125	131	140	151	160	170	180
$10^6X_M{}^{corr}$	2714	2705	2466	2307	2174	2028	1916
$T°, K$	190	200	226	250	273	300	
$10^6X_M{}^{corr}$	1807	1719	1523	1379	1281	1135	

^{a} $X_M{}^{corr} = X_M{}^{exp} - X_D$ was used to compute the values of $X_M{}^{corr}$ with the value of $X_D = -386 \times 10^{-6}$ emu/mole.

TETRAPIVALATOMOLYBDENUM(II)TUNGSTEN(II). Although this compound could never be obtained free of $Mo_2(O_2CC(CH_3)_3)_4$, it could be prepared in concentrations as high as 80 mole % in the solid mixtures. In the most convenient synthetic procedure, the initial W/Mo ratio in the reaction mixture was *ca.* 3. At ratios > 3, net yields were reduced whereas ratios < 3 concentrations of the mixed-metal compound in the isolated product were lower. In a reaction designed to maximize the concentration of mixed-metal compound in the product mixture, 1.06 g (4 mmoles) $Mo(CO)_6$, 8.44 g (24 mmoles) $W(CO)_6$, 5.70 g (56 mmoles) pivalic acid, and 30 ml *o*-dichlorobenzene were placed in a Schlenk flask equipped with a condenser and connected to a gas bubbler. The mixture was refluxed gently (185°C); CO evolution ceased after 5–6 hr, but refluxing was continued an additional 4 hr. When the black solution was cooled to 25°C, the product crystallized as slender yellow needles which were filtered under nitrogen, washed several times with cyclohexane, and then dried under high vacuum. The product is very sensitive to air and to moisture; on exposure to air, it turns black within a few minutes. The only suitable method for storing this compound for a period of days was to seal the material in carefully evacuated and outgassed ampoules. For comparison, $Mo_2(O_2CC(CH_3)_3)_4$ can be stored in air for several days before decomposition becomes noticeable. Analysis for 0.80MoW(O_2CC-$(CH_3)_3)_4$ + 0.20$Mo_2(O_2CC(CH_3)_3)_4$: calcd: Mo 17.65, W 21.49, C 36.14, H 5.46; found: Mo 17.17, W 21.89, C 36.58, H 5.59.

Table III. IR Absorption Frequencies (cm⁻¹)

$MoW(O_2CC(CH_3)_3)_4$	$[MoW(O_2CC(CH_3)_3)_4]I$
314 m	300 w
338 m	
360 m	350 m
420 sh	
440 s	458 s
610 s	620 s
715 w	720 w
765 m	
770 m	780 s
790 s	799 s
890 s	892 s

ᵃ Relative intensities are indicated as s, strong; m, medium; w, weak; and sh,

Mixtures containing 60–80 mole % $MoW(O_2CC(CH_3)_3)_4$ and 40–20 mole % $Mo_2(O_2CC(CH_3)_3)_4$ gave X-ray powder patterns that were essentially identical to that of pure $Mo_2(O_2CC(CH_3)_3)_4$. The two compounds are thus isomorphous and form solid solutions. Both compounds are quite soluble in benzene. IR absorption frequencies for $MoW(O_2-CC(CH_3)_3)_4$ are given in Table III.

TETRAPIVALATOMOLYBDENUM(2.5)TUNGSTEN(2.5) IODIDE, $[MoW(O_2-CC(CH_3)_3)_4]^+I^-$. Because the starting material and the product are both extremely sensitive to air and to moisture, manipulations were performed in a good dry box under purified nitrogen and on the high vacuum manifold. A Soxhlet fritted glass extractor, with 300-ml flasks at each end and Teflon needle valves at the filter by-pass and the vacuum-line connection, was evacuated and outgassed on the high vacuum line. The apparatus was then removed to the dry box where 1.15 grams of a mixture containing 70 mole % (0.81 g) $MoW(O_2CC(CH_3)_3)_4$ and 30 mole % $Mo_2(O_2CC(CH_3)_3)_4$ was placed in the extractor. After evacuation on the vacuum line, 250 ml carefully purified benzene was distilled into the extractor; the starting material dissolved in the benzene to produce a yellow solution. The apparatus was filled with nitrogen, and the quantity of iodine (0.149 g) required to oxidize only the mixed-metal compound was placed on the frit of the extractor. The extractor was then inverted in order to filter the yellow solution into the other 300-ml flask to mix with the iodine. After several minutes of stirring, the gray crystalline product appeared. Stirring was continued for *ca.* 12 hr; then the product was filtered and washed several times with benzene (distilled from the mother liquor). The flask containing the mother liquor was removed under nitrogen, the extractor was stoppered, and the product was dried under high vacuum. Analysis: for $[MoW(O_2CC(CH_3)_3)_4]I$: calcd: Mo 11.83, W 22.66, C 29.61, H 4.41, I 15.64; found: Mo 11.45, W 22.98, C 28.22, H 4.42, I 15.42.

This compound is extremely air-sensitive; indeed it is pyrophoric. Thus reliable carbon and hydrogen analyses were difficult if not impossible to obtain because of problems in sample transfer and weighing. For

for the Mo–W Mixed-Metal Derivatives[a]

$MoW(O_2CC(CH_3)_3)_4$	$[MoW(O_2CC(CH_3)_3)_4]I$
930 w	935 w
1024 m	1030 m
1090 w	1090 w
1215 s	1215 s
1360 s	1360 s
	1375 s
1410 s	1410 s
1475 s	1460 s
1480 s	1490
1505 sh	1520 sh
1540 sh	1558 w
	1564 w

[a] shoulder.

the same reason, the data obtained from magnetic susceptibility measurements had more scatter than usual; these data will be reported elsewhere at a later date. IR absorption frequencies are given in Table III.

BIS(TETRAPROPYLAMMONIUM) NONABROMODITUNGSTATE, $[(C_3H_7)_4N]_2 \cdot W_2Br_9^{2-}$. Into a Schlenk flask equipped with a water-cooled condenser were placed 2.66 g (10 mmoles) $(C_3H_7)_4N^+Br^-$, 3.52 g (10 mmoles) $W(CO)_6$, and 40 ml chlorobenzene. The mixture was brought to reflux (135°C), and gas evolution was monitored. After about 20 min, gas evolution ceased and conversion of $W(CO)_6$ to $W(CO)_5Br^-$ was complete. The golden yellow solution was allowed to cool prior to addition of 20 ml 1,2-dibromoethane. The solution was again brought to reflux, and the color darkened rapidly as CO evolution progressed. After a few minutes, a black crystalline solid began to deposit; the reaction was continued 4–5 hr until CO evolution ceased. During this stage, the evolved gases were sampled and a mass spectrum was obtained. The spectrum revealed that ethylene was a major component of the gas together with CO. When the reaction was complete, the mixture was cooled and filtered; the green-black solid was then washed with chlorobenzene and a small portion of dichloroethane. Finally, the product was recrystallized from acetonitrile (in which it is quite soluble and forms a deep green solution). X-ray powder patterns of the product recovered directly from the reaction mixture and of the recrystallized material demonstrated that they were identical. The initial yield was 90% based on tungsten and 70% after recrystallization. Analysis for $[(C_3H_7)_4N]_2-W_2Br_9$: calcd: W 25.19, Br 49.27, C 19.75, H 3.87; found: W 25.14, Br 48.99, C 19.73, H 4.25.

BIS(TETRABUTYLAMMONIUM) CHLORODECAIODOTETRAMOLYBDATE, $[(C_4H_9)_4N]_2Mo_4I_{10}Cl$. In a typical reaction, 4.89 g (13.2 mmoles) tetrabutyl-ammoniumiodide and 3.50 g (13.2 mmoles) $Mo(CO)_6$ were refluxed (135°C) 15 min in 100 ml chlorobenzene to form a golden solution of $[(C_4H_9)_4N]^+Mo(CO)_5I^-$. This solution was allowed to cool to room temperature prior to addition of iodine (3.37 g, 13.2 mmoles) which rapidly oxidized the molybdenum to $Mo(CO)_4I_3^-$ with liberation of CO.

Thermal decomposition of the carbonyl anion was promoted by refluxing the solution 1 hr; during this time a dark microcrystalline solid precipitated. The black solid was isolated by filtration and placed in a Soxhlet extractor with 1,2-dichloroethane. After two weeks of continuous extraction, virtually all of the solid had passed through the frit; 4.87 grams of recrystallized solid were isolated from the solution in the extraction flask by filtration. Analysis for $[(C_4H_9)_4N]_2Mo_4I_{10}Cl$: calcd: Mo 17.66, I 58.40, Cl 1.63, C 17.68, H 3.34, N 1.29; found: Mo 17.72, I 58.25, Cl 1.53, C 18.08, H 3.32, N 1.46.

Results and Discussion

[Mo$_2$(O$_2$CR)$_4$]I$_3$ Derivatives. The initial objective of work on the oxidation of $Mo_2(O_2CR)_4$ derivatives with halogens was to examine the possibility of achieving addition of halogen across the quadruple bond, as in the reaction:

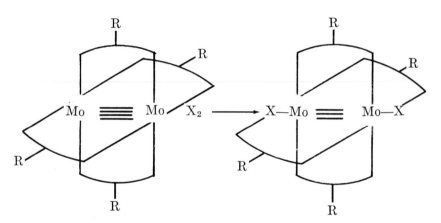

If such reactions could be realized, the resulting products would be expected to have the structure analogous to the known $Re_2(O_2CR)_4X_2$ derivatives (13) except that, upon addition of halogen, the Mo–Mo bond would decrease in bond order from 4 to 3. However, reactions with bromine lead to as yet undefined products, and, in the cases examined, the reactions with iodine lead to triiodide derivatives $Mo_2(O_2CR)_4I_3$.

In the reactions with iodine, it was established that the triiodide was the sole product when the I/Mo ratio was 0–1.5. Thus, neither mono- nor diiodide derivatives could be formed. When the I/Mo ratio was > 1.5, the reaction products approached the composition $Mo_2(O_2CR)_4I_5$ as the I/Mo ratio became very large; these compounds appear to lose iodine readily, and thus they are difficult to obtain in pure form. The reactions reported here lead cleanly to the triiodide derivatives when noncoordinating solvents are used at room temperature. In coordinating solvents such as acetonitrile, or at elevated temperature, the reactions pursue a different course.

All available evidence indicates that the reactions with iodine correspond to a one-electron oxidation of the dimolybdenum species as indicated by Equations 2 and 3. Thermal decomposition of the triiodide

$$Mo_2(O_2CR)_4 + 3/2\ I_2 \rightarrow [Mo_2(O_2CR)_4]^+ + I_3^- \qquad (2)$$

$$Mo_2(O_2CR)_4 + 5/2\ I_2 \rightarrow [Mo_2(O_2CR)_4]^+ + I_5^- \qquad (3)$$

derivatives *in vacuo* at *ca.* 100°C liberates elemental iodine, $Mo_2(O_2CR)_4$, and some unidentified nonvolatile product. Loss of elemental iodine from salts containing the I_3^- anion is expected at elevated temperatures. However, the monoiodide formed in this process is evidently not stable, and it decomposes with formation of $Mo_2(O_2CR)_4$ as by the scheme:

$$[Mo_2(O_2CR)_4]I_3(s) \rightarrow [Mo_2(O_2CR)_4]I(s) + I_2(g)$$
$$2[Mo_2(O_2CR)_4]I(s) \rightarrow Mo_2(O_2CR)_4 + ?$$

The benzoate derivative $[Mo_2(O_2CC_6H_5)_4]I_3$ undergoes a similar reaction in acetonitrile, probably because of insolubility of $Mo_2(O_2CC_6H_5)_4$ which precipitates and is recovered in *ca.* 33% yield based on the initial quantity of triiodide compound used.

The recovery of $Mo_2(O_2CR)_4$ from the thermal decomposition is a strong indication that the basic structure of the dimolybdenum tetracarboxylate (*1, 14*) is retained in the triiodide compounds. This is also indicated convincingly by the IR spectra in Figures 1 and 2 and the data in Table I. The almost perfect band for band matching between the spectra of $Mo_2(O_2CR)_4$ and the corresponding $[Mo_2(O_2CR)_4]I_3$ is most striking in this regard. Particularly for the pivalate derivatives, the data in the 400–700 cm^{-1} region indicate the same molecular framework, since the Mo–O stretching freqeuncies must occur in this region and they should be especially sensitive to any change in molecular structure. The shift in frequency of the bands at 420, 450, and 615 cm^{-1} for the neutral pivalate to 445, 465, and 625 for the triiodide derivative is consistent with the expected increase in such Mo–O stretching freqeuncies upon increase in average oxidation state of the metal atoms.

Finally, the magnetic properties reveal clearly that the triiodide derivatives are species that contain one unpaired electron per dimer as required by oxidation to the cations $[Mo_2(O_2CR)_4]^+$. Magnetic susceptibility data for $[Mo_2(O_2CC(CH_3)_3)_4]I_3$ are presented in Table II. These data provide an excellent fit to the Curie equation and, from the slope, the magnetic moment $\mu = 1.66$ BM was found. The crystalline triiodide

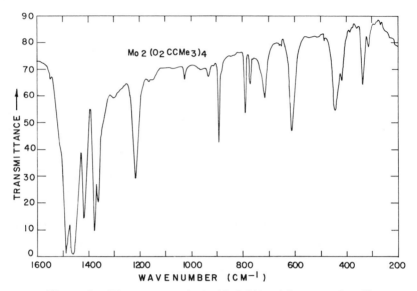

Figure 1.   IR spectrum of $Mo_2(O_2CC(CH_3)_3)_4$ in nujol mull

derivatives all give strong, relatively narrow ESR signals which show little evidence of anisotropy and give g values of 1.93 ± 0.01. From the relation $\mu^2 = g^2s(s+1)$, with $s = \frac{1}{2}$ the calculated moment, $\mu = 1.67$ is obtained; this is in excellent agreement with the moment derived from the susceptibility measurements.

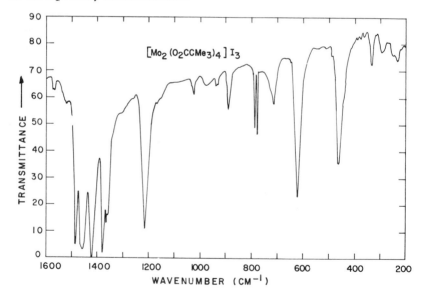

Figure 2.   IR spectrum of $[Mo_2(O_2CC(CH_3)_3)_4]^+I_3^-$ in nujol mull

The findings reported here for the reactions of the dimolybdenum carboxylates with iodine are consistent with the recent observations of Cotton and Pedersen (*15*) during a study of the electrochemical oxidation of $Mo_2(O_2CC_3H_7)_4$ in acetonitrile or ethanol. They found a quasi-reversible one-electron oxidation for the indicated compound by rotating disc polarography and cyclic voltammetry, and $g_{||} = g_\perp = 1.941$ for the oxidized species by ESR studies. Furthermore, the proposed formulation of the triiodide compounds as salts containing the $[Mo_2(O_2CR)_4]^+$ cation with the same molecular structure and eclipsed conformation as that observed for $Mo_2(O_2CCH_3)_4$ is supported by the previously published isolation and structure determination of $K_3Mo_2(SO_4)_4$ (*16*). The latter compound is formed by one-electron oxidation of $K_4Mo_2(SO_4)_4$. The $Mo_2(SO_4)_4^{4-}$ and $Mo_2(SO_4)_4^{3-}$ anions have the same molecular structure (*16, 17*). The lengthening of the Mo–Mo bond from 2.111(1) A in $Mo_2(SO_4)_4^{4-}$ to 2.164(2) A in $Mo_2(SO_4)_4^{3-}$ is evidence that the electron was removed from the δ-bonding orbital upon oxidation from the 4- to 3- ion. It thus appears that the dimolybdenum(II) species may be generally susceptible to one-electron oxidation with retention of structure. The molecular orbital description of the oxidation is best understood in terms of the orbitals localized on the metal atoms: $(\sigma^2)(\pi^4)(\delta^2) \rightarrow (\sigma^2)(\pi^4)(\delta^1)$ (*16*). In these terms, the metal–metal bond order decreases from 4 to 3.5 when one-electron oxidation occurs.

The Mixed–Metal Compounds. The successful synthesis of the compound $MoW(O_2CC(CH_3)_3)_4$ represents the first success in introducing tungsten into a metal–metal quadruple bond, and it is only the second successful formation of any heteronuclear quadruple bond. Garner and Senior (*18*) reported preparation of the Cr–Mo mixed-metal compound $CrMo(O_2CCH_3)_4$. Evidence for the isostructural relation between $Mo_2(O_2CC(CH_3)_3)_4$ and $MoW(O_2CC(CH_3)_3)_4$ comes mainly from mass spectra, IR spectra, and X-ray powder pattern data. IR absorption frequencies for the mixed-metal species are given in Table III; the values should be compared with those for the dimolybdenum pivalate derivatives (Table I). These data indicate clearly a very close relation between the structures of the Mo_2 and the MoW compounds.

Figure 3 is the mass spectrum of the mixture containing 80 mole % $MoW(O_2CC(CH_3)_3)_4$ and 20 mole % $Mo_2(O_2CC(CH_3)_3)_4$. The parent ion peaks for the Mo_2 species are centered at 596 amu, those for the MoW species at 683 amu. For the dimolybdenum species, the parent ion set is in good agreement with that expected according to Hochberg *et al.* (*19*), and the parent ion set for the MoW species agrees with the expected isotope combinations. However, the relative intensities of the two sets are not an accurate reflection of the composition of the mixture.

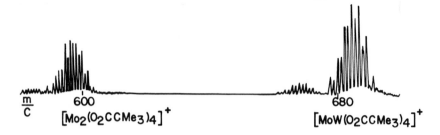

$\dfrac{m}{c}$ 600   680

$[Mo_2(O_2CCMe_3)_4]^+$ $[MoW(O_2CCMe_3)_4]^+$

Figure 3. Parent ion mass spectrum of a mixture containing 80 mole %
$MoW(O_2CC(CH_3)_3)_4$ and 20 mole % $Mo_2(O_2CC(CH_3)_3)_4$

Thus, the relative intensity of the Mo_2 species is greater than that expected from sample analysis.

Because of the extreme reactivity of $MoW(O_2CC(CH_3)_3)_4$, chromatographic methods for sepaarting the mixture were very difficult to
apply. Thus it was not possible to obtain the pure, neutral mixed-metal
compound. However, this greater reactivity could be advantageous in
preparing a pure mixed-metal compound. It was natural to investigate
the reaction with iodine with the expectation that the mixed-metal compound would react preferentially if a limited quantity of oxidant was
available. The success of the method exceeded expectations, but it was
surprising that a simple iodide salt was obtained rather than the triiodide
salt as with the dimolybdenum compounds. This difference probably
facilitates separation since the possibility of solid-solution formation
between oxidized products is diminished.

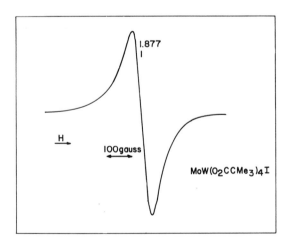

Figure 4. ESR spectrum of polycrystalline
$[MoW(O_2CC(CH_3)_3)_4]I$ at 25°C

The evidence indicates that $[MoW(O_2CC(CH_3)_3)_4]I$ is formed by one-electron oxidation of $MoW(O_2CC(CH_3)_3)_4$ with retention of the molecular structure (*see* IR spectra data in Table III). Again, the shift in the frequency of bands in the 400–700 cm^{-1} region reflects that expected for an increase in average oxidation state of the metal atoms. The ESR spectrum and magnetic susceptibility data also support retention of metal–metal bonding. The ESR spectrum (Figure 4) depicts the narrow, virtually isotropic signal with $g = 1.877$. The lower value of the g factor, lower than that of the dimolybdenum triiodide salts, is expected because of the much larger spin-orbit coupling constant for tungsten. Although difficulty was encountered in obtaining reproducible susceptibility data, an average magnetic moment of 1.58 BM was determined. A value of $\mu = 1.61_5$ BM was calculated from the observed g factor.

The great reactivity of $MoW(O_2CC(CH_3)_3)_4$ and $[MoW(O_2CC(CH_3)_3)_4]I$ suggests that the stability of the metal–metal bond diminishes with introduction of the tungsten atom. If so, corresponding ditungsten species may be surmised to be even less stable—perhaps providing a partial explanation for the lack of success in preparing compounds with the W≡W linkage. Though the great reactivity of the mixed-metal compounds makes synthetic work difficult, it does offer the possibility that new reactions not observed with the Mo_2 species can be developed. Work along this line is in progress.

New Species from Reactions of Halocarbonylmetallate Anions. The basic idea for this aspect of the work was that reactive metal halide fragments might be generated from the halocarbonylmetallate anions under conditions where loss of the CO ligands is promoted in the absence of other coordinating ligands. For example, dinuclear or cluster species might be obtained *via* thermal decomposition of the $M(CO)_4X^{3-}$ anions as in the scheme:

$$M(CO)_4X_3^- \rightarrow [MX_3^-] + 4CO$$
$$n[MX_3^-] \rightarrow M_nX_{3n}^{n-}$$

This scheme also presents the possibility of generating metal atom fragments which might be added to other metal cluster species to build up new clusters. Recently, Matson and Wentworth (*see* Reference 20 and the references cited therein) have been investigating similar reactions between metal halides and halocarbonylmetallate anions in noncoordinating solvents. Their recent synthesis of the mixed-metal anion $CrMoCl_9^{3-}$ and the synthesis of the $W_2Br_9^{2-}$ and $Mo_4I_{10}Cl^{2-}$ salts reported here illustrate the power of this approach.

In the course of investigating the thermal decomposition of $W(CO)_4$-Br_3^-, a search for suitable solvents was necessary. Among the solvents considered was 1,2-dibromoethane. However, a rapid reaction occurs in this solvent before the $W(CO)_4Br_3^-$ anion can be formed. In order to control the reaction, a two-step sequence in chlorobenzene was devised:

$$W(CO)_6 + Br^- \rightarrow W(CO)_5Br^- + CO$$
$$2W(CO)_5Br^- + 7/2\ C_2H_4Br_2 \rightarrow W_2Br_9^{2-} + 7/2\ C_2H_4 + 10CO$$

Although other cations were tried, the tetrapropylammonium ion provided the most easily isolated and crystalline product. The reaction of $W(CO)_5Br^-$ with $C_2H_4Br_2$ proceeded virtually quantitatively with formation of $W_2Br_9^{2-}$ even in the presence of excess dibromoethane. Salts of $W_2Br_9^{2-}$ were not previously reported, although Saillant and Wentworth (21) reported preparation of the $W_2Cl_9^{2-}$ anion.

The structure and properties of $W_2X_9^{2-}$ salts are of considerable interest since, if isostructural, they are related by one-electron oxidation to the salts containing the confacial bioctahedral anions $W_2X_9^{3-}$ (X = Cl, Br). Compounds containing the $W_2X_9^{3-}$ ions have virtually temperature-independent magnetic susceptibilities (22), and the binuclear ions obviously have a spin singlet (S = 0) ground state. The W–W distance (23) observed in $K_3W_2Cl_9$ is only 2.41 A and reflects strong metal–metal bonding that is best described (24) as $(\sigma^2)(\pi^4)$ with bond order $n = 3$. Thus, if $W_2Br_9^{2-}$ in the compound reported here is isostructural, it should show a magnetic moment corresponding to one unpaired electron (S = ½), and the moment should be temperature-independent if the metal–metal bonding remains very strong. Magnetic susceptibility measurements (77°–300°K) for $[(C_3H_7)_4N]_2W_2Br_9$ are listed in Table IV. The magnetic moment derived from the data is indeed independent of temperature and is close to that expected for S = ½. This indicates that the strong metal–metal bonding is retained and that it may be assigned bond order $n = 2.5$.

Table IV. Magnetic Constants[a] of Salts of $W_2Br_9^{2-}$ and $Mo_4I_{10}Cl^{2-}$

Compound	A 10^{-6} emu/ mole	C emu deg/ mole	X_D 10^{-6} emu/ mole	X_{TIP} 10^{-6} emu/ mole	μ, BM
$[(C_3H_7)_4N]_2W_2Br_9$	−452	0.370(3)	−699	247	1.72(1)
$[(C_4H_9)_4N]_2Mo_4I_{10}Cl$	−867	0.385(3)	−1084	217	1.75(1)

[a] Constants were derived from least squares fit to the relation $X_M = A + CT^{-1} = (X_D + X_{TIP}) + CT^{-1}$, and $\mu = 2.828C^{0.5}$.

Preliminary data from the X-ray crystal and molecular structure determination of $[(C_3H_7)_4N]_2W_2Br_9$ can be reported here although a detailed account will be given elsewhere. The crystals are monoclinic, space group $C2/c$, with $a = 36.19$ A, $b = 11.98$ a, $c = 19.53$ a, and $\beta = 96.10°$. A total of 3380 reflections were utilized in refinement, and absorption corrections were applied because of the large linear absorption coefficient (148.4 cm^{-1}) for MoKα radiation. The calculated density was 2.303 g/cm^3 for $Z = 8$, and the observed density was 2.314 g/cm^3. At the present state of refinement ($R = 0.10$), all of the non-hydrogen atom positions have been located, but some of the C and N thermal parameters are abnormally large and the success of further refinement is in doubt. However, the W and Br positions were determined with confidence and they are relatively insensitive to further refinement, as was indicated by the lack of change during the last refinement cycles. The $W_2Br_9{}^{2-}$ anion does show the confacial bioctahedral structure with little distortion from the idealized D_{3h} point group symmetry. Some average bond distances (in Angstroms) and angles (in degrees) in the anion are as follows: $d(\text{W–W})$ 2.58(1), $d(\text{W–Br}_t)$ 2.52(1), $d(\text{W–Br}_b)$ 2.58(1), and the angle W–Br$_b$–W 60.0(1)°. The W–W distance is short; for comparison, $d(\text{Mo–Mo}) = 2.816(9)$ A in the $Mo_2Br_9{}^{3-}$ anion (25) and $d(\text{W–W}) = 2.41$ A (23) in $W_2Cl_9{}^{3-}$. Although the W–W distance in $W_2Br_9{}^{3-}$ is unknown, it should be comparable to that in $W_2Cl_9{}^{3-}$. Thus one-electron oxidation to $W_2Br_9{}^{2-}$ weakens the W–W bond, yet the bonding is still noticeably stronger than in $Mo_2Br_9{}^{3-}$ which formally has a higher bond order. The strength of the metal–metal bonding in the $W_2X_9{}^{n-}$ anions evidently is unique among $M_2X_9{}^{n-}$ species (X = halogen).

The preparation of $[(C_4H_9)_4N]_2Mo_4I_{10}Cl$ is reported here as illustrative of new species which can be prepared by the discussed methods. The structure of this compound is unknown, and, although single crystals can be obtained, none examined to date have given X-ray diffraction data of adequate quality for a structure determination. It was definitely established that the Cl atom is gained during extraction and recrystallization from 1,2-dichloroethane. The product isolated initially from the reaction in chlorobenzene does not contain chlorine, and, even though it does have the same X-ray powder pattern as the material recrystallized from 1,2-dichloroethane, the composition is not well established.

Magnetic data for $[(C_4H_9)_4N]_2Mo_4I_{10}Cl$ are given in Table IV. The constant magnetic moment of 1.75 BM over the 77°–300°K range agrees with that expected for one unpaired electron per tetrameric unit. From the ESR spectrum of a powdered sample, $\bar{g} = 2.027$; using this value, $\mu(\text{calc.}) = 1.755$ BM which is in excellent agreement with the value of $\mu(\text{exp.})$ given above. These data do not prove that the anion is tetrameric, but they are most consistent with that interpretation. Because of the unusual stoichiometry of this compound and the rare occurrence of tetranuclear metal halide clusters, the structure and chemistry of this anion will be interesting to elucidate.

Acknowledgement

We are grateful to Yoshihisa Matsuda for assistance with the ESR measurements and to F. A. Cotton for preprints of work performed in his laboratory on the electrochemical and structural characterization of dimolybdenum(II) derivatives.

Literature Cited

1. Cotton, F. A., *Acc. Chem. Res.* (1969) **2**, 240.
2. Baird, M. C., *Prog. Inorg. Chem.* (1968) **9**, 1–135.
3. Lawton, D., Mason, R., *J. Am. Chem. Soc.* (1965) **87**, 921.
4. Stephenson, T. A., Bannister, E., Wilkinson, G., *J. Chem. Soc.* (1964) 2538.
5. Dubicki, L., Martin, R., *Aust. J. Chem.* (1969) **22**, 1571.
6. Brignole, A. B., Cotton, F. A., *Inorg. Synth.* (1972) **13** ,88.
7. Holste, G., Schäfer, H., *J. Less Common Met.* (1970) **20**, 164.
8. McCarley, R. E., Hoxmeier, R. J., Michel, J., unpublished data.
9. Holste, G., *Z. Anorg. Allg. Chem.* (1973) **398**, 249.
10. Cotton, F. A., Jeremic, M., *Synth. Inorg. Met. Org. Chem.* (1971) **1**, 265.
11. Yagoda, H., Fales, H., *J. Am. Chem. Soc.* (1936) **58** ,1494.
12. Converse, J. G., McCarley, R. E., *Inorg. Chem.* (1970) **9**, 1361.
13. Bennett, M. J., Bratton, W. K., Cotton, F. A., Robinson, W. R., *Inorg. Chem.* (1968) **7**, 1570.
14. Cotton, F. A., Norman, Jr., J. G., *J. Coord. Chem.* (1971) **1**, 161.
15. Cotton, F. A., Pedersen, E., private communication.
16. Cotton, F. A., Frenz, B. A., Webb, T. R., *J. Am. Chem. Soc.* (1973) **95**, 4431.
17. Angell, C. L., Cotton, F. A., Frenz, B. A., Webb, T. R., *J. Chem. Soc. Chem. Commun.* (1973) 399.
18. Garner, C. D., Senior, R. G., *J. Chem. Soc. Chem. Commun.* (1974) 580.
19. Hochberg, E., Walks, P., Abbott, E. H., *Inorg. Chem.* (1974) **13**, 1824.
20. Matson, M. S., Wentworth, R. A. D., *J. Am. Chem. Soc.* (1974) **96** ,7837.
21. Saillant, R., Wentworth, R. A. D., *J. Am. Chem. Soc.* (1969) **91** ,2174.
22. Saillant, R., Wentworth, R. A. D., *Inorg. Chem.* (1969) **8**, 1226.
23. Watson, Jr., W. H., Waser, J., Acta Crystallogr. (1958) **11**, 689.
24. Cotton, F. A., Ucko, D. A., *Inorg. Chim. Acta* (1972) **6**, 161.
25. Saillant, R., Jackson, R. B., Streib, W. E., Folting, K., Wentworth, R. A. D., *Inorg. Chem.* (1971) **10**, 1453.

RECEIVED January 24, 1975.

Structure and Reactivity Patterns of Polyphosphine Ligands and Their Complexes of Rhodium and Cobalt(I)

DEVON W. MEEK, DANIEL L. DuBOIS, and JACK TIETHOF

The Ohio State University, Columbus, Ohio 43210

In the presence of free radicals generated from AIBN, phosphorus–hydrogen and sulfur–hydrogen bonds add cleanly and readily to the carbon–carbon double bonds of vinyl derivatives to produce useful polydentate chelating ligands. Chelating polyphosphine ligands accentuate unusual properties of transition metals. Compared with a monodentate phosphine, a polyphosphine simultaneously provides: (a) more control on the coordination number, stoichiometry, and stereochemistry of the resulting complex; (b) increased basicity (or nucleophilicity) at the metal; and (c) detailed structural and bonding information via metal–phosphorus and phosphorus–phosphorus coupling constants. Alkylation and protonation reactions of RhCl(ttp), ttp = PhP($CH_2CH_2CH_2PPh_2$)$_2$, yield alkyl and hydride complexes that have strikingly different behavior. Several new series of Co(I) complexes of types Co(triphos)H(CO), [Co(triphos)(CO)$_2$]$^+$, and [Co(triphos)(monophos)(CO)]$^+$ were characterized with each of the three triphosphine ligands $CH_3C(CH_2PPh_2)_3$, PhP($CH_2CH_2PPh_2$)$_2$, and PhP($CH_2CH_2CH_2PPh_2$)$_2$.

Unusual properties in transition metals are often induced by *tert*-phosphine ligands. For example, monodentate phosphines have been use extensively for stabilizing both high and low oxidation states of the metal (*e.g.* Ni(IV) and Ni(O) (*1*), for producing a high trans influence (*2, 3*), and for activating small molecules such as H$_2$, O$_2$, and olefins (*4*). In recent years, many investigations of organometallic chemistry and transition metal complexes used chelating phosphines as ligands. In particular, investigations of five-coordinate complexes revealed that the stability, stereochemistry, and magnetic properties depend on a subtle blend of electronic and steric effects (5, 6, 7, 8, 9). Consequently, further development of the coordination chemistry of *tert*-phosphines depends greatly on the concurrent development of synthetic phosphorus chemistry.

A properly designed polyphosphine can accentuate the special effects of phosphine ligands as it accomplishes simultaneously: (a) a more predictable coordination number and stoichiometry in the resulting complexes since the chelate effect minimizes the possibility that one or more phosphino groups will be displaced during a chemical reaction; (b) increased basicity (or nucleophilicity) of the metal atom; and (c) more control over the stereochemistry of the resulting complex (10). In addition, poly(*tert*-phosphine) ligands have tremendous potential in the study of dynamic processes and in obtaining metal–phosphorus and phosphorus–phosphorus NMR coupling constants that would be unobtainable for analogous complexes of monodentate phosphines (11).

This paper is concerned with all these five aspects of polyphosphines. In addition, the methods for synthesizing polyphosphine ligands are discussed briefly.

Poly(tert-phosphine) Ligands

An objective of our research was to synthesize a series of related, flexible polydentate ligands that contain either $-CH_2CH_2-$ or $-CH_2CH_2CH_2-$ connecting units and different types of donor groups so that systematic variations of the catalytic, sterochemical, and spectral properties of the metal could be studied. Until 1971 most polyphosphines were prepared by treating organic polyhalides with alkali metal dialkyl or diaryl phosphides—*e.g.*, the relatively easy preparation of $Ph_2PCH_2CH_2$-PPh_2 (12). However, use of this method for more complicated tri- and tetraphosphines is severely limited by the difficulties in obtaining the appropriate organic polyhalide or in effecting complete reaction with the phosphide reagent (13). The variety of polyphosphines that contain PCH_2CH_2P units was greatly increased by King et al. (14, 15, 16) by the base-catalyzed additions of phosphorus–hydrogen bonds to the carbon–carbon double bonds in various vinyl phosphine derivatives. King and Cloyd recently used the base-catalyzed method to prepare the first extensive series of methylated polyphosphines (17, 18). The syntheses of the methylated phosphine ligands involves the conversion of a P–H bond to a $PCH_2CH_2P(CH_3)_2$ unit *via* the potassium *tert*-butoxide catalyzed addition to $CH_2{=}CHP(S)(CH_3)_2$ followed by desulfurization with $LiAlH_4$ in boiling dioxane (Reactions 1 and 2).

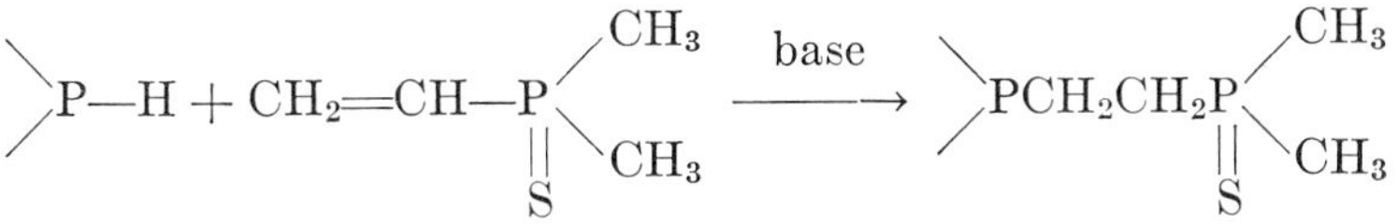

$$\diagdown\hspace{-0.5em}P\!-\!H + CH_2\!\!=\!\!CH\!-\!P\diagdown^{CH_3}_{\underset{\|}{S}CH_3} \xrightarrow{\text{base}} \diagdown\hspace{-0.5em}PCH_2CH_2P\diagdown^{CH_3}_{\underset{\|}{S}CH_3} \quad (1)$$

$$\diagdown\hspace{-0.5em}PCH_2CH_2P\diagdown^{CH_3}_{\underset{\|}{S}CH_3} \xrightarrow{\text{LiAlH}_4} \diagdown\hspace{-0.5em}PCH_2CH_2P(CH_3)_2 \quad (2)$$

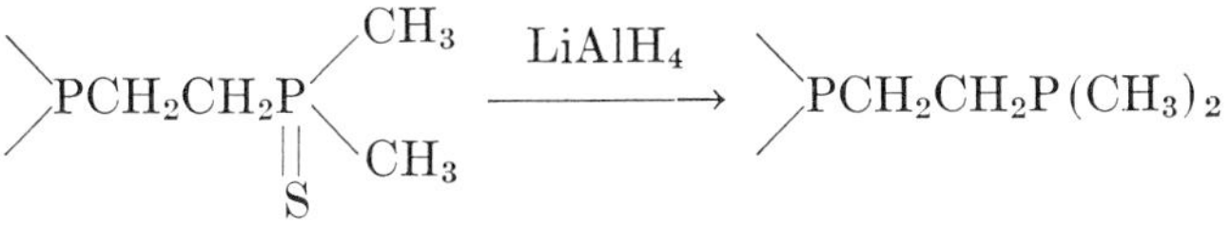

Variations of the base-catalyzed method were used to prepare the mixed alkyl–aryl di(*tert*-phosphine) ($C_6H_5)_2PCH_2CH_2P(CH_3)_2$, the three triphosphines $R'P(CH_2CH_2PR_2)_2$ $R' = CH_3$, $R = CH_3$ or C_6H_5, $R = CH_3$), the mixed aliphatic–aromatic linear tetra(*tert*-phosphine) (I), and the branched penta(*tert*-phosphine) (II) (*17, 18*).

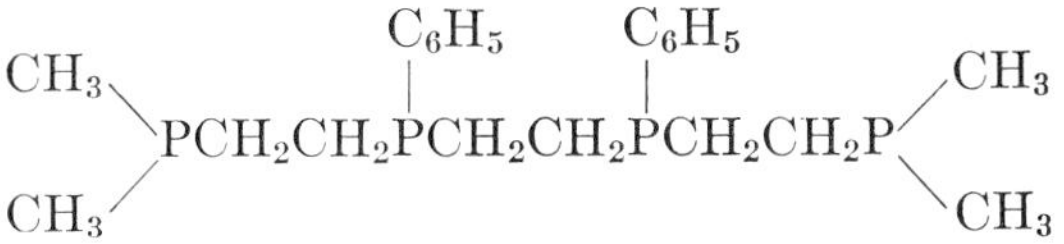

$$CH_3{\diagdown\atop CH_3\diagup}PCH_2CH_2\overset{\overset{\displaystyle C_6H_5}{|}}{P}CH_2CH_2\overset{\overset{\displaystyle C_6H_5}{|}}{P}CH_2CH_2P{\diagup CH_3 \atop \diagdown CH_3}$$

I

$$\begin{array}{c}CH_3{\diagdown\atop CH_3\diagup}PCH_2CH_2{\diagdown}\\[1em]\hspace{4em}PCH_2CH_2\overset{\overset{\displaystyle C_6H_5}{|}}{P}CH_2CH_2P{\diagup CH_3\atop\diagdown CH_3}\\[1em]CH_3{\diagdown\atop CH_3\diagup}PCH_2CH_2{\diagup}\end{array}$$

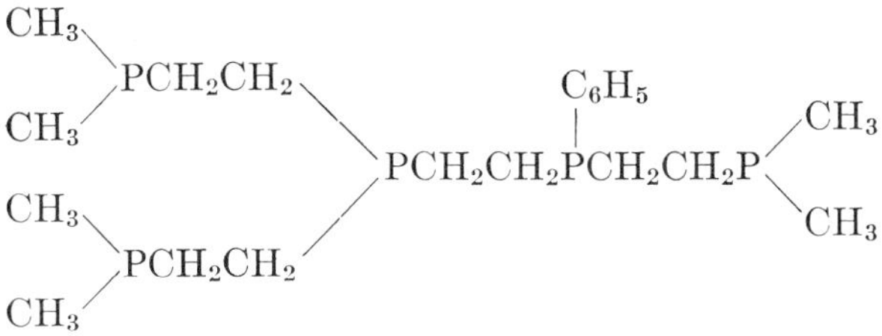

II

Although the base-catalyzed process is a valuable synthetic route to polyphosphines, the method is limited to PCH_2CH_2P connecting linkages and to available vinyl phosphines. Our need for a series of flexible polydentate ligands containing either $-CH_2CH_2-$ or $-CH_2CH_2CH_2-$ units and different types of donor groups led us to investigate alternative routes of ligand synthesis. Thus, we previously prepared the methylated di-

and triphosphines $CH_3{\diagdown\atop CH_3\diagup}PCH_2CH_2CH_2P{\diagup CH_3\atop\diagdown CH_3}$ and $C_6H_5P(CH_2\text{-}$

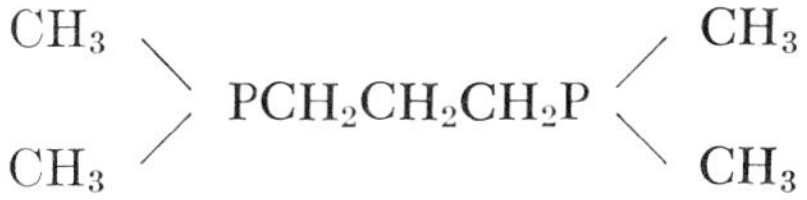

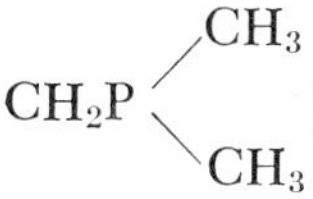

$CH_2P{\diagup CH_3\atop\diagdown CH_3})_2$ by reacting $(CH_3)_2P^-$ with $ClCH_2CH_2CH_2Cl$ and $C_6H_5P\text{-}$

($CH_2CH_2CH_2Cl$)$_2$, respectively (*19, 20*). Recently, we found that the free-radical catalyzed addition of phosphorus–hydrogen or sulfur–hydrogen bonds across carbon–carbon double bonds in vinyl phosphines is a high yield, general reaction for the preparation of polyphosphines, mixed phosphorus–sulfur, and mixed phosphorus–nitrogen compounds (*21*). The general reactions can be summarized by Equations 3, 4, and 5.

$$\text{\textbackslash}P\text{—}H + CH_2\text{=}CH\text{—}P\diagup \longrightarrow \text{\textbackslash}PCH_2CH_2P\diagup \qquad (3)$$

$$\text{—}S\text{—}H + CH_2\text{=}CH\text{—}P\diagup \longrightarrow \text{—}SCH_2CH_2P\diagup \qquad (4)$$

$$\text{\textbackslash}P\text{—}H + CH_2\text{=}CH\text{—}C(O)NH_2 \longrightarrow \text{\textbackslash}PCH_2CH_2C(O)NH_2 \qquad (5)$$

The free-radical catalyzed reactions to produce 1,2-bis(diphenylphosphino)ethane (diphos) and bis(2-diphenylphosphinoethyl)phenyl-

Table I.   Reactants, Products, and Yields Obtained *via* the

Reactants

$Ph_2PCH\text{=}CH_2 + Ph_2PH$
$PhP(CH\text{=}CH_2)_2 + Ph_2PH$

$Ph_2PCH\text{=}CH_2 +$ $\overset{Ph}{\underset{H}{\diagup}}PCH_2CH_2CH_2P\overset{Ph}{\underset{H}{\diagup}}$

$\underset{i\text{-PrO}}{\overset{Ph}{\diagdown}}PCH\text{=}CH_2 + Ph_2PH$
$\underset{O}{\|}$

$Ph_2PCH_2CH_2CH_2P\overset{Ph}{\underset{H}{\diagup}} + Ph_2PCH\text{=}CH_2$

$Ph_2PCH\text{=}CH_2 + PhSH$
$PhP(CH\text{=}CH_2)_2 + PhSH$
$Ph_2PH + CH_2\text{=}CHC\text{—}NH_2$
$\qquad\qquad\qquad \underset{O}{\|}$
$PhPH_2 + CH_2\text{=}CHCH_2NH_2$
$PhPH_2 + CH_2\text{=}CHCH_2OH$

a AIBN is 2,2′-azobis(isobutyronitrile).
b We thank W. H. Myers for this result.
c We thank J. R. Nappier for this result.

phosphine (triphos) yielded products with spectral characteristics that were virtually identical to those of authentic materials. The yields were better than those reported (*14, 15, 16*) for the base-catalyzed process: 88% *vs.* 80% for diphos and 91% *vs.* 50% for triphos. Yields of the other compounds were equally impressive (Table I).

It is interesting that the trimethylene connecting chain of our symmetrical triphosphine ligand $PhP(CH_2CH_2CH_2PPh_2)_2$ (3-3 connecting unit) has a much simpler ^{31}P NMR spectrum than that of the triphosphine $PhP(CH_2CH_2PPh_2)_2$ (2-2 connecting unit). The marked difference is illustrated in Figure 1 where the P–P coupling is < 1 Hz in the 3-3 ligand [$\delta(PhP) = -28.8$ ppm, $\delta(Ph_2P) = -18.2$ ppm] and 29 Hz in the 2-2 ligand [$\delta(PhP) = -16.6$ ppm, $\delta(Ph_2P) = -12.8$ ppm]. The 2-3 ligand $Ph_2PCH_2CH_2PCH_2CH_2CH_2PPh_2$ provides a nice confirmation of the
$$\qquad\qquad\qquad\qquad |$$
$$\qquad\qquad\qquad\qquad Ph$$
effect of the connecting chain on the P–P coupling constant since the P–P

AIBN-Catalyzed Reactions of Vinyl and Allyl Compounds[a]

Product	Yield, %
$Ph_2PCH_2CH_2PPh_2$	88
$Ph_2PCH_2CH_2PCH_2CH_2PPh_2$ $\qquad\qquad\; \mid$ $\qquad\qquad\; Ph$	91
$Ph_2PCH_2CH_2PCH_2CH_2CH_2PCH_2CH_2PPh_2$ $\qquad\qquad\; \mid \qquad\qquad\quad \mid$ $\qquad\qquad\; Ph \qquad\qquad\;\; Ph$	70
$Ph\diagdown$ $\qquad\diagup PCH_2CH_2PPh_2$ $i\text{-}PrO\diagup\; \|$ $\qquad\qquad\; O$	70
$Ph_2PCH_2CH_2CH_2PCH_2CH_2PPh_2$ $\qquad\qquad\qquad\;\; \mid$ $\qquad\qquad\qquad\;\; Ph$	91[b]
$Ph_2PCH_2CH_2SPh$	99
$PhP(CH_2CH_2SPh)_2$	97
$Ph_2PCH_2CH_2C{-}NH_2$ $\qquad\qquad\quad \|$ $\qquad\qquad\quad O$	85[b]
$PhP(CH_2CH_2CH_2NH_2)_2$	75[c]
$PhP(CH_2CH_2CH_2OH)_2$	85

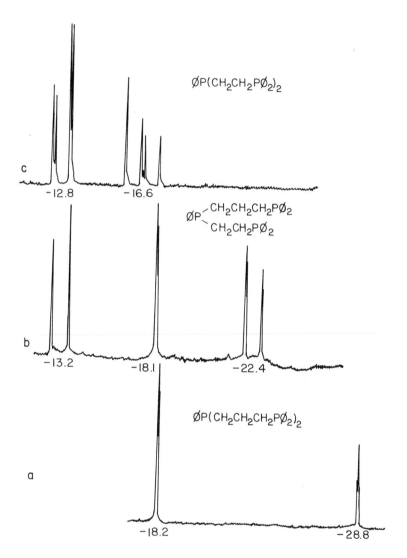

Figure 1. ^{31}P *NMR spectra of* $PhP(CH_2CH_2CH_2PPh_2)_3$ *in* C_6H_6
P

(bottom), of $Ph_2PCH_2CH_2PCH_2CH_2CH_2PPh_2$ *in* C_6H_6 *(center), and
of* $PhP(CH_2CH_2PPh_2)_2$ *in* $CHCl_3$ *(top). The numbers correspond to
the chemical shifts of the peaks relative to 85%* H_3PO_4 *as an external
standard; positive numbers are downfield from the standard.*

coupling through the ethylene unit is 29 Hz whereas the P–P coupling
through the trimethylene unit is ~ 1 Hz (*see* Figure 1).

The free-radical catalyzed method for preparing mixed P- and S-
(and P- and N-) containing ligands is especially noteworthy, as it repre-

sents an easy route to potentially useful ligands. The two P–S compounds prepared in this work are formally analogous to the ligands diphos and triphos, in which one and two PPh_2 groups, respectively, are replaced by

SPh groups. Also, use of compounds of the type $\begin{array}{c}R \\ \diagdown \\ i\text{-}PrO \diagup\end{array} PCH{=}CH_2$

may be exploited for the syntheses of much more complicated mixed

ligands. For example, reduction of the $\begin{array}{c}\diagdown \\ \diagup P{-}OR \\ \| \\ O\end{array}$ group to a secondary

phosphine then provides the capability for additional reactions *via* King's base-catalyzed method or *via* our free-radical catalyzed method.

The chief advantages of the free-radical catalyzed method are: (a) the flexibility of designing several different related polydentate ligands that contain a variety of donor groups with either $-CH_2CH_2-$ or $-CH_2-CH_2CH_2-$ connecting units by simply choosing the appropriate P–H, S–H, and vinyl derivatives; (b) the experimental simplicity of the homogeneous solutions and the one-pot reaction; (c) the faster reaction times as compared with base-catalyzed addition to vinylphosphines; (d) the ease with which the reaction impurities or by-products are removed *in vacuo* to leave a material that is sufficiently pure for use as a ligand in subsequent preparations of complexes; and (e) the high yield (85–99%) that the method routinely gives.

Alkyl and Hydride Rhodium Complexes of $PhP(CH_2CH_2CH_2PPh_2)_2$

Nappier *et al.* (*10*) demonstrated previously that RhCl(ttp), ttp = $PhP(CH_2CH_2CH_2PPh_2)_2$, which is a dissociatively stable analog of Wilkinson's $RhCl(PPh_3)_3$, readily adds a variety of neutral and cationic reagents to give stable, five-coordinate complexes $RhCl(ttp) \cdot A$ (A = BF_3, CO, SO_2, O_2) and $RhCl(ttp) \cdot A'$ (A' = NO^+, N_2Ph^+, CH_3CO^+) respectively. The chelating triphosphine ligand $PhP(CH_2CH_2CH_2PPh_2)_2$ so increases the apparent basicity of the metal center that this rhodium complex mimics the reactions of phosphine complexes of Ir(I)—*e.g.* $IrCl(CO)(PPh_3)_2$ and $IrCl(PPh_3)_3$—more closely than it resembles other rhodium–aryl phosphine complexes. The basic nature of RhCl(ttp) is also demonstrated by the structures of the nitrosyl and phenyldiazo complexes which contain bent Rh–N–O and Rh–N–N angles of 131° and 125° respectively (*10, 22*). Thus, all of the above ttp complexes are formally examples of oxidative–addition conversion of four-coordinate d^8 complexes into five-coordinate d^6 cases.

Alkylation and protonation reactions of RhCl(ttp) yield complexes with spectral properties which indicate that oxidative–addition has converted CH_3^+, $C_2H_5^+$, and H^+ to formally CH_3^-, $C_2H_5^-$, and H^- respectively; this is analogous to the reaction pattern of the cations NO^+ and N_2Ph^+. The alkyl and hydride complexes show striking differences in their tendency to add carbon monoxide or polar solvent molecules. These differences are illustrated by the IR and NMR spectra that are discussed below.

Proton NMR Spectra. The proton NMR spectra of the complexes $[Rh(ttp)HX(L)]^+$ ($L = C_2H_5OH$, CH_3CN, CO) in the region $\tau \sim 24$ appeared as multiplets which are assigned to the Rh–H resonance. First-order coupling of the hydride nucleus to the ^{103}Rh nucleus and three ^{31}P nuclei produces a 12-line pattern (*i.e.* the A portion of an AMX_2Y system), and such a spectrum was observed for $[Rh(ttp)HCl(CH_3CN)]^+$-BF_4 (Figure 2). Similar spectra were observed for $[Rh(ttp)HCl(CO)]$-

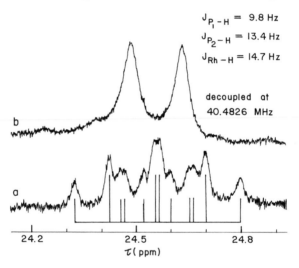

Figure 2. Proton NMR spectrum of $[Rh(ttp)HCl$-$(CH_3CN)]BF_4$ dissolved in CH_3CN (a), and phosphorus nuclei decoupled at 40.4826 MHz (b)

PF_6 and $[Rh(ttp)HBr(CO)]PF_6$. On irradiation of the ^{31}P region (40.48 MHz), the multiplet patterns collapse to doublets from which J_{Rh-H} is determined unambiguously. The remaining doublet coupling is assigned to coupling between the hydride and the unique phosphorus in ttp while coupling to the two equivalent terminal phosphorus atoms produces a triplet in the hydride spectrum. The calculated values of coupling constants and chemical shifts for the hydride complexes are listed in Table II.

The proton NMR spectrum of the methyl group in $[Rh(ttp)(CH_3)-Cl]FSO_3$ and its six-coordinate derivatives $[Rh(ttp)(CH_3)Cl(L)]FSO_3$ ($L = CH_3CN$ or CO) were very similar to those of the analogous hydride

Table II. Proton NMR Data for Hydride Complexes
of the Type [Rh(ttp)HX(L)]⁺

Compound	τ^a	J, Hz		
		$Rh–H$	$P_1–H$	$P_2–H$
[Rh(ttp)HCl(MeCN)]BF₄	24.6	14.7	13.4	9.8
[Rh(ttp)HCl(CO)]BF₄	22.4	15.5	8.9	8.6
[Rh(ttp)HBr(CO)]PF₆	21.3	15.4	8.5	8.5

ᵃ Tau value relative to internal TMS at 10.00.

species except that the magnitudes of the coupling constants were less. As was described previously (*23*), the portion of the spectrum assigned to the rhodium–methyl group overlapped with that of the methylene protons of ttp in the five-coordinate cation; however, because of the trans influence (*2*) of a ligand trans to the methyl group in both [Rh(ttp)CH₃Cl(CO)]FSO₃ and [Rh(ttp)CH₃Cl(CH₃CN)]FSO₃, the

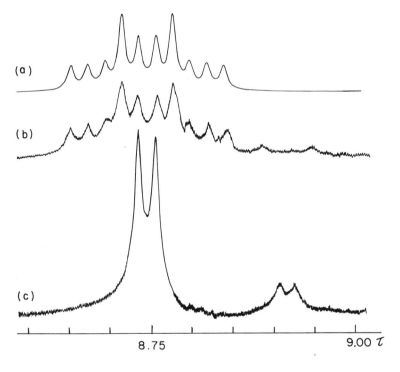

Figure 3. Computer simulation of the proton NMR spectrum of [Rh(ttp)CH₃Cl(CH₃CN)]FSO₃ in CH₃CN based on data in Table III (a), experimental proton spectrum of [Rh(ttp)CH₃Cl-(CH₃CN)]FSO₃ in CH₃CN (b), and phosphorus nuclei decoupled at 40.483 MHz (c)

Table III. **¹H NMR of Rh Alkyls**

Compound	τ^a	Rh–H	H–P₁	H–P₂
			J, Hz	
[Rh(ttp)CH₃Cl]FSO₃	7.40	2.5	3.8	5.5
[Rh(ttp)CH₃Cl(MeCN)]FSO₃	8.75	2.1	4.4	6.2
[Rh(ttp)CH₃Cl(CO)]FSO₃	8.71	1.0	6.0	5.8

ª Tau value of multiplet center relative to internal TMS at 10.00.

rhodium–methyl resonances are shifted significantly up-field from the resonance position of the methylene protons of ttp. Figure 3 shows the 12-line pattern for the A₃ portion of an A₃MX₂Y spectrum that is observed for [Rh(ttp)CH₃Cl(CH₃CN)]⁺, and Table III lists the NMR spectral parameters.

³¹P **Spectra.** The Fourier-transform, proton-decoupled ³¹P NMR spectra of all the hydride and methyl complexes displayed a doublet of

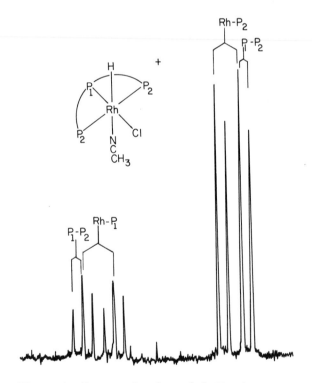

Figure 4. Proton-noise decoupled, Fourier transform ³¹P NMR spectrum of [Rh(ttp)HCl(CH₃CN)]-BF₄ in CD₃CN

triplets and a doublet of doublets (*see* Figure 4) as is expected for an AM_2X coupling pattern.

Stereochemistry of the Rhodium Hydride and Rhodium Methyl Complexes. On the basis of IR and proton and ^{31}P NMR spectra, all of the cationic hydrides $[Rh(ttp)HCl(L)]^+$ (where $L = CO$, CH_3CN, C_2H_5OH) and $[Rh(ttp)HBr(CO)]^+$ are six-coordinate whereas the methyl cation $[Rh(ttp)CH_3Cl]^+$ can be isolated as the five-coordinate species as well as the six-coordinate compounds $[Rh(ttp)CH_3Cl(L)]^+$ ($L = CH_3CN$, CO).

The ^{31}P NMR spectra of the cationic hydrides reveal that the terminal phosphorus nuclei are equivalent; thus, for the six-coordinate complexes $Rh(ttp)XYZ$, the ttp ligand must have a meridional arrangement, *e.g.* III. Consideration of the magnitudes of the $^1J_{Rh-P}$ and $^2J_{P-H}$ coupling

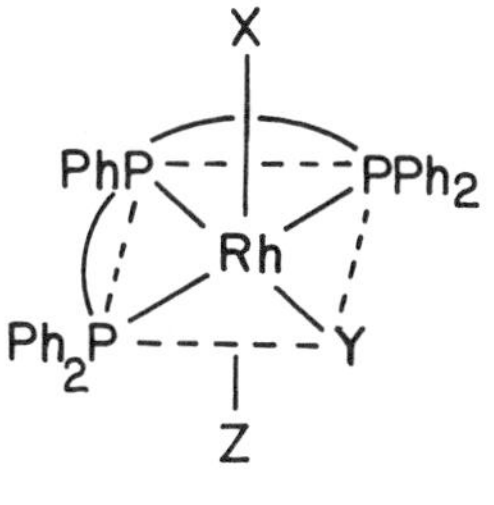

III

constants provides correlations on which to base a structure in the cases of isomers, *e.g.* $[Rh(ttp)HCl(CO)]^+$. In second, and third-row transition metals, the phosphorus–hydride coupling constants are generally much larger for a trans structure than for a cis structure (*24, 25, 26*). (Reviews of ^{31}P NMR data are given in Refs. *24, 25,* and *26*) In all of our six-coordinate rhodium complexes of ttp, the $^2J_{P-H}$ values are relatively small and approximately equal (Table II); consequently, the hydride must be cis to all three phosphorus nuclei which is consistent with both structure IV and structure V.

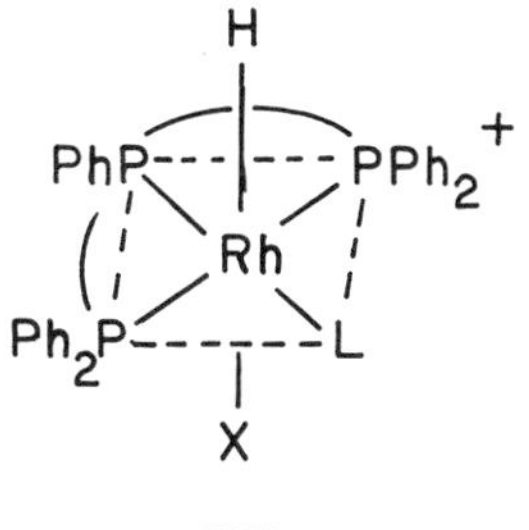

IV 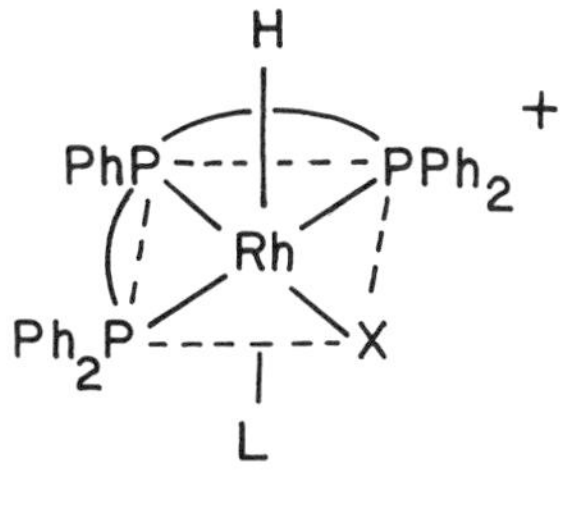   V   X = Cl , Br
L = CO,
CH_3CN,
C_2H_5OH

With the carbonyl complexes, a choice can be made between structures IV and V on the basis of the shift in ν_{CO} on deuteration. Vaska (27) observed that Fermi interaction in complexes having CO trans to a hydride leads to a significant shift in ν_{CO} and to an anomolous ratio of νM–H/νM–D on deuteration. The complexes [Rh(ttp)HCl(CO)]⁺ and [Rh(ttp)DCl(CO)]⁺ have essentially the same ν_{CO} and a normal value of 1.40 for νRh–H/νRh–D. Findings for the bromide complexes were similar. Thus, the cationic hydride–carbonyl complexes have structure VI.

The alkyl complexes [Rh(ttp)Cl(R)]FSO₃ (R = CH₃, C₂H₅) are five-coordinate monomers in contrast to the six-coordinate hydrides. A dimeric cation [Rh(ttp)Cl(CH₃)]₂²⁺ can be excluded on the basis of conductance and molecular weight data; also coordination of the FSO₃⁻ anion in the solid state is eliminated by the IR spectrum. The ³¹P NMR spectra are consistent with the cations [Rh(ttp)ClR]⁺ being square pyramidal, which is analogous to the known structures (10, 22) of [Rh(ttp)Cl(NO)]⁺ and [Rh(ttp)Cl(N₂Ph)]⁺ (Table IV). Thus, the

Table IV. ³¹P NMR Data for the ttp–Rhodium Complexes[a]

Compound	δ,[b] ppm		J, Hz		
	PPh	PPh₂	P–P[c]	P₁–Rh[d]	P₂–Rh[e]
[Rh(ttp)NOCl]PF₆	10.8	2.1	30	139	106
[Rh(ttp)(CH₃)Cl]FSO₃	11.2	1.1	33	115	93
[Rh(ttp)(CH₃)Cl(MeCN)]FSO₃	11.6	−1.0	33	116	92
[Rh(ttp)(C₂H₅)Cl]PF₆	11.7	−1.8	34	120	98
[Rh(ttp)HCl(MeCN)]BF₄	24.0	10.0	37	116	89
[Rh(ttp)HCl(CO)]PF₆	1.1	4.9	38	90	81
[Rh(ttp)(CH₃)Cl(CO)]FSO₃	7.0	−4.9	38	108	84

[a] ttp = PhP(CH₂CH₂CH₂PPh₂)₂.
[b] Value in ppm relative to external 85% H₃PO₄, with positive numbers being downfield from the standard.
[c] Resolution for the coupling constants is ±1 Hz.
[d] P₁ is the unique central phosphorus atom of PhP(CH₂CH₂CH₂PPh₂)₂.
[e] P₂ refers to the two equivalent terminal phosphorus atoms of PhP(CH₂CH₂-CH₂PPh₂)₂.

triphosphine ligand and chloride form the basal plane and the alkyl group occupies the apical position.

The proton NMR data of $[Rh(ttp)HCl(C_2H_5OH)]^+$ in CH_3NO_2 containing added CH_3CN reveal that acetonitrile replaces the labile ethanol molecule quantitatively and that the resulting cation $[Rh(ttp)-HCl(CH_3CN)]^+$ does not exchange CH_3CN rapidly at room temperature. In contrast to the hydride, the methyl complex $[Rh(ttp)Cl(CH_3)]^+$ exchanges CH_3CN rapidly down to $\sim -50°C$. The corresponding ethyl complex does not form a stable CH_3CN or CO adduct; thus, the tendency of the five-coordinate species to add a sixth ligand decreases markedly in the series hydride > methyl > ethyl which probably reflects increasing steric interactions rather than electronic factors. Proton NMR and IR studies also demonstrated that, for a given hydride or alkyl complex, the stability of the six-coordinate complexes decreases in the order CO > CH_3CN > C_2H_5OH.

The ^{31}P NMR spectra of the two methyl cations $[Rh(ttp)Cl(CH_3)L]$ ($L = CH_3CN$, CO) reveal that the terminal phosphino groups are equivalent which requires a planar arrangement of ttp (Figure 5). In addition, the coupling constant $^1J_{Rh-P_1}$ is 24–25 Hz larger for both complexes than that of J_{Rh-P_2}, which strongly suggests that the weaker ligand

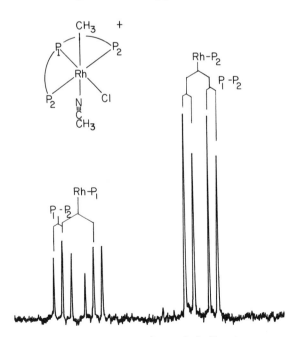

Figure 5. Proton-noise decoupled, Fourier transform ^{31}P NMR spectrum of $[Rh(ttp)CH_3Cl(CH_3-CN)]FSO_3$ in CD_3CN

is trans to P_1. Consequently, structure VII is proposed for [Rh(ttp)Cl-(CH$_3$)(CO)]$^+$; this is a different isomer from that of the analogous hydride, *i.e.* VI.

VII

Cobalt (I) Complexes of Tri(tert-phosphine) Ligands

As a part of our efforts to study systematically the physical and chemical properties of a metal in different stereochemical environments, we synthesized and characterized several new series of Co(I) complexes of the types Co(triphos)H(CO), [Co(triphos)(CO)$_2$]$^+$, and [Co-(triphos)(monophos)(CO)]$^+$ with the three tri(*tert*-phosphine) ligands CH$_3$C(CH$_2$PPh$_2$)$_3$, PhP(CH$_2$CH$_2$PPh$_2$)$_2$, and PhP(CH$_2$CH$_2$CH$_2$PPh$_2$)$_2$ (monophos = the monodentate ligands Ph$_2$PH, (MeO)$_3$P, and Et$_3$P). Table V lists some of the isolated compounds, the most diagnostic IR data for the cobalt hydride and carbonyl stretching frequencies, and the conductivity or mass spectral data for the cationic or the neutral complexes respectively.

The Co(triphos)H(CO) complexes can be prepared by reducing the corresponding Co(triphos)Cl$_2$ complexes with NaBH$_4$ in ethanol while bubbling carbon monoxide through the solution. The yellow, air-sensitive complexes were characterized by elemental analysis and by ^{1}H NMR, IR, and mass spectroscopy. The corresponding deuterated complexes were also prepared in order to verify the IR bands assigned to the Co–H stretching mode. The proton NMR of the hydride region for Co(etp)H(CO) consists of an overlapping doublet of triplets at τ 21.7 with $J_{\phi P\text{-}H} = 65$ Hz and $J_{\phi_2 P\text{-}H} = 44$ Hz (Figure 6). The spectra in Figure 6 show that the two types of phosphorus nuclei can be decoupled selectively. The NMR data indicate that the two terminal diphenylphosphino groups are equivalent and are consistent with either a trigonal bipyramid or a square pyramid containing a plane of symmetry.

Treatment of the hydride complexes with the weak acids NH$_4$BF$_4$ or NH$_4$PF$_6$ in the presence of another ligand provides a general and selective method for preparing the series of complexes [Co(triphos)-

Table V. Diagnostic Characterization Data on the Co(I) Complexes of Three Tri(*tert*-phosphines)[a]

Compound	$\nu_{M-H, M-D}$[b]	ν_{CO}	Parent Ion in Mass Spectrum
HCo(etp)(CO)	1920	1870	
DCo(etp)(CO)	1380	1870	622
HCo(ttp)(CO)	1920(sh)[c]	1880	
DCo(ttp)(CO)	1370	1880	650
HCo(tripod)(CO)	1930	1857	712
DCo(tripod)(CO)	1380	1857	

	ν_{CO}		
	Nujol Mull	CH_2Cl_2 Solution	ΛM[d]
[Co(etp)(CO)$_2$]BF$_4$	1980, 2020	1980, 2024	82
[Co(etp)(P(OMe)$_3$)(CO)]BF$_4$	1945	1950	84
[Co(etp)(Pϕ_2H)(CO)]PF$_6$	1925	1936	67
[Co(etp)(PEt$_3$)(CO)]BF$_4$	1930	1928	76
[Co(ttp)(CO)$_2$]BF$_4$	1940, 2000	1949, 2004	84
[Co(ttp)(P(OMe)$_3$)(CO)]BF$_4$	1920	1932	80
[Co(ttp)(Pϕ_2H)(CO)]BF$_4$	1930	1928	82
[Co(tripod)(CO)$_2$]PF$_6$	1953, 2026	1969, 2026	80
[Co(tripod)(P(OMe)$_3$)(CO)]BF$_4$	1930	1928	84
[Co(tripod)(Pϕ_2H)(CO)]BF$_4$	1905	1915	81
[Co(tripod)(PEt$_3$)(CO)]BF$_4$	1890	1908	82

[a] The three triphosphine ligands are PhP(CH$_2$CH$_2$PPh$_2$)$_2$, etp; PhP(CH$_2$CH$_2$-CH$_2$PPh$_2$)$_2$, ttp; and CH$_3$C(CH$_2$PPh$_2$)$_3$, tripod.
[b] The spectra of the hydride complexes were obtained on Nujol mulls whereas the spectra of the deuterated species were recorded on KBr discs. Hexachlorobutadiene reacts with these hydrides.
[c] sh, shoulder on a more intense absorption.
[d] Molar conductance values on ~10^{-3}M nitromethane solutions.

(CO)$_2$]$^+$ and [Co(triphos)(monophos)(CO)]$^+$ (Equations 6 and 7).

$$\text{HCo(etp)CO} + \text{NH}_4\text{BF}_4 + \text{CO} \xrightarrow[\text{CH}_3\text{OH}]{\Delta}$$
$$[\text{Co(etp)(CO)}_2]\text{BF}_4 + \text{H}_2 + \text{NH}_3 \quad (6)$$

$$\text{HCo(triphos)CO} + \text{NH}_4\text{BF}_4 + \text{PR}_3 \xrightarrow[\text{CH}_3\text{OH}]{\Delta}$$
$$[\text{Co(triphos)(PR}_3)\text{CO}]\text{BF}_4 + \text{H}_2 + \text{NH}_3 \quad (7)$$

This preparative method offers distinct advantages over the more commonly used direct combination of ligand with compounds such as Co$_2$(CO)$_8$. Difficulties such as polymeric complexes are often encoun-

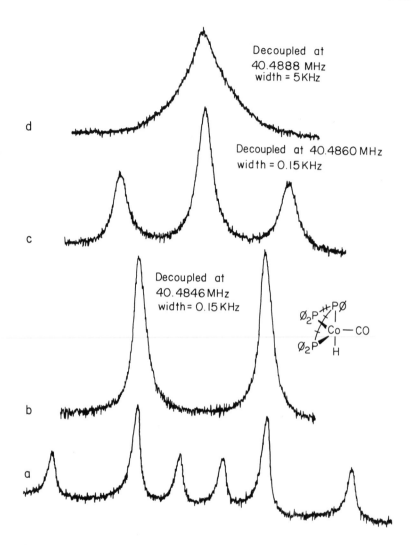

*Figure 6. Proton NMR spectra of Co[PhP(CH₂CH₂PPh₂)₂]H(CO)
in tetrahydrofuran (a), two phosphorus nuclei decoupled at 40.4846
MHz (b), one phosphorus nucleus decoupled at 40.4860 MHz (c), and
all three phosphorus nuclei decoupled at 40.4888 (d). The tentatively
proposed structure has the hydride and carbonyl ligands at axial and
equatorial sites of a trigonal bipyramid, respectively.*

tered by the latter preparative methods (*e.g.* Equation 8) (28). The

$$Co_2(CO)_8 + PhP(CH_2CH_2PPh_2)_2 \xrightarrow[\text{C}_6\text{H}_6]{\Delta \text{ or } h\nu}$$

$$[Co_2(etp)_3(CO)_4]^+BPh_4^- \qquad (8)$$

HCo(triphos)CO + NH₄X route is particularly useful for syntheses of the mixed complexes [Co(triphos)(monophos)(CO)]⁺ since such complexes would be very difficult to obtain from other Co(I) reagents. The [Co(triphos)(CO)₂]⁺ and [Co(triphos)(monophos)(CO)]⁺ complexes are all yellow or orange, air-stable solids, but their solutions are moderately sensitive to oxygen.

The dicarbonyl cations may have either square-pyramidal, trigonal-bipyramidal, or somewhat distorted structures (*29*). IR studies of the

Table VI. Bond Angles of the [Co(triphos)(CO)₂]⁺ Complexes Calculated from the Relative Intensities of the Two Carbonyl Bands[a]

Compound	2θ
[Co(tripod)(CO)₂]PF₆	88°
[Co(etp)(CO)₂]PF₆	96°
[Co(etp)(CO)₂]BF₄	95°
[Co(ttp)(CO)₂]PF₆	115°
[Co(ttp)(CO)₂]BF₄	114°

[a] Calculated according to Cotton and Wilkinson (*30*).

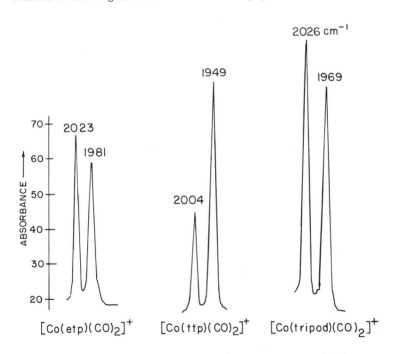

Figure 7.   IR absorption peaks in the ν_{CO} region for dichloromethane solutions of the dicarbonyl cations of PhP(CH₂-CH₂PPh₂)₂ (etp), PhP(CH₂CH₂CH₂PPh₂)₂ (ttp), and CH₃C-(CH₂PPh₂)₃ (tripod)

dicarbonyl compounds indicate that the structural nature of the tridentate ligand determines the manner in which the ligand bonds to the available coordination sites around cobalt(I). Measurement of the relative intensities of the bands corresponding to the symmetric and anti-symmetric stretching frequencies of the dicarbonyls was used to calculate the bond angles between the two carbonyl ligands (30). The values obtained for five complexes are given in Table VI. The four most likely structures for complexes containing the linear tridentate ligands are VIII, IX, X, and XI. For $[Co(ttp)(CO)_2]^+$, only structure VIII is consistent with the bond angles calculated. For $[Co(etp)(CO)_2]^+$, structures IX, X, and XI are

Table VII. ^{31}P NMR Parameters for the

Compound	δ^b	Assignment^c	
$[Co(etp)(CO)_2]BF_4$	122.6	ϕP	t
	78.8	$\phi_2 P$	d
$[Co(etp(P(OMe)_3)(CO)]BF_4$	150.0	$P(OMe)_3$	d-t
	113.3	ϕP	d-t
	68.2	$\phi_2 P$	d-d
$[Co(etp)(P\phi_2H)(CO)]BF_4$	115.4	ϕP	d-t
	66.8	$\phi_2 P$	d-d
	57.2	$\phi_2 PH$	d-t
$[Co(etp)(PEt_3)(CO)]BF_4$	109.6	ϕP	d-t
	64.9	$\phi_2 P$	d-d
	38.7	PEt_3	d-t
$[Co(ttp)(CO)_2]BF_4$	37.0	$\phi_2 P$	d
	−7.3	ϕP	t
$[Co(ttp)(P\phi_2H)(CO)]BF_4$	26.8	$\phi_2 P$	d-d
	3.1	$\phi_2 PH$	d
	−8.2	ϕP	t-d

^{a} The three triphosphine ligands are $PhP(CH_2CH_2PPh_2)_2$, etp; $PhP(CH_2CH_2-CH_2PPh_2)_2$, ttp; $CH_3C(CH_2PPh_2)_3$, tripod.
^{b} Value in ppm relative to 85% H_3PO_4 ±0.2 ppm.

compatible with the calculated bond angles. The steric requirements of the chelating tripod ligand would require an arrangement of the phosphorus atoms similar to that in IX or XI, and the bond angles calculated are in agreement with either of these structures. (The structures of three complexes of $CH_3C(CH_2PPh_2)_3$ have been determined, and the P–M–P angles are fairly constant; *e.g.* mean P–M–P angles in Ni(triphos)-C_2F_4, and Ni(triphos)I are 92.5° (*31*), 89° (*32*), and 95.0° (*33*) respec-

$$[(triphos)Fe{-}H{-}Fe(triphos)]^+,$$

tively.) The solution IR spectra of the three triphosphine complexes are presented in Figure 7. The solid state spectra show no major differences from the solution spectra.

The ^{31}P NMR parameters for some of the $[Co(triphos)(CO)_2]^+$ and $[Co(triphos)(monophos)(CO)]^+$ complexes are listed in Table VII. The ^{31}P spectra of the $[Co(triphos)(monophos)(CO)]^+$ complexes contain extensive and valuable information about the various phosphorus–

Tri(*tert*-phosphine)–Cobalt(I) Complexes[a]

J_{P-P}, Hz	Assignment of Coupled Nuclei	T[e]	Solvent
45[d]	Pϕ_2–Pϕ groups	amb.	CH_3CN
211	ϕP–P(OMe)$_3$	amb.	CH_3CN
88	ϕ_2P–P(OMe)$_3$		
38	ϕP–Pϕ_2		
107	ϕP–Pϕ_2H	amb.	CH_3CN
56	ϕ_2P–Pϕ_2H		
36	ϕ_2P–Pϕ		
364	P–H		
104	ϕP–PEt$_3$	amb.	CH_3CN
53	ϕ_2P–PEt$_3$		
33	ϕP–Pϕ_2		
75	ϕP–Pϕ_2	amb.	CH_3CN
46	ϕP–Pϕ_2H	−62°C	CH_2Cl_2
9	ϕ_2P–Pϕ_2H		
89	ϕ_2P–Pϕ		
317	P–H		

[e] d, doublet; t, triplet; d-t, doublet of triplets, etc.
[d] Coupling constants are accurate to ±2.4 Hz.
[e] Ambient operating temperature of Bruker HX-90 spectrometer = ~33°C.

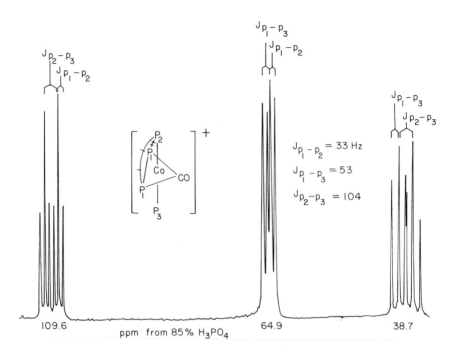

Figure 8. Proton-noise decoupled, Fourier-transform ^{31}P NMR spectrum of $[Co(etp)(CO)P(C_2H_5)_3]BF_4$ in CH_3CN at $-29°C$; etp $= PhP(CH_2CH_2PPh_2)_2$

phosphorus couplings since there are three different types of phosphorus nuclei which generally give well separated resonance patterns. The ^{31}P NMR resonance for the central phosphorus atom of both triphosphine ligands $PhP(CH_2CH_2PPh_2)_2$ and $PhP(CH_2CH_2CH_2PPh_2)_2$ in the [Co-(triphos)(monophos(CO)]$^+$ complexes appears as a doublet of triplets as the result of coupling between the monodentate phosphorus ligand and the two equivalent diphenylphosphino groups. For the etp complexes, the range of values for the P–P coupling constants between the terminal phosphorus nuclei and the central phosphorus atom is from 33 to 45 Hz, which is smaller than the 49–88 Hz range observed for the coupling of the terminal phosphorus atom the the monodentate phosphorus ligand. The magnitude of the coupling constants of the monodentate ligands to the central and terminal phosphorus atoms of the tridentate ligands shows the expected increase for the phosphite complex (24, 25, 26). The ^{31}P spectrum of $[Co(etp)(PEt_3)(CO)]BF_4$ in Figure 8 represents a typical spectrum for this type of complex.

The ^{31}P NMR spectra for the two $[Co(ttp)(monophos)(CO)]^+$ complexes are quite different from the corresponding etp complexes in

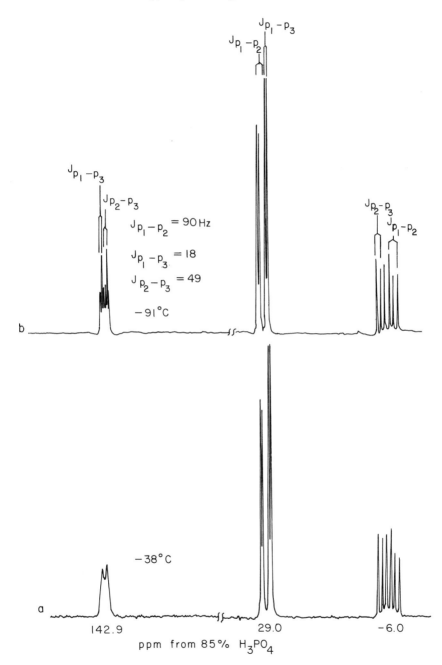

Figure 9. Proton-noise decoupled, Fourier-transform ^{31}P NMR spectra of $[Co(ttp)(CO)P(OCH_3)_3]BF_4$ in dichloromethane at $-38°C$ (a) and at $-91°C$ (b). The coupling constants were determined from the $-91°C$ spectrum.

the magnitudes of both the chemical shifts and the P–P coupling constants. Phosphorus–phosphorus coupling between the central phosphorus atom and the terminal atoms of ttp in the complexes [Co-(ttp)(monophos)(CO)]$^+$ (monophos = Ph$_2$PH and P(OCH$_3$)$_3$) is approximately 90 Hz. The coupling between the central phosphorus atom of ttp and the monodentate ligands Ph$_2$PH and P(OCH$_3$)$_3$ is 46 and 49 Hz respectively. These values are comparable to those found for cis coupling in the etp complexes. However, the coupling between the monodentate ligand and the two diphenylphosphino groups is small, being 9 and 18 Hz for the diphenylphosphine and trimethylphosphite complexes respectively. Two of the spectra of [Co(ttp)(P(OMe)$_3$)(CO)]BF$_4$ are given in Figure 9. The one at −91°C in particular illustrates the need to cool the cobalt samples in order to reduce the broadening effect of the cobalt quadrupole.

Many more coupling constants are observed in the [Co(etp)L-(CO)]X and [Co(ttp)L(CO)]X complexes than are observed in complexes containing only monophosphine ligands. The data listed in Table VII indicate that a large range of phosphorus–phosphorus coupling constants, ∼ 10–210 Hz, can be expected for five-coordinate complexes. Additional structural studies on some of the complexes described here are underway and should provide correlations between the stereochemistry of five-coordinate complexes and the magnitudes of their $^2J_{PP}$ coupling constants.

Acknowledgments

We are grateful to W. H. Myers and T. E. Nappier, Jr. for exploratory studies on the free-radical catalyzed syntheses of ligands and the Rh–ttp complexes, respectively, and to Matthey Bishop Inc. for a loan of rhodium trichloride.

Literature Cited

1. Warren, L. F., Bennett, M. A., J. Am. Chem. Soc. (1974) 96, 3340.
2. Pidcock, A., Richards, R. E., Venanzi, L. M., J. Chem. Soc. A (1966) 1707.
3. Osborn, J. A., Jardine, F. H., Young, J. F., Wilkinson, G., J. Chem. Soc. A (1966) 1711.
4. Muetterties, E. L., Ed., "Transition Metal Hydrides," Marcel Dekker, New York, 1971.
5. Alyea, E. C., Meek, D. W., J. Am. Chem. Soc. (1969) 91, 5761.
6. Cloyd, Jr., J. C., Meek, D. W., Inorg. Chim. Acta (1972) 6, 607.
7. Stalick, J. K., Corfield, P. W. R., Meek, D. W., Inorg. Chem. (1973) 12, 1668.
8. Sacconi, L., J. Chem. Soc. A (1970) 248.
9. Dawson, J., McLennan, T. J., Robinson, W., Merle, A., Dartiguenave, M., Dartiguenave, Y., Gray, H. B., J. Am. Chem. Soc. (1974) 96, 4428.

10. Nappier, Jr., T. E., Meek, D. W., Kirchner, R. M., Ibers, J. A., *J. Am. Chem. Soc.* (1973) **95**, 4194.
11. Chatt, J., Mason, R., Meek, D. W., *J. Am. Chem. Soc.* (1975) **97**, 3826.
12. Chatt, J., Hart, F. A., *J. Chem. Soc.* (1960) 1378.
13. Berglund, D., Meek, D. W., *Inorg. Chem.* (1969) **8**, 2602.
14. King, R. B., Kapoor, P. N., *J. Am. Chem. Soc.* (1971) **93**, 4158.
15. King, R. B., *Acc. Chem. Res.* (1972) **5**, 177.
16. King, R. B., Cloyd, Jr., J. C., Kapoor, P. N., *J. Chem. Soc. Perkin Trans. 1* (1973) 2226.
17. King, R. B., Cloyd, Jr., J. C., *J. Am. Chem. Soc.* (1975) **97**, 46.
18. *Ibid.* (1975) **97**, 53.
19. Kordosky, G., Cook, B. R., Cloyd, Jr., J. C., Meek, D. W., *Inorg. Synth.* (1973) **14**, 14.
20. Siweic, E., M.S. Thesis, The Ohio State University, 1972.
21. DuBois, D. L., Myers, W. H., Meek, D. W., *J. Chem. Soc. Dalton Trans.* (1975) 1011.
22. Caughan, Jr., A. P., Haymore, B. L., Ibers, J. A., Myers, W. H., Nappier, Jr., T. E., Meek, D. W., *J. Am. Chem. Soc.* (1973) **95**, 6859.
23. Peterson, J. L., Nappier, Jr., T. E., Meek, D. W., *J. Am. Chem. Soc.* (1973) **95**, 8195.
24. Nixon, J. F., Pidcock, A., *Annu. Rev. NMR Spectrosc.* (1969) **2**, 345–422.
25. Finer, E. G., Harris, R. K., *Prog. NMR Spectrosc.* (1971) **6**, 61.
26. Verkade, J. G., *Coord. Chem. Rev.* (1972) **9**, 1–106.
27. Vaska, L., *J. Am. Chem. Soc.* (1966) **88**, 4100.
28. Peterson, R. L., Watters, K. L., *Inorg. Chem.* (1973) **12**, 3009.
29. Jesson, J. P., Meakin, P., *J. Am. Chem. Soc.* (1974) **96**, 5760.
30. Cotton, F. A., Wilkinson, G., "Advanced Inorganic Chemistry," 3rd ed., p. 697, Interscience, New York, 1972.
31. Browning, J., Penfold, B. R., *J. Chem. Soc. Chem. Commun.* (1973) 198.
32. Dapporto, P., Fallani, G., Midollini, S., Sacconi, L., *J. Am. Chem. Soc.* (1973) **95**, 2021.
33. Dapporto, P., Fallani, G., Midollini, S., Sacconi, L., *J. Chem. Soc. Chem. Commun.* (1972) 1161.

RECEIVED January 30, 1975. Work supported by the National Science Foundation which also gave an equipment grant that aided in the purchase of the NMR equipment.

29

Rhodium(I) Complexes of Macrocyclic Tetradentate Thioethers

WILLIAM D. LEMKE, KENTON E. TRAVIS, NURHAN E. TAKVORYAN, and DARYLE H. BUSCH

The Ohio State University, Columbus, Ohio 43210

Reduction of the rhodium(III) complexes of the tetradentate macrocyclic thioethers 1,4,8,11-tetrathiacyclotetradecane (TTP) and 13,14-benzo-1,4,8,11-tetrathiapentadecane (TTX) produces four-coordinate, cationic Rh(TTP)$^+$ and Rh(TTX)$^+$. The anions are not coordinated in [Rh(TTP)]X, X = ClO$_4^-$, PF$_6^-$, BF$_4^-$, NCS$^-$, I$^-$, or Br$^-$. Furthermore, Rh(TTP)$^+$ does not react with electron-pair donors including back-bonding ligands like C$_2$H$_4$, CO, and PPh$_3$. Rh(TTP)$^+$ does, however, combine with many elctrophiles including BF$_3$, SO$_2$, NO$^+$, O$_2$, TCNE, and H$^+$. Thus, it is a substantial nucleophile. Rh(TTP)$^+$ undergoes oxidative addition with CH$_3$I, C$_6$H$_5$-CH$_2$Br, and CH$_3$C$\overset{\diagup O}{\underset{\diagdown Cl}{}}$ thereby increasing both the coordination number and the oxidation state of Rh by two; however, it fails to react with H$_2$.

The expectation that, by analogy to phosphines, thioethers should function as π acceptor ligands and thereby stabilize low oxidation state compounds, led several investigators to try to synthesize thioether complexes of rhodium(I). Walton (*1*) treated [Rh(DTH)Cl$_2$]Cl (DTH = CH$_3$SCH$_2$CH$_2$SCH$_3$) with ethanolic potassium hydroxide, a reducing system developed by Chatt and Shaw (*2*), but he failed to obtain a complex of the expected type. Attempts to obtain rhodium(I) derivatives by reducing [Rh(DTH)$_2$Cl$_2$]Cl with sodium borohydride or by electrochemical methods were equally unsuccessful.

Chatt and co-workers (*3*) attempted to produce rhodium(I)–thioether complexes by substitution reactions. No such product was obtained

358

by the reaction of $[Rh(CO)_2Cl]_2$ with C_6H_5SR ($R = Me$, i-Pr, or C_6H_5), and reaction of the same rhodium(I) complex with tridentate thioethers ($RSCH_2CH_2)_2S$ ($R = n$-Pr or C_6H_5) and with $CH_3C(CH_2SEt)_2$ yielded oils. A polymeric material of the formulation $[RhCl(CO)(DTH)]_n$ was obtained from the reaction of $[Rh(CO)_2Cl]_2$ and DTH in acetone. The compound exhibited very low carbonyl frequencies (1830 and 1800 cm^{-1}) which were attributed to bridging carbonyls, and the compound was thought to be a trimer or a higher oligomer. The corresponding product obtained with $CH_3CH_2SCH_2CH_2SCH_2CH_3$ was formulated similarly; however, its IR spectrum had five carbonyl bands, and it was suggested that the trimeric or tetrameric structure involved halogen or sulfur bridges. We report here the first, well characterized monomeric rhodium-(I)–thioether complexes. They were prepared by the reduction of rhodium(III) complexes of macrocyclic tetrathioether ligands.

The rhodium(III) complexes of the macrocyclic tetrathioethers 1,4,8,11-tetrathiacyclotetradecane (TTP, Structure I) and 13,14-benzo-1,4,8,11-tetrathiacyclopentadecane (TTX, Structure II) were recently

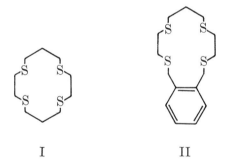

I   II

reported (4). The tetradentate ligands coordinate in a folded fashion yielding *cis*-diacido octahedral complexes, *e.g.* $Rh^{III}(TTP)Cl_2^+$. The many convincing demonstrations that macrocyclic ligands facilitate the stabilization of extreme oxidation states (*e.g.* Ref. 5 and references cited therein) strongly support the possibility that ligands of structures I and II would form stable rhodium(I) derivatives. This is indeed the case, and chemical or electrochemical reduction of the corresponding rhodium-(III) complex leads to the four-coordinate cationic rhodium(I) species $Rh(TTP)^+$ and $Rh(TTX)^+$ which were isolated as the $B(C_6H_5)_4^-$, ClO_4^-, BF_4^-, PF_6^-, NCS^-, I^-, and Br^- salts of TTP and as the PF_6^- salt of TTX. The adduct formation and oxidative addition reactions of $Rh(TTP)^+$ were investigated with a range of potential reactants including O_2, SO_2, CO, BF_3, C_2H_4, tetracyanoethylene, H_2, $P\phi_3$, alkyl, acyl, and arolyl halides, NO^+, and strong Brønsted acids.

Experimental

General Procedure. All manipulations involving complexes of rhodium(I) were done in a vacuum atmospheres glove box equipped with an efficient purification train, or under nitrogen, unless noted otherwise.

Reagents. Standard reagent grade materials were used when possible without further purification. Dichloromethane, benzene, methanol, and spectrograde nitromethane were dried over molecular sieves (Linde). Acetonitrile and diethyl ether were dried over calcium hydride. Any peroxide present in the ether was removed by passing it through a column of alumina; this was followed by testing with KI. In all reactions involving rhodium(I), it is imperative that the solvents be both oxygen- and peroxide-free; therefore all solvents were routinely purged 15–20 min with nitrogen to remove any dissolved oxygen.

Physical Measurements. Analyses were performed by Galbraith Laboratories, Inc., Knoxville, Tenn., and by Alfred Bernhardt Mikroanalytisches Laboratorium, 5251 Elbach über Engelskirchen, West Germany. Conductance was measured on 10^{-3}–$10^{-4}M$ solutions using an Industrial Instruments model RC 16B conductivity bridge. Dried spectrograde nitromethane and dried acetonitrile were the solvents.

IR Spectra. A Perkin–Elmer 337 recording IR spectrophotometer was used routinely to obtain IR spectra at 4000–400 cm^{-1}. For this region, the samples were prepared as Nujol mulls pressed between potassum bromide plates or as solutions in matched liquid cells (0.1-mm path length) with potassium bromide windows. The far-IR spectra were obtained on a Perkin–Elmer 457 grating IR spectrophotometer. Samples for this region were prepared as Nujol mulls and pressed between cesium iodide plates.

Electronic Spectra. Visible, near-IR, and UV absorption spectra were obtained with a Cary model 14R recording spectrophotometer. Solution spectra were run in matching 1-cm quartz cells.

NMR Spectra. The NMR spectra of ligands and complexes were run on a Varian Associates HA-60 or Jeolco MH-100 spectrometer. The solvents used were the deutero derivatives of the solvents needed to dissolve the compounds being studied. Tetramethylsilane was the internal standard.

Mass Spectra. Mass spectra were obtained with an AEI MS-902 mass spectrometer operating at *ca.* 8 or 70 eV.

Electrochemistry. Conventional polarography and cyclic voltammetry were performed with an Indiana Instrument and Chemical model ORNL-1988A controlled-potential and derivative voltammeter. Current–potential curves were recorded on a Hewlett–Packard/Moseley X–Y recorder. All measurements were made in acetonitrile in a vacuum atmospheres glove box under nitrogen with tetra-*n*-butylammonium tetrafluoroborate as supporting electrolyte. Experimental runs were made using a three-compartment H-type polarographic cell. A rotating platinum wire served as the working electrode, and the reference electrode was a chloride-coated silver wire ($+0.30$ V *vs.* Ag/Ag$^+$, 0.1M). For controlled potential electrolysis, a Brinkman–Wenking model 68FF 0.5 electronic potentiostate was used.

Synthesis of Ligands. TETRADENTATE LIGANDS. The ligands 1,4,8,11-tetrathiacyclotetradecane(TTP) and 13,14-benzo-1,4,8,11-tetrathiacyclopentadecane(TTX) were synthesized as reported previously (4).

1,1,1-TRIS(METHYLTHIOMETHYL)ETHANE. In 500 ml of ethanol under nitrogen was dissolved 6.9 g (0.3 mole) freshly cut sodium metal. Methyl mercaptan (Matheson) was bubbled into the solution until no more gas seemed to dissolve (ca. 45 min.). The resulting solution was heated to reflux, and 31 g (0.1 mole) 1,1,1-tris(bromomethyl)ethane prepared by the von Doering and Levy method (6) was added dropwise. The mixture was refluxed an additional 2 hrs, then cooled and filtered. The solvent was removed, leaving a colorless liquid which was distilled (bp 93–95°C/0.5 mm) to yield 15.6 g of the desired product (74%). Anal. calcd for $C_8H_{18}S_3$: C 45.66, H 8.62, S 45.72; found: C 45.62, M 8.61, S 45.60. The mass spectrum has a parent peak $m/e = 210$ (theor., 210). The NMR spectrum has singlets at $\tau = 8.90$ (area = 3), $\tau = 7.85$ (area = 9), and $\tau = 7.33$ (area = 6).

2,6-DITHIAHEPTANE. This compound was previously prepared by another method (7). In 500 ml ethanol under nitrogen was dissolved 13.8 g (0.6 mole) freshly cut sodium metal. To this solution, 32.4 g (0.3 mole) 1,3-propanedithiol (Aldrich) was added dropwise during 10 min, and the mixture was heated to reflux for 30 min. The resulting solution was cooled to room temperature, and 85.1 g (0.6 mole) methyl iodide (Baker) was added dropwise during 20 min. The mixture was then heated to reflux and was refluxed 12 hrs. After cooling to room temperature, the ethanol was removed on a flash evaporator. The resulting oil was filtered to remove the sodium iodide, and it was then distilled (bp 77–78°C/0.5 mm) to yield 20.0 g of the desired product (49%). The mass spectrum has a parent peak at $m/e = 136$ (theor., 136). The NMR spectrum has a quintet at $\tau = 8.90$ (area = 2), a singlet at $\tau = 7.90$ (area = 6), and a triplet at $\tau = 7.42$ (area = 4).

TRICHLORO[1,1,1-TRIS(METHYLTHIOMETHYL)ETHANE]RHODIUM(III). One gram of RhCl$_3$ · 3H$_2$O dissolved in 100 ml hot ethanol was added dropwise for 1 hr to a rapidly refluxing solution of 2 g 1,1,1-tris(methylthiomethyl)ethane in 300 ml ethanol. By the end of this time, the product had begun to precipitate as a light yellow powder. Approximately 200 ml ethanol was removed on a flash evaporator, and diethyl ether was added to complete precipitation of the product. The product was collected on a filter frit, washed with diethyl ether, and dried in a vacuum oven overnight at 60°C. The molar conductance (CH$_3$NO$_2$) is 11 cm^2/ohm mole which is indicative of a nonelectrolyte. The complex is sparingly soluble in methanol, ethanol, acetonitrile, and nitromethane.

trans-DICHLOROBIS(2,6-DITHIAHEPTANE)RHODIUM(III) CHLORIDE. A hot ethanolic solution of 1 g RhCl$_3$ · 3H$_2$O was added dropwise during 1 hr to a refluxing solution of 1.5 g 2,6-dithiaheptane in 200 ml ethanol. The resulting bright orange solution was concentrated to approximately 30 ml; cold diethyl ether was then added until the solution became cloudy. After storage overnight in the refrigerator at 0°C, bright orange crystals of the complex were obtained. The product was collected, washed with diethyl ether, and dried in a vacuum desiccator at room temperature. Crystals of the complex contain water as was demonstrated by the strong OH–OH stretching band in the IR spectrum.

trans-Dichlorobis(2,5-dithiahexane)rhodium(III) Chloride, [Rh-(DTH)$_2$Cl$_2$]Cl. This complex was prepared by the Walton method (*1*) by adding excess 2,5-dithiahexane (K and K) to a refluxing solution of RhCl$_3$ · 3H$_2$O. The initial yellow precipitate of the dimer dissolved leaving a bright yellow solution from which the bright yellow product was isolated by flashing off the ethanol and inducing crystallization by adding diethyl ether. The IR spectrum of the product is identical to that described by Walton.

PREPARATION OF THE RHODIUM(I) COMPLEXES OF 1,4,8,11-TETRATHIACYCLOTETRADECANE. A solution of 0.5 g [Rh(TTP)Cl$_2$]Cl in 50 ml dried, deaerated methanol was prepared in an inert atmosphere glove box or under a purge of nitrogen. To this solution was added 0.25 g (excess) sodium borohydride dissolved in ~ 5 ml methanol. The solution turned immediately from bright yellow to reddish gold. A solution of 2–3 g (excess) of the appropriate salt (LiClO$_4$, LiBF$_4$, NH$_4$PF$_6$, NH$_4$I, NH$_4$SCN or LiBr) dissolved in methanol was then added. With ClO$_4^-$, BF$_4^-$, PF$_6^-$, and I$^-$, bright golden crystals of the complex formed; for the SCN$^-$ and Br$^-$ derivatives, the volume was reduced to about 15 ml and crystallization was induced by the addition of diethyl ether. The product was collected on a filter frit; it was washed with a small amount of dry, peroxide-free, deaerated diethyl ether and dried by pumping on a vacuum line in the glove box.

(13,14-Benzo-1,4,8,11-Tetrathiacyclopentadecane)rhodium(I) Hexafluorophosphate, [Rh(TTX)]PF$_6$. To a solution of 0.5 g [Rh-(TTX)Cl$_2$]Cl · 3.5H$_2$O in 50 ml dry, deaerated methanol was added 0.15 g sodium borohydride dissolved in 5 ml methanol. Excess (2–3 g) NH$_4$PF$_6$ dissolved in 10 ml methanol was then added to the gold solution. The brown-gold product crystallized out of solution and was collected on a filter frit and dried on a vacuum line in the glove box.

Reaction of [Rh(TTP)]$^+$ with Small Covalent Molecules. SULFUR DIOXIDE (1,4,8,11-TETRATHIACYCLOTETRADECANE)RHODIUM(I) PERCHLORATE, [Rh(TTP)SO$_2$]ClO$_4$. To a solution of 0.3 g [Rh(TTP)Cl$_2$]Cl in 30 ml methanol, 0.15 g sodium borohydride in 5 ml methanol was added under a nitrogen purge to give a gold solution containing [Rh(TTP)]$^+$. Sulfur dioxide (C. P. Matheson) was then bubbled through the solution together with the nitrogen purge for 3–5 min. To the resulting bright yellow solution was then added 2 g LiClO$_4$ dissolved in 10 ml methanol. The bright yellow solid which formed immediately was collected on a filter frit under nitrogen and dried in a vacuum desiccator at room temperature.

REACTION OF [Rh(TTP)]$^+$ WITH OXYGEN. A solution of [Rh(TTP)] in methanol was prepared as described above by the reduction of [Rh-(TTP)Cl$_2$]Cl in methanol with sodium borohydride. Upon exposure to air for 3–5 min, the gold solution turned deep blue initially, and then it rapidly became deep red-brown. At this point, LiClO$_4$ dissolved in methanol was added to precipitate the product which was obtained as a brown amorphous solid. The molar conductance (CH$_3$NO$_2$) is 83 cm^2/ohm mole.

TETRACYANOETHYLENE(1,4,8,11-TETRATHIACYCLOTETRADECANE)RHODIUM(I) HEXAFLUOROPHOSPHATE, [Rh(TTP)C$_6$N$_4$]PF$_6$. To a solution of 0.35 g [Rh(TTP)]PF$_6$ in 50 ml of acetonitrile was added 0.1 g tetracyanoethylene (Aldrich) dissolved in 10 ml ocetonitrile. The solution turned immediately from gold to bright orange-yellow, and a bright orange-yellow solid began forming. The volume of the solution was reduced to 20 ml to complete precipitation of the product which was collected, washed with diethyl ether, and dried.

BORON TRIFLUORIDE(1,4,8,11-TETRATHIACYCLOTETRADECANE)RHODIUM(I) HEXAFLUOROPHOSPHATE, [BF$_3$Rh(TTP)]PF$_6 \cdot$ CH$_3$CN. To a solution of 0.3 g [Rh(TTP)]PF$_6$ in 40 ml acetonitrile was added 0.15 ml freshly distilled and deaerated boron trifluoride etherate (Baker). As the solution was stirred 5 min, its color changed from gold to pale yellow. When the volume was reduced to 15 ml, a pale yellow solid began precipitating, and small amounts of diethyl ether were added for 10–15 min to complete precipitation. The product was isolated as a light yellow amorphous powder.

NITROSYL(1,4,8,11-TETRATHIACYCLOTETRADECANE)RHODIUM(III) TETRA-FLUOROBORATE, [Rh(TTP)NO](BF$_4$)$_2$. To a solution of 0.25 g [Rh(TTP)]-BF$_4$ in 40 ml acetonitrile, a solution of 0.1 g (excess) nitrosonium tetra-fluoroborate (Aldrich) dissolved in 5 ml acetonitrile was added drop-wise. During the addition, the solution turned from the golden color of [Rh(TTP)]$^+$ to a bright green. The volume of the solution was then reduced to approximately 10–15 ml, and small aliquots of diethyl ether were added for 30 min. The bright green microcrystalline product was collected, washed with diethyl ether, and dried by pumping on a vacuum line.

ACETYLCHLORO(1,4,8,11-TETRATHIACYCLOTETRADECANE)RHODIUM(III) HEXAFLUOROPHOSPHATE, [CH$_3$CO)Rh(TTP)Cl]PF$_6$. To a solution of 0.3 g [Rh(TTP)]PF$_6$ in 50 ml acetonitrile was added 0.1 ml (0.09 g) acetyl chloride (Mallinckrodt) (excess). The solution turned immediately from gold to pale yellow. The procedure described above was used to isolate a pale yellow microcrystalline solid.

IODOMETHYL(1,4,8,11-TETRATHIACYCLOTETRADECANE)RHODIUM(III) PERCHLORATE, [CH$_3$Rh(TTP)I]ClO$_4$. The addition of 0.15 g sodium borohydride in 5 ml methanol to a solution of 0.3 g [Rh(TTP)Cl$_2$]Cl in 40 ml under nitrogen gave a gold solution containing [Rh(TTP)]$^+$ When 2 ml (excess) freshly distilled methyl iodide (Baker) was added, the solution became yellowish gold; the addition of 2 g LiClO$_4$ dissolved in 10 ml methanol clouded the solution. As the volume was reduced, the product crystallized out as a yellowish gold solid; it was collected on a filter frit and dried in a vacuum desiccator at room temperature.

HYDRIDO(1,4,8,11-TETRATHIACYCLOTETRADECANE)RHODIUM(III) TETRA-FLUOROBORATE, [HRh(TTP)](BR$_4$)$_2$. To a solution of 0.35 g [Rh-(TTP)]BF$_4$ in 40 ml nitromethane was added 0.2 ml 48% aqueous HBF$_4$ (Alfa) from a fresh sample. The solution turned slowly from gold to greenish yellow; its volume was reduced to 10 ml, and small amounts of benzene were added for 15 min to induce crystallization. The product was isolated as a greenish yellow microcrystalline solid. The IR spectrum (Nujol) has a weak band at 2220 cm^{-1} which is assigned to the Rh–H stretching frequency. The molor conductance (CH$_3$CN) is 279 cm^2/ohm mole. (Note: when aqueous HBF$_4$ of indeterminate age was used, ill-defined products were obtained which may have resulted from the oxidation of the rhodium(I) by contaminants such as HF).

σ-ALLYLCHLORO-(OR BROMO)-(1,4,8,11-TETRATHIACYCLOTETRADECANE)-RHODIUM(III) HEXAFLUOROPHOSPHATE, [(σ-C₃H₅)Rh(TTP)Cl]PF₆. To a solution of 0.4 g [Rh(TTP)]PF₆ in 50 ml acetonitrile was added 0.1 ml allyl chloride (or allyl bromide). The solution turned immediately from gold to a very pale yellow. The volume was reduced to 15 ml, and small amounts of diethyl ether were added for 15 min. The product was obtained as an off-white powder which could be recrystallized from acetonitrile–ether solutions.

trans-DICHLORO(1,4,8,11-TETRATHIACYCLOTETRADECANE)RHODIUM(III) HEXAFLUOROPHOSPHATE, [Rh(TTP)Cl₂]PF₆. An excess of a saturated solution of chlorine in acetonitrile was added to a solution of 0.35 g [Rh(TTP)]PF₆ in 50 ml acetonitrile. As the solution was stirred at room temperature for 45 min, its color changed from gold to bright yellow. The excess chlorine was removed by pumping on a vacuum line, and the volume of the solution was reduced to 15 ml. Bright yellow microcrystals of the product formed when diethyl ether was added slowly. The product was collected, washed with diethyl ether, and dried by pumping on a vacuum line.

BENZOYLCHLORO(1,4,8,11-TETRATHIACYCLOTETRADECANE)RHODIUM(III) HEXAFLUOROPHOSPHATE, [C₆H₅CO)Rh(TTP)Cl]PF₆. Freshly distilled benzoyl chloride (Baker) (0.1 ml, 0.11 g) was added dropwise to a solution of 0.3 g [Rh(TTP)]PF₆ in 50 ml acetonitrile. As the solution was stirred for 5 min, its color changed from gold to pale yellow. After the volume was reduced to 15 ml, small amounts of diethyl ether were added to crystallize the product. The product was obtained as a pale yellow amorphous powder which could be recrystallized from acetonitrile by the slow addition of diethyl ether.

trans-DIIODO(1,4,8,11-TETRATHIACYCLOTETRADECANE)RHODIUM(III) HEXAFLUOROPHOSPHATE, [Rh(TTP)I₂]PF₆. To a solution of 0.4 g [Rh-(TTP)]PF₆ in 40 ml acetonitrile was added 0.1 g iodine dissolved in 10 ml acetonitrile. The solution was stirred 6 hrs at room temperature while its color turned gradually from gold to bright orange. Reducing the volume of the solution to 10–15 ml and adding small aliquots of diethyl ether for 30 min gave bright orange crystals of the desired product. The product was collected, washed with diethyl ether, and dried as described previously.

BENZYLBROMO(1,4,8,11-TETRATHIACYCLOTETRADECANE)RHODIUM(III) HEXAFLUOROPHOSPHATE, [C₆H₅CH₂Rh(TTP)Br]PF₆. Benzyl bromide (0.11 ml, 0.15 g) was added dropwise to a solution of 0.35 g [Rh(TTP)]-PF₆ in 40 ml acetonitrile. The color of the solution changed immediately from gold to bright yellow. Reducing the volume to 10 ml and slowly adding small portions of diethyl ether yielded bright yellow crystals of the desired complex which were collected, washed with diethyl ether, and dried as described previously.

REACTION OF [Rh(TTP)PF₆ WITH CH₃SO₂Cl. Methane sulfonyl chloride (Baker) was freshly distilled before use. To a solution of 0.3 g [Rh(TTP)]PF₆ in 50 ml acetonitrile was added 0.2 ml CH₃SO₂Cl. The color of the solution changed immediately from gold to a very deep red. The volume was reduced to 10 ml, and small amounts of benzene were added until the solution became cloudy. When the solution was allowed to sit 20 min, a small amount of a dark red-brown powder formed. The powder was collected, washed with benzene, and dried on a vacuum line. The IR spectrum of the powder has no bands attributable to the SO_2 group. The ESR spectrum of the initial deep red solution indicates no paramagnetic species. Analysis of the product is most consistent with a dichloro-rhodium(III) complex. Anal.: calcd for $[Rh(C_{10}H_{20}S_4)Cl_2]$-$PF_6CH_3CN$: C 22.93, H 3.68, S 20.41, Rh 16.38; found: C 22.85, H 3.94, S 20.56, Rh 16.38.

REACTION OF $[Rh(TTP)]^+$ WITH H_2. Hydrogen (C. P. Burdett) was bubbled vigorouly through a solution of 0.45 g $[Rh(TTP)]PF_6$ in 50 ml acetonitrile for 30 min; there was no color change. The IR spectrum of the solution has no bands that are attributable to a Rh–H stretching frequency. When hydrogen was bubbled through a solution of $[Rh-(TTP)]PF_6$ for 4–5 hrs, the color gradually became bluish because of the formation of the Rh(II) species, $[Rh(TTP)]^{+2}$, presumably as the result of traces of oxygen impurities in the hydrogen. This reaction was not investigated further. The Rh(II) species are the subject of a separate report.

CATALYTIC EXPERIMENTS WITH $[Rh(TTP)]^+$. An attempt was made to catalyze the hydrogenation of maleic acid with $[Rh(TTP)]BF_4$ using reported (literature) conditions. A solution of 0.171 g $[Rh(TTP)]BF_4$ (0.0075M) and 0.290 g maleic acid (0.05M) in 50 ml dimethylacetamide was heated to 60°C, and then hydrogen was bubbled vigorously through the solution 1 hr at either 60° or *ca.* 25°C. During this time, no color change was observed. The volume was reduced to 10 ml by distilling off the DMA, and an NMR spectrum was obtained; no traces of succinic acid are detected in the NMR spectrum.

Results and Discussion

Formation and Characterization of the Rh(I) Complexes. The extreme difference between the behavior of macrocyclic ligand derivatives $[Rh(TTP)Cl_2]Cl$ and $[Rh(TTX)Cl_2]Cl$ and that of rhodium complexes containing acyclic chelate ligands deserves additional emphasis. We confirmed Walton's observations that both chemical (sodium borohydride) and electrochemical reduction of $[Rh(DTH)_2Cl_2]Cl$ invariably leads to mixtures of the free ligand, some colloidal rhodium metal, and an unidentified dark brown species. To obviate the possibility that the ethylene bridge in DTH provides too small a bite for rhodium, thereby producing only weak chelation, similar experiments were conducted with CH_3SCH_2-$CH_2CH_2SCH_3$. However, the reduction of the corresponding Rh(III) complex with sodium borohydride in methanol immediately produced a black precipitate of rhodium metal and free ligand. The result was the same when reduction was attempted with $[Rh\{1,1,1$-tris(methylthio-methyl)ethane$\}Cl_3]$. In contrast, $[Rh(TTP)Cl_2]Cl$ and $[Rh(TTX)Cl_2]$-Cl can be reduced either electrochemically or chemically (sodium borohydride), forming the rhodium(I) species $Rh(TTP)^+$ and $Rh(TTX)^+$, respectively. With the macrocyclic ligand derivatives, we conclude that the kinetic inertness toward substitution of the Rh–S bonds that derives

Table I. Elemental Analyses

Complex	C	H	S	X
[Rh(C$_8$H$_{18}$S$_3$)Cl$_3$]	22.89	4.32	22.92	
[Rh(C$_5$H$_{12}$S$_2$)$_2$Cl$_2$]Cl · 2.5H$_2$O	22.79	5.17	24.35	
[Rh(TTP)]ClO$_4$	25.51	4.28	27.24	7.53(Cl)
[Rh(TTP)]BF$_4$	26.21	4.40	27.99	
[Rh(TTP)]PF$_6$	23.26	3.90	24.84	
[Rh(TTP)]I	24.10	4.05	25.74	25.47(I)
[Rh(TTP)]SCN	30.76	4.69	37.33	
[Rh(TTP)]Br	26.61	4.47	28.42	
[Rh(TTX)]PF$_6$	31.14	3.83	22.17	
[Rh(TTP)SO$_2$]ClO$_4$	22.45	3.77	29.97	17.95(O)
[Rh(TTP)O$_2$]ClO$_4$	23.83	4.01	25.50	19.09(O)
[Rh(TTP)C$_6$N$_4$]PF$_6$	29.83	3.13	19.90	8.70(N)
[BF$_3$Rh(TTP)]PF$_6$ · CH$_3$CN	23.05	3.71	20.51	
[Rh(TTP)NO](BF$_4$)$_2$	20.88	3.51	22.30	2.43(N)
[(CH$_3$CO)Rh(TTP)Cl]PF$_6$	24.23	3.90	21.56	5.96(Cl)
[CH$_3$Rh(TTP)I]ClO$_4$	21.56	3.78	20.93	
[HRh(TTP)](BF$_4$)$_2$	22.00	3.88	23.49	18.85(Rh)
[C$_3$H$_5$Rh(TTP)Br]PF$_6$	24.50	3.95	20.12	12.54(Br)
[C$_3$H$_5$Rh(TTP)Cl]PF$_6$	26.24	4.37	21.56	5.96(Cl)
[Rh(TTP)Cl$_2$]PF$_6$	20.44	3.43	21.84	12.07(Cl)
[C$_6$H$_5$CH$_2$Rh(TTP)Br]PF$_6$	29.70	3.96	18.66	11.62(Br)
[C$_6$H$_5$CORh(TTP)Cl]PF$_6$	31.08	3.84	19.52	5.40(Cl)
[Rh(TTP)I$_2$]PF$_6$	15.59	2.62	16.65	32.96(I)

from the structure of the ligand (8, 9) is responsible for the fact that these Rh(I) complexes can be isolated.

Synthesis of the rhodium(I) complexes of TTP and TTX is not as straightforward as, for example, that of the Rh(I) complexes of phosphines. Substitution reactions, such as the refluxing of [Rh(1,5-cyclooctadiene)$_2$Cl]$_2$ with TTP, generally failed. Also the Rh(III) in RhCl$_3$ · 3H$_2$O is not simultaneously reduced and chelated by TTP (10). The only successful routes that we found involve the reduction of the previously characterized Rh(III) complexes. Voltammetry in acetonitrile solution on [Rh(TTP)Cl$_2$]Cl showed a single cathodic process at −0.72 V (vs. Ag$^+$/AgCl) which represents a two-electron reduction. Small samples of the Rh(I) salt were prepared by electrolysis, but reduction with sodium borohydride in methanol was used to prepare most of the salts reported here.

The addition of various salts to the reduced solutions led to isolation of the series of compounds Rh(TTP)X (where X = B(C$_6$H$_5$)$_4^-$, PF$_6^-$, ClO$_4^-$, Br$^-$, I$^-$, and SCN$^-$) and Rh(TTX)PF$_6$. Elemental analyses are reported in Table I. The molar conductances are all consistent with

of the New Rhodium Complexes

Found, %

	C	H	S	X	Yield, %
	22.86	4.24	22.46		76
	22.87	4.80	24.39		68
a)	25.40	4.24	27.16		60
b)	25.35	4.19	26.99	7.55 (Cl)	
	25.98	4.20	27.89		55
	23.18	3.95	24.57		65
	23.85	3.86	25.52	25.35 (I)	42
	30.83	4.75	36.97		31
	26.11	4.32	27.99		30
	30.88	3.79	21.95		51
	22.38	3.74	30.91	17.84 (O)	63
	23.39	3.93	24.31	20.21 (O)	
	30.03	2.96	20.05	8.81 (N)	71
	23.39	4.08	20.70		55
	20.51	3.59	22.39	2.27 (N)	
	24.14	3.72	21.63	5.99 (Cl)	
	21.35	3.54	21.58		
	21.97	3.77	23.32	19.09 (Rh)	
	24.49	3.98	20.07	12.59 (Br)	
	26.51	4.53	22.42	5.95 (Cl)	
	20.41	3.36	21.63	11.99 (Cl)	
	29.66	3.94	18.52	11.69 (Br)	
	31.76	3.97	19.80	5.36 (Cl)	
	15.46	2.54	16.55	33.24 (I)	

formulation as uni-univalent electrolytes (Table II) (*11, 12*). The IR spectra of the salts [Rh(TTP)]X and [Rh(TTX)]PF$_6$ resemble those of the Rh(III) complexes (*4*), which confirms the presence of the essentially unaltered ligand. The main variations are attributable to the different anions present and to the fact that the anions remain uncoordinated. Uncoordinated BF$_4^-$ has strong bands at 1060 and 520 cm^{-1}, ClO$_4^-$ at 1086 and 620 cm^{-1}, and PF$_6^-$ at 850 and 555 cm^{-1}. The spectrum of [Rh-(TTP)]SCN shows ν_{CN} at 2040 cm^{-1} which is near that for KSCN (2053

Table II. Molar Conductances of the Rh(I) Complexes

Complex	Solvent	λ_M, cm^2/ohm mole
[Rh(TTP)][B(C$_6$H$_5$)$_4$]	CH$_3$NO$_2$	86
[Rh(TTP)]BF$_4$	CH$_3$CN	135
[Rh(TTP)]PF$_6$	CH$_3$CN	140
[Rh(TTP)]ClO$_4$	CH$_3$CN	147
[Rh(TTP)]Br	CH$_3$CN	132
[Rh(TTP)]I	CH$_3$CN	130
[Rh(TTP)]SCN	CH$_3$CN	165
[Rh(TTX)]PF$_6$	CH$_3$CN	150

cm^{-1}) and farther in frequency from N-bonded NCS ($\sim$ 2065–2100 cm^{-1}) and S-bonded NCS ($\sim$ 2100–2127 cm^{-1}). This supports the conclusion that $Rh(TTP)^+$ does not tend to expand its coordination number to five by coordinating to its counter ion.

The electronic spectra of the Rh(I) complexes are listed in Table III together with those of two phosphine complexes. According to the interpretations of Gray and co-workers (13), the pertinent ordering of energy

Table III. Electronic Spectra of Some Square-Planar Rhodium(I) Complexes

Complex	Solvent	λ_{max}, nm	ϵ
[Rh(TTP)]PF$_6$	CH$_3$CN	355 sh	1077
		318 sh	3736
		293 sh	2418
[Rh(TTP)]I	CH$_3$CN	352 sh	985
		315 sh	3360
		292	6512
[Rh(TTX)]PF$_6$	CH$_3$CN	357 sh	907
		324 sh	3174
		295	6009
[Rh(2=phos)$_2$]B(C$_6$H$_5$)$_4$[a]	chlorobenzene	405	6160
		341	6650
		313	10,900
[Rh(2=phos)$_2$]Cl[a]	EPA[c]	474	250
		405	4900
		341	4900
		309	8400
[Rh(2-phos)$_2$]Cl[b]	EPA	470	150
		405	5150
		314	9300
		296	8400

[a] 2=phos is cis-[(C$_6$H$_5$)$_2$PCH=CHP(C$_6$H$_5$)$_2$] (14).
[b] 2-phos is (C$_6$H$_5$)$_2$PCH$_2$CH$_2$P(C$_6$H$_5$)$_2$ (13).
[c] EPA is a 5:5:2 mixture of diethyl ether–isopentane–ethyl alcohol.

levels is probably $b_{2_g(xy)} < e_{g(xz,yz)} < a_{1_g(z^2)} < a_{2_u}\pi < b_{1_g(x^2-y^2)}$, and the ground state configuration is $b_{2_g}{}^2 e_g{}^4 a_{1_g}{}^2$. The lowest vacant orbital is $a_{2_u}\pi$ which is derived from vacant ligand orbitals with π symmetry. Thus, the spectral bands represent metal-to-ligand charge transfer, and it is not surprising that the lowest energy bands for $Rh(TTP)^+$ occur at energies higher than those for the Rh(I) phosphine complexes.

Using a different ordering of d levels, Vaska et al. (14) tentatively assigned the first transition observed in the spectrum of [Rh(2=phos)$_2$]-B(C$_6$H$_5$)$_4$ at 405 nm as the $xy \rightarrow x^2 - y^2$ transition. The spectra of the analogous complexes [Co(2=phos)$_2$]B(C$_6$H$_5$)$_4$ and [Ir(2=phos)$_2$]B-(C$_6$H$_5$)$_4$ have this band at 730 and 525 nm, respectively. This transition is equal to 10 Dq, and thus the energy of the transition is a measure of the ligand field stabilization energy. Since the reactivity of the Co(I), Rh(I), and Ir(I) complexes toward oxygen and hydrogen decreased as the energy of this band increased, Vaska suggested that a correlation exists and that the electronic spectra of such square-planar d^8 complexes may aid in predicting their related reactivities. However, this model would assign a very high ligand field strength to TTP and predict that thioether complexes of Rh(I) would be more stable and less reactive than the corresponding phosphine complexes. This is contrary to experience.

Reactivity of Rh(TTP)⁺ toward Nucleophiles. The new Rh(I) complexes show no tendency to expand their coordination numbers by adding Lewis bases. Nitrogen bases, triphenyl phosphine, carbon monoxide, and ethylene all fail to react with Rh(TTP)⁺. With ethylene, the absence of interaction was confirmed by solution IR spectral measurements. Although CO and C_2H_4 owe much of their coordinating abilities to back bonding, they must still be considered to be moderate σ donors. The failure of electron-pair donors to interact with Rh(TTP)⁺ is a reflection of the very great electron density residing on the Rh(I) atom, a structural property that is well illustrated by the contrasting great reactivity of Rh(TTP)⁺ toward electrophiles.

Reactions of Rh(TTP)⁺ with Electrophiles. The most characteristic reaction of Rh(TTP)⁺ is its addition to electrophiles. It combines with H⁺, NO⁺, BF_3, SO_2, tetracyanoethylene, and O_2 to form a series of well characterized products. The variations in reactivity toward such molecules as SO_2, O_2, and BF_3 that is observed among the many known square-planar d^8 complexes were attributed to the following factors (15): (a) the electron affinity of the covalent molecule, (b) the nucleophilicity of the metal in the complex, and (c) the ability of d orbitals on the metal to overlap effectively with suitable orbitals on the electrophile. Molecules such as O_2 and SO_2 are most appropriately viewed as π acids of considerable electron affinity, although they do have some ability to act as σ donors, and also as Lewis (σ-bonding) acids.

Solutions of [Rh(TTP)]PF₆ in acetonitrile react readily with boron trifluoride etherate to give the adduct [F₃BRh(TTP)]PF₆ · CH₃CN which was isolated as a light yellow solid. In addition to the usual bands for [Rh(TTP)]PF₆, the IR spectrum of the adduct has bands at 1055 (strong), 520 (moderate), and 2290 cm⁻¹ (weak). The bands at 1055 and 520 cm⁻¹ are consistent with the existence of the F₃BRh structural unit (16). The absorption at 2290 cm⁻¹ is assignable to ν_{CN} for CH₃CN. Since this band occurs at 2278 cm⁻¹ in pure acetonitrile and at lower frequencies among its complexes, the CH₃CN is assumed to be uncoordinated. The molar conductance in acetonitrile of [F₃BRh(TTP)]PF₆ is 132 cm²/ohm mole which indicates that the substance is a uni-univalent electrolyte (11). Thus the adduct is assigned a five-coordinate structure, and it probabbly exists as a distorted tetragonal pyramid with the sulfur atoms of the TTP forming the base and the BF₃ group the apex.

The formation of the stable adduct with the Lewis acid BF₃ established the enhanced basicity of the Rh(I) in Rh(TTP)⁺ over that of the previously known Rh(I)–phosphine complexes. Although [Ir(PPh₃)-(CO)Cl] adds BF₃, the rhodium analog does not (17). A stronger Lewis acid, e.g. BBr₃ or BCl₃, is required for an observable interaction with [Rh(PPh₃)(CO)Cl] (18). Indeed, the only other Rh(I) complex known to form a stable BF₃ adduct is chloro-bis(3-diphenylphosphino-propyl)phenylphosphine rhodium(I) (19). The enhanced nucleophilicity of the rhodium in [Rh(TTP)]⁺ is considered as evidence of the poor π-acceptor qualities of the sulfur atoms in the thioether ligand as compared with those of the phosphorus atoms in their similar complexes.

The bright yellow complex [Rh(TTP)SO₂]ClO₄ was isolated by bubbling SO₂ through a solution of Rh(TTP)⁺ in methanol, followed by the addition of LiClO₄. Its molar conductance (CH₃NO₂) is 90 cm²/ohm mole which indicates that the salt is a uni-univalent electrolyte. The symmetric and antisymmetric vibrational frequencies for the S-bonded ligand in this complex occur at ν_{sym}, 1032 cm⁻¹ and ν_{asym} 1165 cm⁻¹. These agree well with those observed for other S-bonded SO₂ complexes (19, 20, 21). The Lewis acid character of SO₂ in its adducts of this class is evident in the x-ray structure of [IrCl(SO₂)(CO)(PPh₃)₂] as determined by LaPlaca and Ibers (22). It reveals that the SO₂ ligand is bonded to the metal via the sulfur atom. The metal is coordinated in a tetragonal pyramidal fashion with the SO₂ at the apex. The geometry about the sulfur atom of the SO₂ is tetrahedral and may be ascribed to the donation by the metal of a pair of electrons to an approximately sp^3 hybridized vacant sulfur orbital.

Tetracyanoethylene (TCNE) is a very strong π acid so it is not surprising that the addition of TCNE to an acetonitrile solution of Rh(TTP)⁺ leads to the very rapid formation of the orange-yellow adduct [(TCNE)Rh(TTP)]⁺. The molar conductance of its PF₆⁻ salt in CH₃NO₂ is 93 cm²/ohm mole (uni-univalent electrolyte), and its IR spectrum has a sharp band at 2210 cm⁻¹ with a shoulder at 2195 cm⁻¹ that is attributable to the C≡N groups of TCNE. There is a similar band in the IR spectrum of [(TCNE)Ir(PPh₃)₂(CO)Cl] (23, 24). It is again assumed that the electrophilic ligand binds in the apical site in the pseudotetragonal-pyramidal array of donors about the Rh(I).

The light green nitrosyl complex [Rh(TTP)NO](BF₄)₂ is readily prepared by adding an acetonitrile solution of NOBF₄ to [Rh(TTP)]BF₄ dissolved in acetonitrile. The molar conductance of this compound (245 cm^2/ohm mole in CH$_3$CN) confirms that it is a di-univalent electrolyte. Its IR spectrum has an NO stretching mode at 1750 cm^{-1} which is comparable to that of other related nitrosyl complexes—*e.g.* ν_{NO} = 1680 cm^{-1} for [IrCl(CO)(PPh$_3$)$_2$(NO)]$^+$ (*25*), 1699 cm^{-1} for [Rh(NO){PhP-[(CH$_2$)$_3$PPh$_3$]$_2$}Cl]BF$_4$ (*19*), and 1560 cm^{-1} for [IrCl$_2$(NO)(PPh$_3$)$_2$] (*26*). These three compounds all have bent M–N–O groupings (*26, 27*). The compounds Fe(CO)$_2$(NO)$_2$ (ν_{NO}=1810, 1766) and Mn(CO)(NO)$_3$ (ν_{NO} = 1823, 1734) have linear M–N–O groupings (*28*). It is suggested that [Rh(TPP)]$^+$ donates an electron pair to NO$^+$ and forms a complex in which the bent NO group occupies the axial metal ion site.

The Lewis base behavior of Rh(I) in Rh(TTP)$^+$ suggests that the complex might combine with the proton; this was demonstrated. The addition of a 48% aqueous solution of HBF$_4$ from a freshly opened bottle to a solution of [Rh(TTP)]BF$_4$ in nitromethane produced the five-coordinate adduct [HRh(TTP)]$^{2+}$ which was isolated as the yellow-green tetrafluoroborate salt [HRh(TTP)](BF$_4$)$_2$. The IR spectrum of the salt has a weak band at 2220 cm^{-1} which is assignable to the Rh–H stretching vibration. The Rh–H mode in [HRhCl$_2$(PPh$_3$)$_3$]O · 5CH$_2$Cl$_2$ occurs at 2105 cm^{-1} (*29*), that in [HRh{PhP[(CH$_2$)$_3$PPh$_2$]$_2$}Cl$_2$] at 2195 cm^{-1} (*19*). The molar conductance of the protonated material in CH$_3$CN is 279 cm^2/ohm mole which is indicative of a di-univalent electrolyte.

The most widely studied adducts of square-planar d^8 complexes are those with the oxygen molecule O$_2$ (*see* References *30* and *31* for review). In these derivatives, the two oxygen atoms are bound equivalently to the metal atom, and the bonding scheme is often compared to that proposed for ethylene adducts. Presumably, the O$_2$ donates a pair of electrons to the bond, but the metal also donates strongly to the bond. When the metal atom can back donate very strongly, irreversible O$_2$ binding is expected.

When freshly reduced solutions containing Rh(TTP)$^+$ are exposed to air, the bronze-gold color changes to brown. The addition of sodium tetraphenylborate leads to the isolation of a brown diamagnetic salt with the stoichiometry [Rh(TTP)O$_2$](BPh$_4$). Careful study of the IR spectra of the product and related materials leads to the assignment of a band at 845 cm^{-1} to the RhO$_2$ group. As others suggested (*32, 33*), the band is probably not purely an O–O stretching mode, but rather it is combined with an M–O stretching motion. As is true of many of the O$_2$ adducts, [Rh(TTP)O$_2$]$^+$ can be reduced back to Rh(TTP)$^+$; however, the O$_2$ is not removed merely by heating or lowering the partial pressure of O$_2$. (In the course of studies of Rh(II) complexes formed with the ligands of interest here, we discovered that Rh^{II}(TTP)$^+$ is formed as an intermediate in the O$_2$ oxidation of Rh(TTP)$^+$. Still more remarkably, Rh^{II}(TTP)$^+$ forms a 1:1 O$_2$ adduct that is closely analogous to the cobalt(II) adducts with O_2. These matters are the subject of a separate publication.)

Oxidative Addition Reactions of Rh(TTP)⁺. This material is so organized that the term oxidative addition may be used in a somewhat restricted sense. Inasmuch as both nucleophilic and electrophilic addition were already considered, the reactions that were reserved for this section on oxidative addition involve a simultaneous real change in coordination number from four to six and a formal change in the oxidation state of Rh from I to III.

Rh(TTP)⁺ reacts with alkyl halides, acyl halides, aroyl halides, and sulfonyl halides, but it shows no evidence of reaction with molecular hydrogen. These observations further emphasize the fact that Rh(TTP)⁺ is essentially a nucleophile and it therefore reacts with those reagents RX that can oxidatively add by nucleophilic attack (34). Rh(TTP)⁺ does not react with H_2, and H_2 seems always to add to d^8 complexes *via* a concerted mechanism (35). It appears that Rh(TTP)⁺ has very little diradical character, *i.e.* it is not a good analog of a carbene (35). It is possible that this unreactivity may be associated with the stereochemistry of chelation by the macrocyclic ligand. Earlier studies on the oxidative addition reactions of Rh(I) complex with a tetraaza macrocycle revealed that the Rh(I) had strong nucleophilic properties but the activation of molecular H_2 was not reported (36, 37). This possibility is supported by reports that dialkyl sulfide complexes of rhodium chloride catalyze the hydrogenation of olefins (38).

In some ways, the simplest and least interesting oxidative addition reactions are those with halogens. However, there was a rather interesting and useful finding in the reaction of Rh(TTP)⁺ with Cl_2—the product was *trans*-Rh(TTP)Cl_2^+. Earlier studies (4), in which the Rh(III) complexes of TTP were prepared by substitution reactions, yielded only *cis*-Rh(TTP)Cl_2^+. Thus, as the result of two very different preparative routes, the cis and trans isomers are both available. The NMR spectra are second order and difficult to obtain because of solubility problems. They are sufficiently well resolved to show that the isomers are different (Figure 1). The far-IR and electronic spectra of the compounds provide support for the isomeric structural assignments.

The far-IR spectrum of *cis*-[Rh(TTP)Cl_2]PF_6 has two moderately strong bands at 307 and 290 cm⁻¹ that are assignable to the Rh–Cl stretching vibrations predicted for a *cis*-diacido-ML_4X_2 complex (4). The far-IR spectrum of the product obtained by oxidative addition of chlorine to Rh(TTP)⁺ has a single strong band at 360 cm⁻¹ that is assignable to an Rh–Cl stretching vibration. Similarly, a single band at 362 cm⁻¹ was assigned to the lone Rh–Cl stretch, and Walton (1) considered this evidence of the trans structure of [Rh(DTH)$_2$Cl$_2$]Cl.

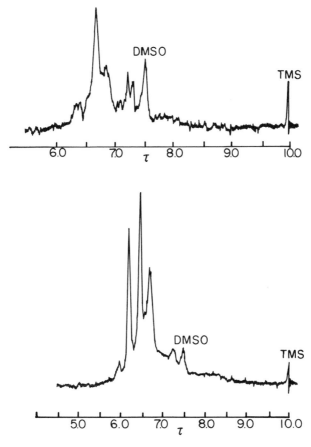

Figure 1. Proton NMR spectra of cis-[Rh(TTP)-
Cl$_2$]PF$_6$ *(top) and of* trans-[Rh(TTP)Cl$_2$]PF$_6$ *(bottom)*

As was mentioned in the report on *cis*-[Rh(TTP)Cl$_2$]Cl (4), the
electronic spectra of *trans*-ML$_4$X$_2$ isomers exhibit transitions at lower
energies and lesser intensities than do those of the cis isomers because of
a greater distortion in the ligand field of the cis isomers. The visible
spectra of *cis*- and *trans*-[Rh(TTP)Cl$_2$]PF$_6$ are presented in Figure 2.
Thus, the electronic spectrum of *trans*-[Rh(TTP)Cl$_2$]PF$_6$ has absorptions
at 426 ($\epsilon = 291$), 340 (sh, $\epsilon = 582$), and 260 nm ($\epsilon = 36,683$) whereas
trans-[Rh(cyclam)Cl$_2$]$^+$ (cyclam = 1,4,8,11-tetraazacyclotetradecane)
has absorptions at 406 ($\epsilon = 78$), 310 (sh, $\epsilon = 80$), 242 (sh, $\epsilon = 3300$),
and 204 nm ($\epsilon = 37,100$) (39). With the symmetry approximated as D_{4h},
the first two bands can be assigned to the *d–d* transitions $^1E_g \leftarrow {}^1A_1$ and
$^1A_2 \leftarrow {}^1A_1$; the positions of the second bands reveal that the ligand field
strength of TTP is noticeably but not a great deal smaller than that of

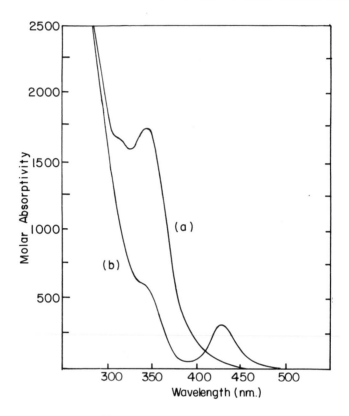

Figure 2.   Visible spectra of (a) cis-$[Rh(TTP)Cl_2]PF_6$ *and
(b)* trans-$[Rh(TTP)Cl_2]PF_6$

cyclam. In contrast, the electronic spectrum of *cis*-[Rh(TTP)Cl$_2$]PF$_6$ has bands at 352 ($\epsilon = 1741$), 315 (sh, $\epsilon = 1692$), and 248 nm ($\epsilon = 18,582$) whereas the bands of *cis*-[Rh(cyclam)Cl$_2$]$^+$ are at 354 ($\epsilon = 223$), 299 ($\epsilon = 308$), and 207 nm ($\epsilon = 33,900$) (*39*). Thus, the generalization applies for each pair of cis or trans isomers.

trans-[Rh(TTP)I$_2$]PF$_6$ was also prepared by oxidative addition. It is a bright orange solid that has the conductance of a 1:1 electrolyte (152 cm^2/ohm mole in CH$_3$CN). Its electronic spectrum has bands at 450 ($\epsilon = 537$), 330 (sh, $\epsilon = 2150$), and 258 nm ($\epsilon = 37,000$) whereas that of *trans*-[Rh(cyclam)I$_2$]$^+$ has absorptions at 515 (sh, $\epsilon = 64$), 466 ($\epsilon = 204$), 353 ($\epsilon = 13,100$), 275 ($\epsilon = 34,500$), and 266 nm ($\epsilon = 22,800$) (*39*). The previously reported *cis*-[Rh(TTP)I$_2$]$^+$ has absorptions at 405 ($\epsilon = 2460$), 320 ($\epsilon = 8550$), and 245 nm ($\epsilon = 22,500$) (*4*).

Both four- and five-coordinate d^8 complexes react with alkyl and acyl halides, but only the most reactive alkylating agents and d^8 complexes generally form stable adducts (*40*). Thus, Rh(TTP)$^+$ and CH$_3$I

react in methanol solution to form the expected adduct $[CH_3Rh(TTP)I]$-
ClO_4. The molar conductance (169 cm²/ohm mole) reveals that the
substance is a 1:1 electrolyte and confirms the coordination of the I⁻ as
well as the CH_3 group. The proton NMR spectrum of this compound in
d_6-DMSO (Figure 3) has a doublet at 8.95 τ that is assignable to the
methyl group bound to rhodium with $J_{103Rh-H} = 2.0$ Hz. A similar doublet
was observed for $RhCl(I)(CH_3)(PPh_3)_2CH_3I$ at 7.12 τ and $J_{103Rh-H} = 2.4$
Hz. The portion of the NMR spectrum of $[CH_3Rh(TTP)I]^+$ that is associated with the TTP protons resembles more closely the corresponding
spectrum for *trans*-$[Rh(TTP)Cl_2]^+$ than that of *cis*-$[Rh(TTP)Cl_2]^+$; this
therefore suggests that the CH_3 and I are trans in this complex.

   Allyl bromide reacts with $Rh(TTP)^+$ in acetonitrile to form bright
yellow $[(\sigma-C_3H_5)Rh(TTP)Br]PF_6$. Its molar conductance (160 cm²/
ohm mole) confirms the electrolyte type (1:1), and the IR spectrum has
a weak but well defined band at 1618 cm⁻¹ that is assignable (*41*) to the
C=C stretching vibration of a σ-bonded allyl group. In addition, the
band typical of π allylic structures in the 500–520 cm⁻¹ region is not
observed. Limited solubility prevented NMR studies.

   The reaction of $Rh(TTP)^+$ with benzyl bromide in acetonitrile yields
the bright yellow 1:1 electrolyte $[(C_6H_5CH_2)Rh(TTP)Br]PF_6$ ($\lambda_M =$
148 cm²/ohm mole, CH_3CN). The presence of the phenyl group is
apparent in the IR spectrum. The proton NMR spectrum has, in addition

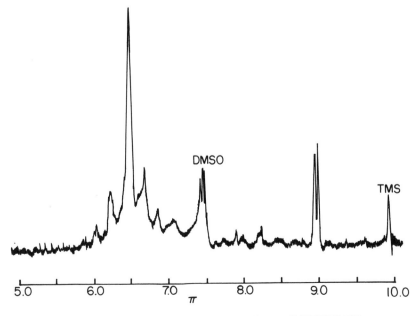

Figure 3.   Proton NMR spectrum of $[CH_3Rh(TTP)I]ClO_4$

to the usual pattern attributable to TTP protons in the 5.8–7.2 τ region, a sharp apparent singlet with some fine structure at 2.59 τ that is assignable to the phenyl protons and a poorly resolved resonance at 8.05 τ that is attributable to the CH_2 protons. The reaction, at reflux, of benzyl chloride with $Rh(PPh_3)_3Cl$ in dichloromethane leads to a π allylic structure (III) (42). Structure III was assigned on the basis of the complexity of

$$Ph_3P\!\!-\!\!-\!\!-\!\!Rh\!\!-\!\!\cdots$$

with Cl above and Cl below, and a phenyl ring to the right.

III

the phenyl proton resonances. The relatively simple spectrum of $[(C_6H_5\text{-}CH_2)Rh(TTP)Br]PF_6$ eliminates III as a possible structure for this compound.

Pale yellow $[CH_3C\!\!\overset{O}{\diagup}\!\!-Rh(TTP)Cl]PF_6$ is formed by the reaction

of $Rh(TTP)^+$ with $CH_3C\overset{\diagup O}{\diagdown Cl}$ ($\lambda_M = 155$ cm^2/ohm mole, CH_3CN, 1:1

electrolyte). The IR spectrum has the strong C=O absorption at 1681 cm^{-1} with a shoulder at 1655 cm^{-1}. Similar bands appear in the IR spectra

of related compounds—$[CH_3C\!\!\overset{O}{\diagup}\!\!-IrCl_2(CO)(PEt_2Ph)_2]$, $\nu_{C=O} = 1639$

cm^{-1} (43) and $[CH_3C\!\!\overset{O}{\diagup}\!\!-Rh\{PhP[(CH_2)_3PPh_2]_2\}Cl_2]$, $\lambda_{C=O} = 1704$ cm^{-1}

(19). In addition, there is a band of medium intensity in the IR spectrum

of $[CH_3C\!\!\overset{O}{\diagup}\!\!-Rh(TTP)Cl]PF_6$ at 585 cm^{-1}. Such bands were tentatively assigned to an M–C–O bending mode. In addition to the usual TTP proton resonances, the NMR spectrum of this compound has a sharp

singlet at 7.48 τ that is assignable to the methyl protons of the $CH_3C\overset{\diagup O}{\diagdown}$

group. For the complexes $[CH_3C\!\!\overset{O}{\diagup}\!\!-IrXY(CO)(PMe_2Ph)_2]$ where X and

Y may be Cl and/or Br, this resonance occurs as a singlet at 8.29–8.42 τ (*44*).

Benzoyl chloride gives a similar product $[C_6H_5C\overset{O}{=}Rh(TTP)Cl]PF_6$ (pale yellow solid; $\lambda_M = 131$ cm²/ohm mole; 1:1 electrolyte; $\nu_{C=O} = 1655$, sh at 1635 cm⁻¹). Related compounds have $\nu_{C=O}$ at 1650–1700 cm⁻¹ (*29, 44*). In addition, a band at 653 cm⁻¹ is assignable to an M–C–O mode (*45*). The NMR spectrum has poorly resolved phenyl resonances in the 2.65–2.9 τ region in addition to the TTP resonances.

Attempts to obtain CH_3SO_2 adducts by the reaction of $Rh(TTP)^+$ with CH_3SO_2Cl indicated that the chlorine atom had been abstracted to form $Rh(TTP)Cl_2^+$. Collman and co-workers (*46*) observed this kind of reaction earlier, and they found that Cl · abstraction by the metal liberated RSO_2 · radicals that dimerized to form a disulfone. $Rh(TTP)^+$ tends to undergo one-electron oxidation under the influence of a range of reagents, thereby yielding an intricate series of rhodium(II) complexes. These are the subject of a separate study.

Acknowledgment

The mass spectra were obtained by Richard Weisenberger.

Literature Cited

1. Walton, R. A., *J. Chem. Soc. A* (1967) 1852.
2. Chatt, J., Shaw, B. L., *J. Chem. Soc. A* (1966) 1437.
3. Chatt, J., Leigh, G. J., Storace, A. P., Squire, D. A., Starkey, B. J., *J. Chem. Soc. A* (1971) 899.
4. Travis, K., Busch, D. H., *Inorg. Chem.* (1974) **13**, 2591.
5. Lovecchio, F. V., Gore, E. S., Busch, D. H., *J. Am. Chem. Soc.* (1974) **96**, 3109.
6. von Doering, W. E., Levy, L. K., *J. Am. Chem. Soc.* (1955) **77**, 509.
7. Mathias, S., *Bol. Fac. Filos. Cienc. Letras Univ. Sao Paulo* (1942) **14**, 75.
8. Busch, D. H., Farmery, K., Goedken, V., Katovic, V., Melnyk, A. C., Sperati, C. R., Tokel, N., ADV. CHEM. SER. (1971) **100**, 44.
9. Hinz, F. P., Margerum, D. W., *J. Am. Chem. Soc.* (1974) **96**, 4993.
10. Evans, D., Osborn, J. A., Wilkinson, G., *Inorg. Synth.* (1968) **11**, 99.
11. Cianpoli, M., Paoletti, P., *Inorg. Chem.* (1967) **6**, 1261.
12. Dubois, T. D., Meek, D. W., *Inorg. Chem.* (1969) **8**, 146.
13. Geoffrey, G. L., Wrighton, M. S., Hammond, G. S., Gray, H. B., *J. Am. Chem. Soc.* (1974) **96**, 3105.
14. Vaska, L., Chen, L. S., Miller, W. V., *J. Am. Chem. Soc.* (1971) **93**, 6671.
15. Vaska, L., *Acc. Chem. Res.* (1968) **1**, 335.
16. Shriver, D. F., Jackovitz, J. F., Beollas, M. J., *Spectrochim. Acta* (1968) **24A**, 1469.
17. Scott, R. N., Shriver, D. F., Vaska, L., *J. Am. Chem. Soc.* (1968) **90**, 1079.
18. Powell, P., Noth, H., *Chem. Commun.* (1966) 637.
19. Nappier, T. E., Thesis, The Ohio State University, 1972.
20. Vaska, L., Bath, S. S., *J. Am. Chem. Soc.* (1966) **88**, 1333.

21. Brage, E. H., Hübel, W., *Angew. Chem.* (1963) **75**, 345.
22. La Placa, S. J., Ibers, J. A., *Inorg. Chem.* (1966) **5**, 405.
23. Baddley, W. H., Venanzi, L. M., *Inorg. Chem.* (1966) **5**, 33.
24. Baddley, W. H., *J. Am. Chem. Soc.* (1966) **88**, 4545.
25. Hodgson, D. J., Payne, N. C., McGinnety, J. A., Pearson, R. G., Ibers, J. A., *J. Am. Chem. Soc.* (1968) **90**, 4486.
26. Mingos, D. M. P., Ibers, J. A., *Inorg. Chem.* (1971) **10**, 1035.
27. Mingos, D. M. P., Robinson, W. T., Ibers, J. A., *Inorg. Chem.* (1971) **10**, 1043.
28. Johnson, B. F. G., McCleverty, J. A., *Prog. Inorg. Chem.* (1966) **7**, 277.
29. Baird, M. C., Mague, J. T., Osborn, J. A., Wilkinson, G., *J. Chem. Soc. A* (1967) 1347.
30. Choy, V. J., O'Connor, C. J., *Coord. Chem. Rev.* (1972) **9**, 145.
31. Valentine, J. S., *Chem. Rev.* (1973) **73**, 235.
32. McGinnety, J. A., Doedens, R. J., Ibers, J. A., *Inorg. Chem.* (1967) **6**, 2243.
33. Otsuka, S., Nakamura, A., Tatsumo, Y., *J. Am. Chem. Soc.* (1969) **91**, 6994.
34. Shriver, D. F., *Acc. Chem. Res.* (1970) **3**, 231.
35. Halpern, J., ADV. CHEM. SER. (1968) **70**, 1.
36. Collman, J. P., Murphy, D. W., Dolcetti, G., *J. Am. Chem. Soc.* (1973) **95**, 2687.
37. Collman, J. P., private communication.
38. James, R. B., Ng, F. T. T., *J. Chem. Soc. Dalton Trans.* (1972) 335.
39. Bounsall, E. J., Koprich, S. R., *Can. J. Chem.* (1970) **48**, 1481.
40. Collman, J. P., Roper, W. R., *Adv. Organomet. Chem.* (1968) **7**, 53.
41. Lawson, D. N., Osborn, J. A., Wilkinson, G., *J. Chem. Soc. A* (1966) 1733.
42. O'Connor, C. D., *J. Inorg. Nucl. Chem.* (1970) **32**, 2299.
43. Chatt, J., Johnson, N. P., Shaw, B. L., *J. Chem. Soc. A* (1967) 604.
44. Deeming, A. J., Shaw, B. L., *J. Chem. Soc. A* (1969) 1128.
45. Darensbourg, M. Y., Darensbourg, D. J., *Inorg. Chem.* (1970) **9**, 32.
46. Dolcetti, G., Hoffman, N. N., Collman, J. P., *Inorg. Chim. Acta* (1972) **6**, 532.

RECEIVED January 24, 1975. Work supported by the National Science Foundation.

30

Synthesis and Study of Octa-aza Annulene Complexes With Unusual Properties

VIRGIL L. GOEDKEN and SHIE-MING PENG

University of Chicago, Chicago, Ill. 60637

Prototype complexes, [M(C$_{10}$H$_{20}$N$_8$)](ClO$_4$)$_2$, derived from the metal template condensation of 2,3-butanedihydrazone and formaldehyde, undergo further ligand oxidation with various oxidants to form molecular complexes of type M(C$_{10}$H$_{14}$N$_8$) containing completely conjugated 16π-electron systems. The nickel complex is an eclipsed cofacial dimer, [Ni(C$_{10}$H$_{14}$N$_8$)]$_2$, with a Ni–Ni bond length of 2.784(2) A. The dimers are stacked in the lattice with the Ni atoms forming a chain down the c crystal axis. The Co(III)–alkyl complexes have unusual NMR spectra with the proton resonances of the macrocyclic ligand displaying abnormal up-field shifts while the axial ligand protons are observed downfield from their normal positions. The temperature dependence and solvent dependence, together with shifts on chemical alteration of the complex, indicate that the abnormal shifts may be attributed to a thermally populated paramagnetic state at ambient temperatures.

The study of synthetic macrocyclic ligand complexes has considerable merit. In general, these complexes have a robust constitution and are not easily destroyed, not even when subjected to strongly acidic, basic, oxidative, or reductive media. These aspects have been advocated and utilized, principally by Busch, in a lengthy series of far reaching investigations (*see* Reference *1* and references cited therein, as well as Reference *2*). This paper reports the synthesis of some bis-α-diimine macrocyclic complexes and their chemical transformation to yield a series of completely conjugated 16π-electron ligand complexes. A number of these complexes have some uncommon properties which are discussed in detail.

Our initial objective was to synthesize a completely conjugated ligand system which, because of abnormal shortening of the metal–ligand distances, might interact strongly enough with the metal to create complexes with unusual chemical and physical properties. It is well known that a compression of the electronic energy level occurs in complexes of soft or polarizable ligands such as dithiolene and dithiolato-type ligands (3). Busch and co-workers also demonstrated that the redox potentials of metals of macrocyclic ligands depend strongly on the degree and type of conjugation present in the ligand as well as on the ring size (4, 5).

The macrocyclic ligand system chosen for this study, $M(C_{10}H_{14}N_8)$, is that obtained by the metal template condensation of 2,3-butanedihydrazone with aldehydes, followed by four-electron ligand oxidation. The simplicity of the syntheses and the inexpensive starting materials make the complexes available for a wide variety of studies. Although the Cu, Ni, Co, and Fe complexes of this ligand have been prepared, those of Ni and Co appear to be most unusual and they will be discussed in greatest detail.

Syntheses

$[Cu(C_{10}H_{20}N_8)Cl(H_2O)](ClO_4)$. Formaldehyde, 3.22 g (38% in aqueous solution), was added to a solution containing 3.41 g $CuCl_2 \cdot 2H_2O$ in 50 ml water. Then 4.56 g 2,3-butanedihydrazone (prepared by the Busch and Bailar method; see Reference 6) was added as a solid, and the solution was stirred 10 min. The solution turned dark green as the reaction proceeded. An aqueous solution of 5 g $NaClO_4$ and 0.1 ml $HClO_4$ was added to precipitate the complex which separates as shiny green plates. The complex was filtered, washed with water, and then dried in the air.

$[Ni(C_{10}H_{20}N_8)](ClO_4)_2$. Formaldehyde, 1.61 g (38% in water), was added to a solution containing 3.65 g $Ni(ClO_4) \cdot 6H_2O$ in 30 ml CH_3CN. To this blue solution, 2.28 g 2,3-butanedihydrazone was added. The mixture was allowed to stand 2 hr; then 50 ml diethyl ether was added to induce precipitation of the product. The brown product was filtered, and then recrystallized from acetonitrile.

$[Co(C_{10}H_{20}N_8)](ClO_4)_2$. The procedure is the same as for the nickel complex, but the reactions must be carried out under an inert atmosphere.

$[Fe(C_{10}H_{20}N_8)(CH_3CN)_2](ClO_4)_2$. A solution containing 3.22 g 38% formaldehyde (in water) and 3.44 g 2,3-butanedione was added to a solution of 7.27 g $Fe(ClO_4)_2 \cdot 6H_2O$ in 100 ml CH_3CN. Then 2.6 g anhydrous hydrazine was added dropwise, and the red solution was filtered to remove insoluble residues. The solution was then stoppered and refrigerated 12 hr during which time the product crystallized. The product was filtered, washed with ethanol, and then air dried.

$[Cu(C_{10}H_{14}N_8)]$. A solution of 200 mg $[Cu(C_{10}H_{20}N_8)Cl(H_2O)](ClO_4)$ and 1 ml triethylamine was prepared in 10 ml acetonitrile. Molecular oxygen was then bubbled through the solution 5 min. The product, which precipitates as lustrous flaky green crystals, was filtered, washed with CH_3CN, and then dried *in vacuo* (yield about 30%).

[Ni($C_{10}H_{14}N_8$)]₂. A solution of 200 mg [Ni($C_{10}H_{20}N_8$)](ClO_4)₂ in 30 ml CH_3CN was prepared, and 0.5 ml pyridine was added. Molecular oxygen was then bubbled through the solution 10 min. A lustrous, black, finely divided precipitate formed, which was filtered, washed with CH_3CN, and then dried *in vacuo*. Note: the desired product reacts slowly with more molecular oxygen, going back into solution to give an intense green solution which we were unable to isolate and characterize. In order to avoid this difficulty, other oxidants may be used with equal facility. Four equivalents of either I_2 or [Fe(acac)₃], acac = acetyl-acetonato ligand, dissolved in CH_3CN may be used as an oxidant in place of molecular oxygen. Yields of the [Ni($C_{10}H_{14}N_8$)]₂ dimer approach 30%. Other oxidants, such as *o*-chloranil or 2,3-dichloro-5,6-dicyano 1,4-benzoquinone, are capable of introducing double bonds into the macrocyclic ligand, but the products are charge transfer adducts that contain some form of the oxidant.

[Co($C_{10}H_{14}N_8$) (C_5H_5N)₂]I₃. A 5-ml solution containing 1.0 g I_2 in CH_3CN was added to a solution of 200 mg [Co($C_{10}H_{20}N_8$)(CH_3CN)₂]-(ClO_4)₂ and 0.5 ml pyridine dissolved in 15 ml CH_3CN. The black, crystalline product precipitates within a few minutes of mixing the reagents. The product was filtered, washed with CH_3CN, and then dried *in vacuo*.

[Co($C_{10}H_{14}N_8$)(axial base)(R)]. In these compounds base = pyridine, 3-picoline, 4-picoline, acetonitrile, or methylhydrazine and R = $-CH_3$, $-CH_2CH_3$, or $-C_6H_5$. These organo–Co(III) complexes were prepared using a general scheme previously reported in which an organihydrazine undergoes oxidative deamination with the organo-fragment being transferred to the cobalt complex (7). Under the more basic conditions used here, ligand oxidation also occurs, giving the desired octa-aza annulene complex. A solution of 200 mg [Co($C_{10}H_{20}N_8$)-(CH_3CN)₂](ClO_4)₂, 0.5 ml of the appropriate organohydrazine, and 1 ml of the desired axial base was prepared in 15 ml CH_3CN. To this solution, 482 mg potassium *tert*-butoxide was added with vigorous stirring. The solutions became a deep green-black color. The solutions were filtered to remove the $KClO_4$ and other insoluble residues. Air was then bubbled through the solution for several minutes. An exothermic reaction ensues during which N_2 is given off and the product crystallizes from the solution. The product was filtered, washed with CH_3CN, and then dried *in vacuo*.

[Fe($C_{10}H_{14}N_8$)(C_5H_5N)₂]. A solution was prepared by dissolving 200 mg [Fe($C_{10}H_{20}N_8$)(CH_3CN)₂](ClO_4)₂, 0.5 ml pyridine, and 160 mg (4 equivalents) potassium *tert*-butoxide in 20 ml acetonitrile under N_2 atmosphere. The solution darkens considerably to brownish black on mixing. The solution was centrifuged to remove $KClO_4$ and other insoluble residues. A solution of 482 mg (4 equivalents) [Fe(acac)₃], dissolved in a minimum of CH_3CN, was added to the solution containing the macrocyclic complex. The product begins crystallizing on the sides of the reaction vessel after several minutes; however, maximum yields (15%) were obtained when the solution was chilled in an ice box 12 hr.

The product was then filtered under N_2, washed with ethanol, and dried *in vacuo*.

Results and Discussion

The reaction of 2,3-butane dihydrazone with formaldehyde and certain divalent transition metal perchlorates leads to the formation of bis-α-diimine macrocyclic complexes, **II**. Under suitable conditions, these complexes undergo a four-electron ligand oxidation, with two additional double bonds being introduced into the macrocyclic ligands. Deprotonation also occurs in each of the six-membered chelate rings resulting in the formation of completely conjugated 16π-electron macrocyclic ligand complexes (*see* the Scheme). Oxidation of the simple macrocyclic complexes **II** occurs under mild conditions if a suitable base is present to remove a proton from the hydrazine linkage, thus facilitating ligand oxidation. These hydrazine protons can sometimes be removed by rela-

tively weak bases such as pyridine; the ease of proton removal depends greatly on the central metal. At least one proton is removed from the nickel(II) ionization in the presence of pyridine. A much stronger base, *tert*-butoxide ion, is needed to effect proton removal and to facilitate ligand oxidation.

A few comments concerning the possibility of isomerization among these octaza complexes is appropriate. The nitrogen atoms of the hydrazine linkage are ambidentate, and coordination to either nitrogen may occur, depending on whether the 2,3-butanedihydrazone has the *syn-anti* or *anti-anti* configuration. Three isomers are possible for each formula given: two 5-6-5-6 chelate systems (one of D_{2h} symmetry, **IIIa**, and one of C_{2h} symmetry, **IIIb** and one 5-5-6-6 chelate system, **IIIc**). Each depicted double bond arrangement represents only one of many valence tautomers. Completely conjugated 16π-electron ligand complexes of C_{2h} symmetry, **IIIb,** have been prepared by two methods. In the first, reported by Baldwin and co-workers (8), the free ligand (which has C_{2h} symmetry) is formed first and then the metal ion is inserted. Alternatively, it is possible to obtain some C_{2h} complexes in small yield simply by mixing the biacetyl,

IIIb   IIIc

metal perchlorate, hydrazine, and aldehyde together in a solution of acetonitrile (9, 10). The properties of complexes with D_{2h} symmetry are quite different from those of complexes with C_{2h} symmetry.

The NMR spectrum of 2,3-butanedihydrazone confirms that, as isolated, it has the *anti-anti* configuration as shown in Structure I. Since α-diimine ligands are good chelating ligands because of enhanced bonding attributable to the π-acceptor ability of the α-diimine moiety, it is not surprising that the resultant complexes have the symmetrical D_{2h} structure. In general, this overall donor atom arrangement is maintained when the bis-α-diimine complexes are further oxidized. The copper complex appears to be an exception.

The identities of the oxidized products were established by various spectroscopic and crystallographic techniques. The IR spectra of complexes containing the $C_{10}H_{14}N_8$ ligand were devoid of any absorptions in the range normally ascribed to N–H vibrations, and they were also

Table I.   Properties of Oxidized Complexes

Complex	Magnetic Moment, BM	Parent Peak Observed in Mass Spectrum	
$[Cu(C_{10}H_{14}N_8)]$	1.91	309	(^{63}Cu)
$[Ni(C_{10}H_{14}N_8)]_2$	diamagnetic	608	(^{58}Ni)
$[Co(C_{10}H_{14}N_8)(C_5H_5N)_2]I_3$	diamagnetic	—	
$[Fe(C_{10}H_{14}N_8)(C_5H_5N)_2]$	diamagnetic	302	(^{56}Fe)

devoid of any absorptions attributable to perchlorate anions. Mass spectra of the molecular species also established that the macrocyclic ligand had a weight of 246 which is consistent with the state of oxidation visualized in the Scheme. Some properties of these oxidized complexes are given in Table I.

The IR spectra of these complexes and of the organo–cobalt(III) complexes in Table II were of three types. Those of the cobalt and iron complexes were very similar which suggests that the macrocyclic ligands were of the same geometrical isomer and contained a similar pattern of delocalization. The spectra of the Ni and Cu complexes differed significantly from each other and also from those of the Fe and Co complexes. Furthermore, the spectrum of the Cu complex was identical to that of the $[Ni(C_{10}H_{14}N_8)]$ of C_{2h} symmetry, which has been structurally characterized (11), which suggests that isomerization from a structure with

Table II.   NMR Data for the New

Complex	Solvent	T, °C
$[Co(C_{10}H_{14}N_8)\,(py)\,(CH_3)]^0$	$CDCl_3$	
	d_5-py $CDCl_3$	$+50\,°C$
	$CDCl_3$	$-50\,°C$
$[Co(C_{10}H_{14}N_8)\,(py)\,(C_2H_5)]^0$	$CDCl_3$	
$[Co(C_{10}H_{14}N_8)\,(py)\,(C_6H_5)]^0$	$CDCl_3$	
$[Co(C_{10}H_{14}N_8)\,(4\text{-picoline})\,(CH_3)]^0$	$CDCl_3$	
$[Co(C_{10}H_{14}N_8)\,(3\text{-picoline})\,(CH_3)]^0$	$CDCl_3$	
$[Co(C_{10}H_{14}N_8)\,(NH_2NHCH_3)\,(CH_3)]^0$	d_6-DMSO	
$[Co(C_{12}H_{18}N_8)\,(py)\,(CH_3)]^{0\,d}$	$CDCl_3$	
$[Co(C_{22}H_{22}N_8)\,(py)\,(CH_3)]^{0\,d}$	$CDCl_3$	
$[Co(C_{10}H_{16}N_8)\,(CH_3CN)]\,(ClO_4)_2\,^e$	CD_3NO_2	
$[Co(C_{10}H_{14}N_8)\,(CN)\,(CH_3)]^{-f}$	d_5-py & D_2O	

[a] The numbers in parentheses refer to integrated intensity.
[b] The phenyl and pyridine resonances overlap and cannot be distinguished.
[e] Broad peak, probably NH of CH_3NHNH_2.
[d] The ligand was modified by placing methyls on C(1) and C(4) (Figure 1).

D_{2h} symmetry to one with C_{2h} symmetry had occurred. This was essentially confirmed by comparing the X-ray powder patterns of the Cu and the Ni complexes; the pattern of the Cu complex was virtually identical to that of the triclinic form of the Ni octa-aza annulene complex of C_{2h} symmetry. (The C_{2h} complex of Ni occurs in two crystalline forms, triclinic and orthorhombic.) The Cu complex was then oxidized to the corresponding Cu(III) complex which gave a species with a low spin d^8 configuration that was suitable for NMR studies. The methyl region of the ¹H NMR consisted of two singlets at 2.43 and 2.75 ppm (δ), which further substantiated our contention that isomerization had occurred.

The nickel complexes of stoichiometry $NiC_{10}H_{14}N_8$ have either D_{2h} or C_{2h} symmetry (Structures **IIIa** and **IIIb**). The C_{2h} structure is a monomer; it has a well defined NMR spectrum and properties that are generally consistent with other four-coordinate square planar nickel(II)

Organo–Cobalt (III) Complexes [a]

Assignments—Chemical Shifts (δ)

CH_3 (ligand)	$C-R$ ($C-R$)	$Co-R$	Axial Ligand
0.62 (12,s), 4.32 (2,s)		8.04 (3,s)	8.54–8.81 (3,m) ; 13.77–14.07 (2,m)
0.49 (12,s), 4.22 (2,s)		9.00 (3,s)	
0.57 (12,s), 4.23 (2,s)		8.16 (3,s)	8.56–8.85 (3,m) ; 13.84–14.10 (2,m)
0.74 (12,s), 4.57 (2,s)		7.59 (3,s)	8.49–7.74 (3,m) ; 13.62–13.87 (2,m)
0.59 (12,s), 4.33 (2,s)		8.98 (2,q,J = 4 cps) 4.10 (3,t,J = 7 cps)	8.30–8.67 (3,m) ; 13.47–13.97 (2,m)
0.53 (12,s), 4.42 (2,s)		8.16–8.83 (6,m)	12.75–13.08 (4,m) [b]
0.61 (12,s), 4.34 (2,s)		8.01 (2,s)	3.06 (3,s), 8.33–8.55 (2,m) ; 13.63–13.86 (2,m)
0.60 (12,s), 4.34 (2,s)		8.03 (3,s)	3.15 (3,s), 8.35–8.60 (2,m) ; 13.56–13.90 (2,m)
0.60 (12,s), 4.40 (2,s)		7.82 (3,s)	2.63 (3,s), 4.33 (?) [c]
0.91 (12,s), 0.78 (6,s)		6.56 (3,s)	8.21–8.37 (3,m) ; 12.52–12.77 (2,m)
0.83 (12,s), 6.83–7.50 (10,m)		7.07 (3,s)	8.30–8.70 (3,m) ; 12.80–13.30 (2,m)
2.15 (12,s), 6.62 (2,s)		3.73 (3,s)	2.37 (3,s)
0.95 (12,s), 4.95 (2,s)		6.94 (3,s)	

[e] The ligand was protonated on nitrogen.
[f] $[Co(C_{10}H_{14}N_8)(py)(CH_3)]^0$ was dissolved in d_5-py, then saturated NaCN in H_2O was added.

complexes. On the other hand, the D_{2h} isomer is actually a dimer that contains a Ni–Ni bond and gives a weak parent peak in the mass spectrum at 608. The compound is virtually insoluble in all solvents, giving only a faint color in $CHCl_3$ or CH_2Cl_2. Furthermore, the solution spectra are different from those obtained in the solid state; this suggests some possible dimer–dimer interaction in the solid state. The solutions are unstable; they react moderately rapidly with molecular oxygen and decompose slowly even in rigorously degassed chlorocarbon solvents. The unusual implications of these observations prompted us to determine the crystal structure of the complex.

Exhaustive efforts to grow suitable crystals afforded only a tiny specimen, $0.024 \times 0.054 \times 0.23$ mm, which was obtained by the slow diffusion of oxygen into a 10:1 acetonitrile–pyridine solution of [Ni-$(C_{10}H_{20}N_8)$](ClO_4)$_2$. Crystal data: space group $C2/c$, $a = 15.497(9)$ A, $b = 18.534(9)$ A, $c = 13.123(7)$ A, $\cos \beta = -0.8121(2)$, $\rho_{calcd} = 1.718$, $\rho_{exp} = 1.70$ g/cm^3 for which $Z = 8$. The tiny crystal size and consequent weak intensity of the diffractions necessitated the collection of three quadrants of data. Data were collected with Mo Kα radiation to $\sin \theta/\lambda = 0.5946$. The equivalent data were merged, and the processed data from all 2077 independent reflections measured were used in the solution and refinement of the structure. The structure was solved by the heavy

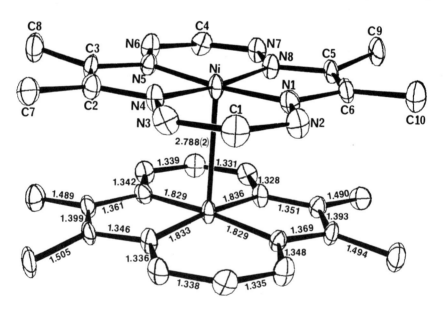

Figure 1.   Molecular structure and interatomic distances of the [Ni($C_{10}H_{14}N_8$)] dimer. The estimated standard deviations are as follows: Ni–N distances, 0.005 A; N–N, C–N, and C–C distances, 0.006–0.008 A.

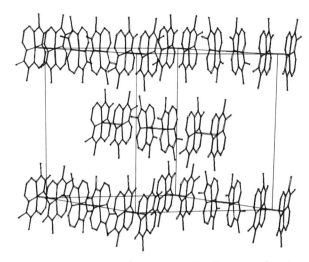

Figure 2. Packing diagram of the dimers indicating
the stacking of the molecules in the unit cell

atom method and refined with anisotropic thermal parameters for all nonhydrogen atoms. All hydrogen atoms were located on a difference Fourier, and idealized coordinates were redetermined assuming standard geometry and C–H distances of 0.95 A. These were included at fixed positions in the final refinement. At convergence, $R_1 = 10.4$ and $R_2 = 6.1\%$ where $R_1 = \Sigma|Fo - Fc|/\Sigma|Fo|$ and $R_2 = [\Sigma w(|Fo| - |Fc|)^2/\Sigma w|Fo|^2]^{0.5}$.

The structure consists of a dimeric macrocyclic complex containing a Ni–Ni bond with the dimeric units stacked along the c crystal axis (Figures 1 and 2). There are a number of unusual features associated with this structure. First, the dimers contain a Ni–Ni bond of length 2.784(2) A. Although Ni–Ni bonds are not common, an increasing number have been found with bridging N-donor or S-donor ligands in recent years with Ni–Ni bonds ranging in length from 2.38 to 2.81 A (*12, 13, 14, 15, 16, 17*). That most closely related to ours contains a dianionic quadridentate N-donor ligand and has a metal–metal distance of 2.81 A. The distances in the latter structure and in the Ni dimer are comparable to the Co–Co distance, 2.794 A, in the $[Co(CN)_5]^{3-}$ dimer which contains a single σ bond (*18, 19*).

However, attractive interaction between the two macrocyclic moieties in addition to a Ni–Ni σ bond must be present to account for the eclipsed conformation of the two units forming the dimer. Although highly delocalized systems such as porphyrin and chlorophyl-type molecules have attractive π–π interactions leading to aggregation in solution, a staggered arrangement of the atoms that minimizes nonbonding repulsions is preferred. The most likely explanation for the eclipsed arrangement of our dimer is that, even at the relatively large Ni–Ni separation, some δ bonding is present between the two metal ions. The fact that the Ni atoms are displaced 0.104 A from the plane of four nitrogen atoms towards one another lends some credence to this viewpoint. The interplanar separation of the least squares plane defined by the two macrocyclic rings is 3.00 A and nonbonding repulsions may prevent closer approach of the two metal atoms.

The double bonds of the macrocyclic ligand are essentially completely delocalized; the average C–C and C–N distances in the five-membered rings, 1,396 and 1.357 A, are very close to the accepted C–C distances of benzene and the aromatic C–N distance of pyridine, 1.35 A. The N–N and C–N distances of the six-membered chelate rings are also intermediate between those expected for single and double bonds. This degree of delocalization is much more extensive than that observed in the C_{2h} structure (which has an alternating double bond arrangement) or in the Co(III) complexes of D_{2h} symmetry (see below). This difference in delocalization accounts for the differences in the IR spectra of the various octa-aza annulene complexes which was commented upon earlier.

The packing arrangement of the dimer molecules is significant. The adjacent dimers are related by a crystallographic two-fold axis and are stacked along the c crystal axis with a Ni–Ni separation of 3.800 A. This stacking, and possible metal–metal interaction of adjacent dimers, may account for the limited solubility of the complex. One other interesting point is that a 0.50-A movement of the Ni atoms from one another in the dimer would lead to a new dimer with equivalent Ni–Ni separation but with the macrocyclic ligand rotated 90° relative to one another. This would appear to be a more favorable arrangement than the one found because it would minimize most of the repulsive interactions.

Truex and Holm (20) observed monomer–dimer equilibrium in solution for their cationic species, [Ni(MeHMe-2,9-diene)]⁺; in view of our results, their dimeric species may also contain a Ni–Ni bond. Although our dimer complex is diamagnetic (i.e., repelled by a magnetic field) in the solid state, all samples investigated gave a moderately strong isotropic ESR signal, $g = 2.001$, indicating some unpaired spin density in the π-ligand system. Since this signal is very similar to that described by Truex and Holm for their cationic 15π system, the signal we observe may be caused by a small impurity of an analogous cationic species in our samples.

In the cobalt complexes there are again remarkable differences in the properties of the C_{2h} and D_{2h} isomers. The Co(III) complex, $[Co(C_{10}H_{14}N_8)(C_5H_5N)_2]I_3$ (tri-iodide anion), is diamagnetic, but all attempts to observe its ^{1}H NMR spectrum failed, even when it was scanned over the range of 100 to -100 ppm on a Bruker 270 spectrometer. However, the organo–Co(III) complexes gave well resolved NMR spectra, but the resonances were significantly shifted from their expected positions. In general, the spectra were characterized by large downfield shifts for the ligands occupying axial sites and high field shifts for those protons located in the equatorial plane of the macrocyclic ligand (Table II). For the $[Co(C_{10}H_{14}N_8)(C_5H_5N)(CH_3)]$ complex, the Co–CH$_3$ resonance is observed at 8.04 ppm (δ), the pyridine resonances at 8.7 and 13.9 ppm, and the ligand methyl groups at 0.62 ppm. In contrast, the Co(III)–alkyl complexes with the C_{2h} form of the ligand are five-coordinate, with typical NMR spectra Co–CH$_3$ at 0.04 ppm and the ligand –CH$_3$'s at 2.69 and 2.42 ppm (*21*). Whereas the organo–Co(III) complexes of the C_{2h} form of the ligand are five-coordinate with "well behaved" magnetic properties, the organo–Co(III) complexes of the D_{2h} form of the ligand are six-coordinate with unusual magnetic properties. The ^{1}H NMR spectra depend markedly on (a) changes of the axial ligand, (b) minor changes in the macrocyclic ligand, (c) solvent, and (d) temperature. These observations have led us to believe that the anomalous shifts are the result of paramagnetic contact shifts arising from the thermal population of a low-lying triplet state.

The spectra depend on the nature of the axial base, but not to the extent that we had hoped. Changing one nitrogen donor for another led only to minor changes in the spectra. However, exchanging pyridine for

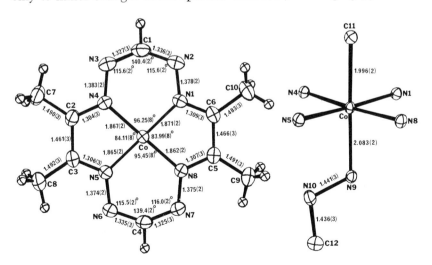

Figure 3.   Two views of [Co(C$_{10}$H$_{14}$N$_8$)(CH$_3$)(NH$_2$NHCH$_3$)] with selected interatomic distances and angles. Left: view perpendicular to the plane of the molecule, and right: a simplified side view

a π-acceptor such as cyanide lowers the Co–CH_3 absorption from 8.04 to 6.93 ppm. The alkyl groups produce a strong σ-trans effect; thus the effect of the axial ligands (trans to the alkyl group) on the electronic structure of the complex is less than might be expected.

The NMR spectra are also sensitive to minor changes in the macrocyclic ligand. Substituting methyl groups for the hydrogen atoms of C(1) and C(4) (see Figure 3), a position relatively isolated from the coordination sphere, shifts the absorption of the Co(III)–CH_3 group from 8.04 to 6.56 ppm. Protonation of the ligand by adding perchloric acid to the molecular complex produces $[Co(C_{10}H_{16}N_8)(CH_3CN)(CH_3)](ClO_4)_2$, a species with a more normal NMR spectrum—the Co(III)–CH_3 resonance appears at 3.78 ppm. Substituting a polar solvent, d-pyridine, for $CDCl_3$ results in a downfield shift for Co(III)–CH_3 of almost.1 ppm to a value of 9.0 ppm for the $[Co(C_{10}H_{14}N_8)(C_5H_5N)(CH_3)]$ complex.

The strongest evidence supporting our contention that the anomalous NMR spectra are due to paramagnetic contact shifts arising from a thermally populated triplet state is derived from the temperature dependence of the NMR spectra (Table II). Cooling the sample from $+50°$ to $-50°C$ decreases the population of the triplet state and shifts the absorption maxima of Co(III)–CH_3 from 8.16 ppm to 7.59 ppm in the direction expected for an increase in the population of the singlet ground state. The NMR spectra of the complexes are independent of concentration and reproducible from preparation to preparation, thus eliminating the possibility that contact shifts result from rapid electron exchange between the cobalt(III)–alkyl complexes and some paramagnetic impurity.

The $[Co(C_{10}H_{14}N_8)(C_5H_5N)_2]I_3$ complex and the organo–Co(III) complexes all give an isotropic ESR signal at room temperature, $g = 2.001$ (half-width $\approx$ 40–80 gauss), which is indicative of unpaired spin density in the π-ligand system. The intensity of the signal from the organo–Co(III) complexes appears to be weaker by a factor of ten than the signal from the bispyridine complex.

The crystal structure of one of the organo–Co(III) complexes was determined in order to define unambiguously the double bond arrangement and the extent of delocalization in the macrocyclic ligand, and to shed some light on the magnetic peculiarities of these organo–cobalt(III) complexes. Crystals of the hydrazine complex, $[Co(C_{10}H_{14}N_8),NH_2-NHCH_3)(CH_3)]$, were suitable for X-ray structural determination. Crystals of the compound belong to space group $P2_1/c$, with $a = 6.961(1)$ A, $b = 23.377(3)$ A, $c = 10.061(2)$ A, $\beta = 103.44(1)°$, $\rho_{calcd} = 1.526$, and $\rho_{exp} = 1.51(2)$ for which $Z = 4$. A total of 5294 data were measured on a Picker FACS-1 diffractometer with Mo $K\alpha$ radiation; 4106 independent data with F values $\geq$ 2 were used in the structural refinement. The structure was solved using the heavy atom method, and it was refined by full matrix least squares techniques. Refinement using anisotropic thermal parameters for all nonhydrogen atoms (with contributions for all the hydrogen atoms located from a difference Fourier and placed in assuming standard distances and geometry) converged to $R_1 = 4.2\%$ and $R_2 = 4.2\%$. The macrocyclic ligand is flat and contains an α-diimine function in each of the five-membered chelate rings, and a three-atom, N–C–N, delocalized system in each of the six-membered chelate rings (Figure 3). The angles defined by N(2)–C(1)–N(3) and N(6)–C(4)–N(7), 140.2(2)° and 139.4(2)° respectively, are unusually wide. This is apparently attributable to the preference of the N–N–C linkage for an angle less than 120°.

The parameters of the inner coordination polyhedron are totally consistent with those reported for a number of organo–cobalt(III) complexes (*21*) with one exception. The average cobalt–nitrogen (planar) distance, 1.866 A, is shorter, although only slightly, than in other organo–cobalt(III) complexes. The crystal structures of three cobaloxime complexes, which are the most closely related to our organo–cobalt complexes, have mean Co–N (planar) distances which vary between 1.88 and 1.90 A (*22, 23, 24*).

An alternative explanation for the abnormal NMR shifts is that the 16π dianionic ligand is behaving as an antiaromatic ring system. The observed shifts are close in magnitude to those observed for the dianion of 15,16-dimethyldihydropyrene; this demonstrates antiaromatic character (*25*); viz., the met hyl protons on the inside of the dihydropyrene ring appear at very low field, $\delta = 21.0$ ppm, compared with the high field position, -4.25 ppm, for the parent, uncharged aromatic ring. However, it is difficult to see how this interpretation could account for the temperature variation, solvent dependence, and ESR signals that are observed for the macrocyclic Co(III) complexes discussed.

Furthermore, the results of the crystallographic analysis are not compatible with those needed to fulfill the antiaromatic requirement, for example, and the alternating double bond arrangement. Although the double bonds are localized, they are localized in the form of two bis-α-diimine chelate linkages and two 3-atom delocalized systems, each containing a negative charge.

Conclusions

A number of geometrical and valance isomers of the dihydro-octaza-annulene metal complexes have been isolated and characterized. The findings are summarized in Table III. It is noteworthy that different

Table III.   Summary of Ligand Structures

Complex[a]	Ligand Structure	Comments
[NiL]	C_{2h}; **IIIb**	localized double bonds
[Ni$_2$L$_2$]	D_{2h}; **IIIa**	delocalized double bonds; Ni–Ni bond
[CuL]	C_{2h}; **IIIb**	localized double bonds
[Cu^{III}L]$^{+1}$	C_{2h}; **IIIb**	localized double bonds
[CoL(NH$_2$NHCH$_3$)CH$_3$]	D_{2h}; **IIIa**	α-diimine chelate rings; allylic type of delocalization in six-membered chelate rings
[CoL(CH$_3$)]	C_{2h}; **IIIb**	localized double bonds

[a] L = C$_{10}$H$_{14}$N$_8$.

metal ions tend to stabilize different geometrical structures, and furthermore that, for a given geometrical isomer, different metal ions result in differing extent of double bond delocalization throughout the ligand. The variation in double bond delocalization for these systems is far greater than found with aromatic nitrogen donor ligands. The crystal structures of a large number of phenanthroline complexes have been determined; there is very little change in the C–N bond lengths within the five-membered chelate rings on going from metals with strong π-backbonding tendencies (which might be expected to stabilize the α-diimine form of phenanthroline) to metal ions with negligible backbonding tendencies (26).

Literature Cited

1. Busch, D. H., "Alfred Werner Commemoration Volume," p. 174, Verlag Helv. Chimica Acta, Basel, 1967.
2. Busch, D. H., Farmery, K., Goedken, V., Katovic, V., Melnyk, A., Sperati, C., Tokel, N., Adv. Chem. Ser. (1971) 100, 44.
3. Schrauzer, G. N., Acc. Chem. Res. (1969) 2, 72.
4. Lovecchio, F., Gore, E., Busch, D., J. Am. Chem. Soc. (1972) 94, 4529.
5. Takvoryan, N., Farmery, K., Katovic, V., Lovecchio, F., Gore, E., Anderson, L., Busch, D., J. Am. Chem. Soc. (1974) 96, 731.
6. Busch, D. H., Bailar, J., J. Am. Chem. Soc. (1956) 78, 1137.
7. Goedken, V., Peng, S.-M., Park, Y.-A., J. Am. Chem. Soc. (1974) 96, 284.
8. Baldwin, J., Holm, R., Harper, R., Huff, J., Koch, S., Truex, T., Inorg. Nucl. Chem. Lett. (1972) 8, 393.
9. Goedken, V., Peng, S.-M., unpublished data.
10. Goedken, V., Peng, S.-M., Chem. Commun. (1973) 62.
11. Goedken, V., Peng, S.-M., J. Am. Chem. Soc. (1973) 95, 5773.
12. Bonamico, M., Dessy, G., Fares, V., Chem. Commun. (1969) 697.
13. Jarcow, O., Schulz, H., Nast, R., Angew. Chem. (1970) 82, 43.
14. Corbett, M., Hoskins, B., Chem. Commun. (1969) 1602.
15. Bonamico, M., Dessy, G., Fares, V., Chem. Commun. (1969) 1106.
16. Bailey, N., James, T., McCleverty, J., McKenzie, E., Moore, R., Worthington, J., Chem. Commun. (1972) 681.

17. Sacconi, L., Mealli, C., Gatteschi, D., *Inorg. Chem.* (1974) **13**, 1985.
18. Brown, L., Raymond, K., Goldbert, S., *J. Am. Chem. Soc.* (1972) **94**, 7664.
19. Simon, G., Adamson, A., Dahl, L., *J. Am. Chem. Soc.* (1972) **94**, 7654.
20. Truex, T., Holm R., *J. Am. Chem. Soc.* (1972) **94**, 4529.
21. Dodd, D., Johnson, M., *J. Organomet. Chem.* (1973) **52**, 34.
22. Crumbliss, A., Bowman, J., Gaus, P., McPhail, A., *Chem. Commun.* (1973) 415.
23. McFadden, D., McPhail, A., *J. Chem. Soc. Dalton Trans.* (1974) 363.
24. Lenhert, P., *Chem. Commun.* (1967) 980.
25. Mitchell, R., Klopfenstein, C., Boekelheide, V., *J. Am. Chem. Soc.* (1969) **91**, 4931.
26. Frenz, B., Ibers, J., *Inorg. Chem.* (1972) **11**, 1109.

RECEIVED January 24, 1975. Work supported by the National Institutes of Health grant HL14827 and the Materials Research Laboratory sponsored by the National Science Foundation.

31

Metal Complexes of Ligands Derived from Carbon Disulfide

JOHN P. FACKLER, JR.

Case Western Reserve University, Cleveland, Ohio 44106

Carbon disulfide reacts with a wide range of bases to pro-duce ligands which readily coordinate to metals. New synthetic work including the formation of aryl xanthates, disulfide products, and oxidized metal dithiolates is re-viewed. The sulfur addition and abstraction reactions are discussed, along with the photobleaching reaction of [Ni(n-butyldtc)₃]⁺. Bridged mercaptide complexes of nickel triad elements are also described.

Carbon disulfide, CS_2, is a remarkably versatile electrophile that reacts with a wide range of bases (1) to produce dithiols or deprotonated dithioacids (Reaction 1). Perhaps the most extensively studied dithiols

$$CS_2 + B^{z-} \rightleftharpoons S_2CB^{z-} \qquad (1)$$

are the dithiocarbamates that were first prepared by Delepine (2) in 1907. Table I lists the names and formulas of the more common 1,1-dithiolate ligands.

Coucouvanis (3) reviewed the major developments in the reaction chemistry of metal 1,1-dithiolates up to 1969, and Eisenberg (4) de-scribed the structural relationships known at that same time. It is my purpose to discuss here primarily what was discovered by our group, and by a few groups elsewhere, since 1970. Figure 1 is a pictorial sum-mary of the wealth of reaction chemistry of these compounds that was found since 1965. My emphasis in this paper is on recent synthetic achievements and on some new intra- and intermolecular reactions. At least part of our continued interest in these species stems from their biological characteristics (5) (as fungicides and other enzyme inhibitors), their vulcanization capabilities (6), and in general their major significant interest as industrial chemicals (7).

Table I. Derivatives of CS$_2$

Base	Formula	Name
Monobasic		
NR$_2$$^-$	S$_2$CNR$_2$$^-$	dithiocarbamate
OR$^-$	S$_2$COR$^-$	xanthate
SR$^-$	S$_2$CSR$^-$	thioxanthate[a]
R$^-$ or Ar$^-$	S$_2$CR$^-$ or S$_2$CAR$^-$	dithioalkylate or arylate
Dibasic		
NR^{z-}	S$_2$CNR^{2-}	dithiocarbimate
O^{2-}	S$_2$CO^{2-}	dithiocarbonate
S^{2-}	S$_2$CS^{2-}	trithiocarbonate
CR$_2$$^{z-}$	S$_2$CCR$_2$$^{2-}$	1,1-dithiolate

[a] Also alkyltrithiocarbonate.

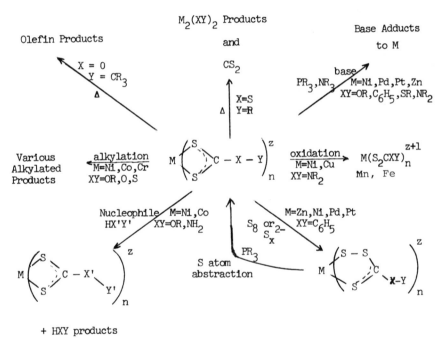

Figure 1. Reaction chemistry of metal 1,1-dithiolates

Synthesis

Aryl Xanthates. It is 165 years since Zeise (8) reported the formation of xanthates from alcohol and CS$_2$ on reaction with base. Yet only three papers, a 1948 Italian patent (9), a 1970 report by Lee (10), and a recent (1971) report by Reichle (11), relate to the synthesis of aryl xanthates. We established (12) that the Italian patent is incorrect.

Phenols in the presence of CS_2 and base with alcohol solvents produce aliphatic, not aromatic, xanthates. (*See* Ref. *48*.)

Lee (*10*) first synthesized an aryl xanthate by reacting thallous phenoxide with CS_2 in benzene (Reaction 2). Reichle (*11*) synthesized

$$TlOC_6H_5 + CS_2 \xrightarrow{\text{benzene}} TlS_2COC_6H_5 \qquad (2)$$

a copper(I) derivative, which he reported to be $[P(C_6H_5)_3]_2CuOC_6H_5 \cdot CS_2$, by reacting CS_2 with the bisphosphine phenoxide of copper(I). We prepared the same material by reacting $[P(C_6H_5)_3]_3CuBr$ with $TlS_2COC_6H_5$ or $TlS_2COC_6H_4$-p-CH_3 (Reaction 3).

$$L_3CuBr + TlS_2COAr \rightarrow L_2CuS_2COAr \qquad (3)$$

The general synthesis of aryl xanthates is described by Schussler *et al.* (*12*). It follows from the ability of thallium(I) xanthates to undergo metathetical reactions with metal halides in absolute ethanol (Reaction 4). The thallium(I) precursors could be obtained with phenol

$$MX_2 + 2TlS_2COAr \xrightarrow[C_2H_5OH]{\text{absolute}} M(S_2COAr)_2 + 2TlX \qquad (4)$$

and the p-CH_3, p-O-n-butyl, p-*tert*-butyl, and 3,5-dimethyl derivatives. With p-bromophenol, the thallium(I) complex would not form although Reichle's method led to the copper(I) derivative. The p-nitrophenol product could not be obtained by either technique. Thus the synthesis clearly depends on the acidity of the parent phenol. The metal ion also is important in stabilizing the product. With sodium or potassium halides, reaction with the thallium(I) complex gave TlX and the corresponding alkali metal phenoxide, thereby eliminating CS_2. Table II lists some of the aryl xanthates studied so far.

The physical properties of these aryl xanthates resemble closely those of their alkyl counterparts. The thallium compounds are heat- and light-sensitive, and they decompose in acetone or THF to form Tl_2S, Tl^0,

Table II.   Aryl Xanthates[a]

TlS_2COAr	$Ni,S_2COAr)_2$	$Co(S_2COAr)_3$
$[(C_6H_5)_3P]_2CuS_2COAR$	$Pd(S_2COAr)_2$	
	$Pt(S_2COAr)_2$	
$(S_2COAr)_2$		

[a] OAr = phenoxide and the p-methyl-, 3,5-dimethyl-, p-butoxy-, and p-*tert*-butyl derivatives of phenoxide.

and phenol. The odor of COS is detected. The solids change color from pale yellow to orange after about two days at room temperature in a stoppered container. In air, thallous phenoxide is formed nearly completely in a few days. The other metallic derivatives that were prepared are more stable, but they do decompose in the presence of water.

The structure of the nickel(II) phenoxide derivative in preliminary refinement (*13*) has the typical NiS_4 planar geometry of 1,1-dithiolates. The phenyl rings, however, are perpendicular to this plane, which indicates that no π conjugation occurs beyond the oxygen atom. The $^1A_{1g} \rightarrow {}_1A_{2g}$ transition of the nickel complex occurs at 15.6 kK and is rather insensitive to ring substitutions. The d orbital splitting produced is slightly larger than that of the dithiophosphates and the alkyl xanthates, but it is not as great as that of the dithiocarbamates. Biological testing of these and related materials is in progress.

Sulfur Addition. Since the initial studies of sulfur addition to anionic 1,1-dithiolate derivatives (Reaction 5) by Coucouvanis (*14, 15*)

$$M(S_2C\!\!=\!\!X)_2{}^{2-} + \frac{2}{x} S_x \rightarrow M(S_3C\!\!=\!\!X)_2{}^{2-} \qquad (5)$$

in our group in the mid-1960's and the later work by Fetchin and co-workers (*16, 17*) on sulfur addition to derivatives of dithioacids (Reactions 6), only a modest effort was expended on synthesis. Furlani and

$$M(S_2CAr)_2 + S_8 \qquad \rightarrow M(S_3CAr)(S_2CAr)$$
$$Zn(S_3CAr)_2 + MK_2 \qquad \rightarrow M(S_3CAr)_2 \qquad + ZnX_2 \qquad (6)$$
$$M(S_3CAr)_2 + P(C_6H_5)_3 \rightarrow M(S_3CAr)(S_2CAr) + SP(C_6H_5)_3$$

Luciani (*18*) synthesized some perthiocarboxylato derivatives of pivalic acid which have properties similar (*19*) to those of the perthioaryl acid complexes. Perhaps parenthetically, it should be noted that vibrational analyses and other physical measurements (*20, 21*) have now established that the complexes synthesized near the turn of the last century by Hofmann and identified as $Ni(NH_3)_3CS_3$ and $Pt(NH_3)_2CS_3 \cdot H_2O$ are salts of the type $[Ni(NH_3)_6][Ni(CS_3)_2]$ and $[Pt(NH_3)_4][Pt(CS_3)_2]$ which contain coordinated trithiocarbonates. Recently (*22*) we synthesized some cadmium derivatives of the perthiocarboxylates to further the understanding of catenated sulfur complexes of this element.

Oxidized Metal Dithiocarbamates. One of the most intriguing developments in recent years in the chemistry of CS_2 derivatives is the exceptional ability of the *N,N'*-dialkyldithiocarbamate ligand to stabilize complexes in which the metal is in a high formal oxidation state. The inorganic chemistry researchers at Nijmegen including J. A. Cras, J. J. Steggerda, and their students H. C. Brinkhoff and J. Willemse deserve substantial credit for this discovery. Their first paper (23), which received very little notice, was a report of the synthesis and structure of $Br_2Cu^{III}S_2CN(n\text{-butyl})_2$. In 1969 Brinkhoff *et al.* (24) reported, among other compounds, the synthesis of $Ni^{IV}[S_2CN(n\text{-butyl})_2]_3Br$. Our interest in oxidized nickel group dithiolates prompted us to examine this material thoroughly. The crystal structure and photochemical bleaching–debleaching properties (25, 26) were shown to be entirely consistent with the ligand stabilized nickel(IV) description. In fact, the short C—N thioureide bond length of 1.318 A (Figure 2) and the associated

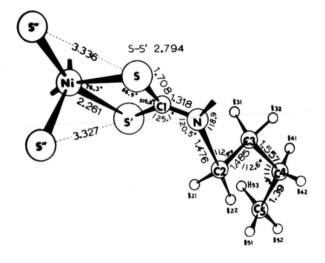

Journal of the American Chemical Society

Figure 2.   Structure of $Ni^{IV}[S_2CN(n\text{-}butyl)_2]_3Br$

shift from 1505 cm^{-1} for the stretching frequency of this bond in the nickel(II) complex to 1545 cm^{-1} in the nickel(IV) species are consistent with the ability of the ligand to distribute positive charge away from the metal ion effectively (1). Oxidation of the ligand system through S···S

1

bonding as depicted by the resonance structures in Reaction 7 also could contribute to the stability of these complexes. However, other CS_2 derivatives capable of ready dimerization to disulfides might also be expected to stabilize high oxidation states of metal ions if this were the dominant factor. To date, only dithiocarbamates (or their selenium analogs) and 1,1-dithiolates have demonstrated this ability. The x-ray crystal structure of the selenium complex $Ni[Se_2CN(n\text{-butyl})]_3Br$ was reported (27).

In the last few years, iron(IV), manganese(IV), palladium(IV), and platinum(IV) complexes were characterized (28, 29, 30, 31, 32). A few other studies such as the reports of cobalt(IV) by Salek and Straub (33), by Gahan and O'Connor (34), and also by Nigo et al. (35) and the report of the ruthenium(IV) and rhodium(IV) complexes by Gahan and O'Connor (34) appear to be in error. These latter materials are synthesized by reacting BF_3 with solutions of the trisdithiocarbamates. Hendrickson and Martin (36) demonstrated by classical methods that "$Co(S_2CN(CH_3)_2)_3BF_4$" is $Co_2[S_2CN(CH_3)_2]_5BF_4$ and Pignolet and Mattson reported (37) the crystal structure (Figure 3) of $Ru_2[S_2CN-(C_2H_5)_2]_5BF_4$. The ability of dithiolate ligands to bond to metals in various ways, while still remaining bidentate, is nicely demonstrated by this structure.

Our attempts to prepare metal complexes in high oxidation states stabilized by other ligands capable of distributing the metal charge effectively have not yet been very successful. Complexes with cyanodithioformate (Reaction 8) are rather insoluble (38). Complexes of *p*-hydroxy-

(7)

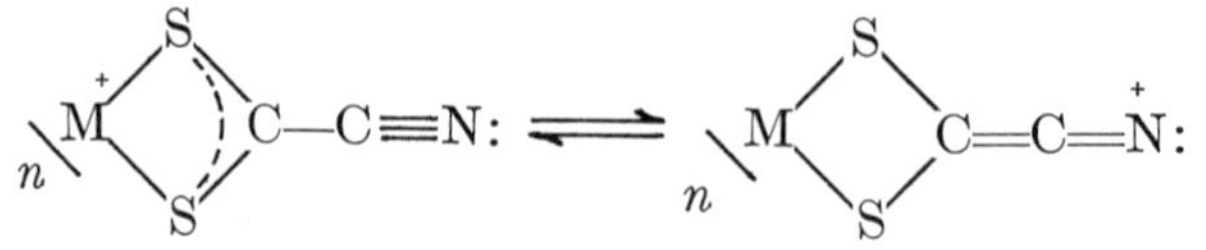

$$(8)$$

dithiobenzoate (**2**) were synthesized recently in our laboratory. Their

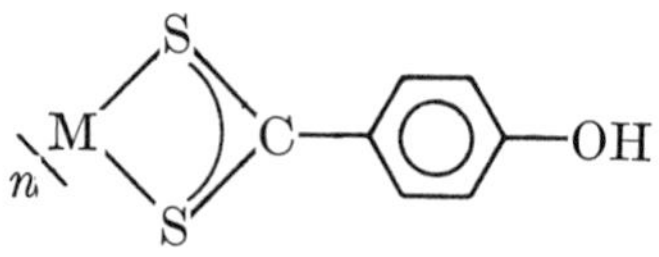

2

potential to stabilize high formal oxidation states by positive charge removal from the metal center (Reaction 9) is currently being evaluated.

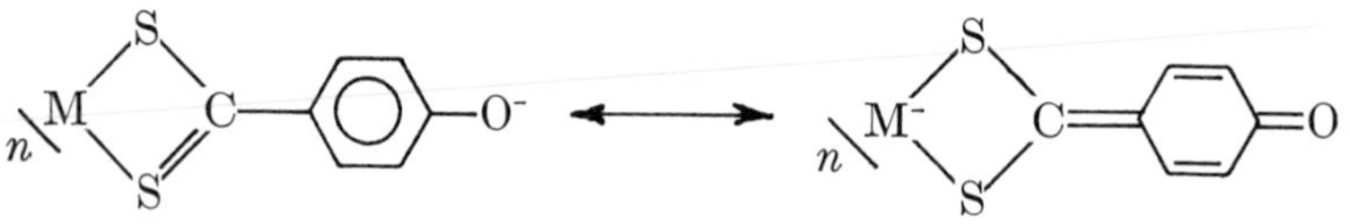

$$(9)$$

Other Miscellaneous Synthetic Information

Since the synthesis and structural analysis of $Cu_8(S_2CC(CN)_2)_6^{4-}$ was reported (*39*), Hollander and Coucouvanis (*40*) prepared and reported the structures of the Cu_8 clusters with the ligand 1,2-dithiosquarate (**3**) and 1,1-dicarboethoxyethylene-2,2-dithiolate (**4**). Each of

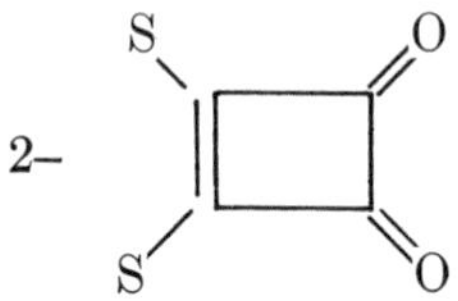

3

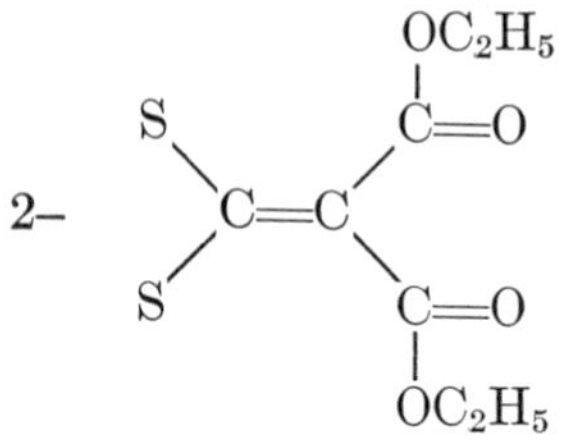

4

these compounds is basically a cube of copper atoms surrounded by a dodecahedron of sulfur atoms. Interestingly the Cu–Cu distance in the two compounds differs very little: 2.822–2.906 A. Molecular orbital cal-

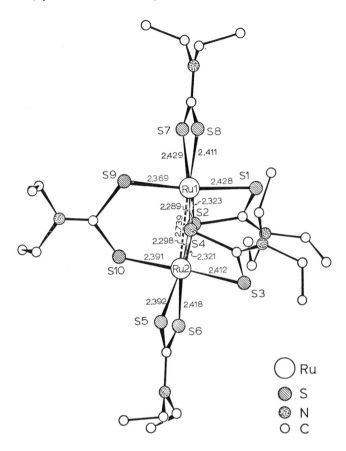

Figure 3.   Crystal structure of $Ru_2[S_2CN(C_2H_5)_2]_5BF_4$

culations of the modified Wolfsberg–Helmholtz (MWH) variety were made by our group (*41*) whereas Messmer (*42*) used Slater–Johnson SCF-αX scattering theory; both methods led to the same general conclusion. The d orbital manifold is not split extensively in Cu_8 and it is completely overlapped by the band of s and p orbitals. The lowest unfilled orbital in either theory is predominantly $4s4p$. The four low lying electronic transitions found between 10.2 kK and 26.0 kK are ligand-to-metal according to the MWH theory and $d \rightarrow sp$ according to the SCF-αX theory. The bonding picture which emerges is that of a small cube of slightly oxidized metallic copper surrounded by a dodecahedral sheath of sulfide ions. The MWH parameterized charge on each copper is $+0.49$ with a $3d^{9.9974}s^{0.4874}p^{0.022}$ configuration. Changes in the Cu–Cu distance (to 2.38 A) have little effect on this bonding description in the MWH model.

Reactions

Sulfur Addition and Abstraction. By using mass spectral intensities of the $C_6H_5CS^+$ fragment upon addition of sulfur enriched in ^{34}S, it was possible (26) to demonstrate that the photochemically initiated sulfur addition reaction (Reaction 10) to planar nickel(II)dithiobenzoate is strikingly specific. The results are consistent with early studies on the

$$ \underset{2}{} \text{Ni} \Big(\!\!\! \underset{S}{\overset{S}{\diagdown}} \!\!\! \Big) C\!-\!C_6H_5 + \tfrac{1}{8}\,{}^*S_8 \xrightarrow[\text{CHCl}_3]{h\nu} \text{Ni} \left(\!\!\! \begin{array}{c} S\!-\!S^* \\ \diagdown \!\!\!\! \diagup \\ S \end{array} C\!-\!C_6H_5 \!\!\right)\!\!\left(\!\!\! \begin{array}{c} S \\ \diagdown \\ S \end{array} C\!-\!C_6H_5 \!\!\right) \quad (10) $$

trithiocarbonate complex that used radioactive ^{33}S tagging. Some scrambling occurs with heating. With zinc, however, total sulfur atom scrambling results. Radiochromatography had been used, notably by Russian investigators in the early 1960's (43), to demonstrate that ^{35}S incorporation into the metal dithiocarbamates was significantly slower for the nickel(II) complex than for the zinc(II) species.

NMR studies of solutions that contain equal mixtures of perthiocumate of zinc(II) (5) and dithiocumate of zinc(II) (6) can be used

             5   **6**

to evaluate the approximate rate of sulfur atom exchange in these species (32). Although a concentration dependence is found which mitigates against a simple mechanistic explanation, the activation energy is small, $\sim 5\,\text{kcal/mole}$, with an extrapolated room temperature exchange lifetime of $\sim 10^{-5}$ sec at infinite dilution. The sulfur-rich zinc complexes are efficient sulfur atom transfer reagents. It appears that zinc dithiolates activate sulfur (vulcanization catalysis) by forming sulfur-rich complexes which have very labile disulfide sulfur atoms.

Oxidized Dithiocarbamates. The ability of thiuram disulfide (7) to

7

undergo ready exchange with dithiocarbamates has been known for more than a decade (43) although it was realized only recently that an oxidized metal complex might be an intermediate. The nickel(IV) complex also displays some interesting reaction chemistry (26). It reacts with the nickel(II)dithiocarbamate to produce a paramagnetic species which, in the case of the di-*n*-butyldithiocarbamate with iodide as the counter ion, was isolated as $Ni(dtc)_2I$ by Willemse. We reported (32) that in 50% $CHCl_3$–toluene glass a 1:1 mixture of the nickel(IV) (bromide) and the nickel(II) complexes exhibits a three-line EPR spectrum (at $-140°C$) with g values of 2.212, 2.190, and 2.165. The powder spectrum of $Ni(dtc)_2I$ shows (32) principal g values of 2.260, 2.215, and 2.027. A distorted square pyramidal geometry could account for these findings.

Solutions of $Ni(n\text{-butyl }dtc)_3Br$ in basic solvents are bleached by visible light to nickel(II) complexes and thiuramdisulfide. The reaction is reversible in CH_3CN, and the rate of return to the nickel(IV) cations can be studied conveniently with spectrophotometry. The experimental data (26) are consistent with a rate law (Equation 11) in which thiuram disulfide–L_2 and halide both play a role.

$$\frac{d\text{Ni}^{IV}}{dt} = -\frac{d\text{Ni}^{II}}{dt} = \frac{k_1[\text{Ni}^{II}][\text{L}_2]}{k_2[\text{Br}^-] + k_3[\text{L}_2]} \tag{11}$$

Intra- and Intermolecular Rearrangements. Pignolet and co-workers (44) described a variety of intramolecular rearrangements of the tris(dithiocarbamates) and suggested that with iron(II) as the metal ion the process involves a trigonal prismatic intermediate. With the nickel(IV) complex, only a broad PMR spectrum was observed with no resolution

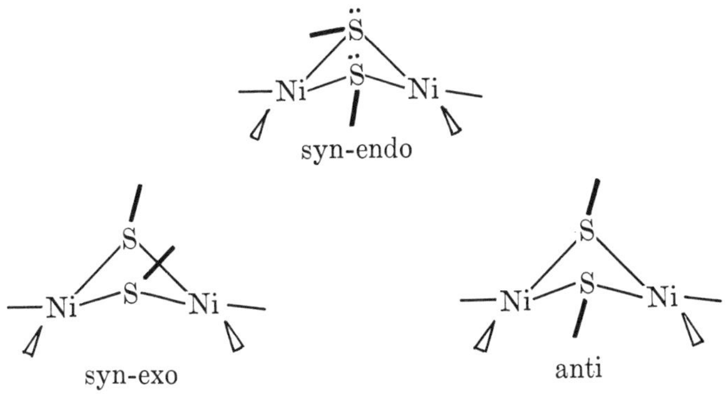

syn-endo

syn-exo

anti

nonplanar

8

of the methylene protons. Similar findings were obtained by Willemse (32) for the tris paladium(IV) and platinum(IV) species. Perhaps the magnetic asymmetry is insufficient to resolve the methylene protons in these nickel triad complexes.

Some NMR studies in our laboratories (45) are designed to bear directly on the so-called Ni–Ni bond (46) in $Ni_2(S_2CRS)_2(SR)_2$ complexes. Both Villa et al. (46) and we (47) reported that these bridged mercaptide complexes exist in folded syn-endo or anti configurations, 8. The short 2.8 A Ni–Ni distance was thought by Villa et al. (46) to imply Ni–Ni bonding. Since the chelate can be exchanged to maintain the dimer, we are carefully examining the proton and ^{13}C NMR spectra of the Ni and Pd complexes (9) which, if folded, should display diastereo-

9

topic methylene protons. While kinetic data have not yet been obtained, at $-50°C$ the Pd complex in CD_2Cl_2 shows a significant broadening of the external methylene groups while the bridging methylene groups remain sharp. A rapid flip-flop, butterfly fashion across the fold of the M_2S_2 rhombus could account for these results. In C_{2v} average symmetry (folded), the bridge methylenes are equivalent, but the external ones are non-equivalent and become equivalent only when a D_{2h} (planar) intermediate occurs. Since this happens when temperatures are as low as $-40°C$, any M–M bond forcing a folded structure must be very weak (less than 10 kcal/mole).

Acknowledgments

Several colleagues were responsible for the studies described here. Their names are noted in the references.

Literature Cited

1. Reid, E. E., "Organic Chemistry of Bivalent Sulfur," Vol. IV, Chemical Publishing, New York, 1962.

2. Delepine, M., *C.R. Acad. Sci.* (1907) **144**, 1125.
3. Coucouvanis, D., *Prog. Inorg. Chem.* (1970) **11**, 233.
4. Eisenberg, R., *Prog. Inorg. Chem.* (1970) **12**, 295.
5. Thorn, G. D., Ludwig, R. A., "The Dithiocarbamates and Related Compounds," Elsevier, New York, 1962.
6. Hofmann, W., "Vulcanization and Vulcanizing Agent," Palmerton, New York, 1967.
7. Rao, S. R., "Xanthates and Related Compounds," Marcel Dekker, New York, 1971.
8. Zeise, W. C., *Rec. Meas. Acad. Roy. Sci. Copenhagen* (1915) **1**, 1.
9. Saccardi, P., Italian Patent **432,082** (1948); *Chem. Abstr.* (1949) **43**, 8604g.
10. Lee, A. G., *J. Chem. Soc. A* (1970) 467.
11. Reichle, W. T., *Inorg. Chem. Acta* (1971) **5**, 325.
12. Schussler, D. P., Fackler, Jr., J. P., unpublished data.
13. Schussler, D. P., Chen, H. W., Fackler, Jr., J. P., unpublished data.
14. Coucouvanis, D., Fackler, Jr., J. P., *J. Am. Chem. Soc.* (1967) **89**, 1346.
15. *Ibid.* (1967) **89**, 1745.
16. Fackler, Jr., J. P., Fetchin, J. A., Smith, J. A., *J. Am. Chem. Soc.* (1970) **92**, 2910, 2912.
17. Fackler, Jr., J. P., Fetchin, J. A., Fries, D. C., *J. Am. Chem. Soc.* (1972) **94**, 7323.
18. Furlani, C., Luciani, M. L., *Inorg. Chem.* (1968) **7**, 1586.
19. Giuliani, A. M., *Inorg. Nucl. Chem. Lett.* (1971) 1001.
20. Burke, J. M., Fackler, Jr., J. P., *Inorg. Chem.* (1972) **11**, 2744.
21. Müller, A., Christophliemk, P., Tossidis, I., Jorgensen, C. K., *Z. Anorg. Allg. Chem.* (1973) **401**, 274.
22. Reynolds, L. W., Fackler, Jr., J. P., unpublished data.
23. Beurskens, P. T., Cras, J. A., Steggerda, J. J., *Inorg. Chem.* (1968) **7**, 810.
24. Brinkhoff, H. C., Cras, J. A., Steggerda, J. J., *Rev. Trav. Chim.* (1969) **88**, 633.
25. Avdeef, A., Fackler, Jr., J. P., Fischer, Jr., R. G., *J. Am. Chem. Soc.* (1970) **92**, 6972.
26. *Ibid.* (1973) **95**, 774.
27. Beurskens, P. T., Cras, J. A., *J. Cryst. Mol. Struct.* (1971) **1**, 63.
28. Pasek, E. A., Straub, D. K., *Inorg. Chem.* (1972) **11**, 259.
29. Brown, K. L., *Cryst. Struct. Commun.* (1974) **3**, 493.
30. Golding, R. M., Harris, C. M., Jessop, K. J., Tennant, W. C., *Aust. J. Chem.* (1972) **25**, 2567.
31. Martin, R. L., Rohde, N. M., Robertson, G. B., Taylor, D., *J. Am. Chem. Soc.* (1974) **96**, 3647.
32. Willemse, J., Thesis, Wijmegen, 1974.
33. Salek, R. Y., Straub, D. K., *Inorg. Chem.* (1974) **13**, 3017.
34. Gahan, L. R., O'Connor, M. J., *J. Chem. Soc. Chem. Commun.* (1974) 68.
35. Nigo, Y., Masuda, I., Shiura, K., *Chem. Commun.* (1970) 476.
36. Hendrickson, A. R., Martin, R. L., *J. Chem. Soc. D* (1974) 873.
37. Pignolet, L. H., Mattson, R. M., unpublished data.
38. Simmons, H. E., Blomstron, D. C., Vest, R. D., *J. Am. Chem. Soc.* (1962) **84**, 4756.
39. McCandlish, L. R., Bissell, R. C., Coucouvanis, D., Fackler, Jr., J. P., *J. Am. Chem. Soc.* (1968) **90**, 7357.
40. Hollander, F. J., Coucouvanis, D., *J. Am. Chem. Soc.* (1974) **96**, 5646.
41. Avdeef, A., Ph.D. Thesis, Case Western Reserve University, 1973.
42. Messmer, R. P., as reported by Slater, J. C., Johnson, K. H., *Phys. Today* (Oct. 1974) 34.
43. Khodzhaeva, I. V., Kissin, Y. V., *Russ. J. Phys. Chem.* (1963) **37**, 412.

44. Pignolet, L. H., Weiher, J., Patterson, G. S., Holm, R. H., *Inorg. Chem.* (1974) **13**, 1263.
45. Lin, I. J. B., Fackler, Jr., J. P., unpublished data.
46. Villa, A. G., Manfredotti, A. G., Nardelli, M., Pelizzi, C., *Chem. Commun.* (1970) 1322.
47. Fackler, Jr., J. P., Zegarski, W. J., *J. Am. Chem. Soc.* (1973) **95**, 8566.
48. McKay, A. F., Garmaise, D. L., Paris, G. Y., Gelblum, S., Ranz, R. J., *Can. J. Chem.* (1960) **38**, 2042. These authors apparently synthesized $KS_2COC_6H_5$ in dimethylformamide. Although they did not isolate the salt, the methyl ester was obtained in a 42% yield by adding methyl chloride to the reaction mixture. The potassium salts of both 2,5- and 2,6-dimethylphenylxanthate have been isolated from dioxane (J. P. Fackler, Jr. and H. W. Chen).

RECEIVED January 24, 1975. Work supported by the National Science Foundation, the National Institutes of Health, the Petroleum Research Fund, and the General Electric Foundation.

Binuclear and Mixed Metal Binuclear Chelates of Schiff-Base Derivates of 1, 3, 5-Triketones

RICHARD L. LINTVEDT, BARBARA TOMLONOVIC, DAVID E. FENTON, and MILTON D. GLICK

Wayne State University, Detroit, Mich. 48202

Some new binucleating, Schiff-base ligands were prepared by reacting the 1,3,5-triketones 1-phenyl-1,3,5-hexanetrione and 2,2-dimethyl-3,5,7-octanetrione with ethylenediamine in 1:1 molar ratios. The products, abbreviated $H_4(BAA)_2en$ and $H_4(PAA)_2en$ respectively, contain two different coordination sites which enable them to bind selectively to two different metal ions. Several heteronuclear complexes were prepared and characterized; these included $NiZn(BAA)_2en$, $NiCu(BAA)_2en$, $NiVO(BAA)_2en$, and $NiUO_2(BAA)_2en$. A single-crystal x-ray structure was determined for NiZn-$(BAA)_2en$. The Cu(II) chelates of these ligands have unusually strong antiferromagnetism; $Cu_2(PAA)_2en$ is diamagnetic at room temperature, has no EPR signal, and gives a high resolution NMR spectrum. $Cu_2(PAA)_2en$ undergoes electrochemical reversible reduction vs. SCE at -0.61 V in a one-step, two-electron reduction per molecule.

There has been considerable interest recently in the physical properties of polynuclear transition metal complexes. This interest covers a wide range of areas including metalloenzymes, homogeneous catalysis, electrical conductivity, and magnetic exchange interactions. Until now much of the research involved investigating an isolated example of an interesting polynuclear complex. Although many compounds with interesting and important properties were discovered in this manner, it is not a systematic method of either preparing new materials or understanding the physical principles responsible for their properties.

Ideally, one would like to prepare many closely related complexes, each one designed to vary a parameter related to the properties under investigation. It is probable that only through such molecular design will a thorough understanding of the properties and reactivity of poly-nuclear complexes be achieved. A surprisingly little studied class of polynuclear transition metal complexes which presents an excellent op-portunity for molecular design of hundreds of new compounds consists of the β-polyketonate chelates and their derivatives. The generalized struc-ture of the β-polyketone ligands may be depicted as a homologous series in which the well known 1,3-diketones are the simplest members. The

1,3-diketone   1,3,5-triketone

1,3,5,7-tetraketone β-polyketone

organic molecules may be altered in various ways including varying R and R', replacing hydrogens on the backbone carbons with other groups, replacing the ketonic oxygens with other electron pair donors such as N, S, Se, and P, and increasing the number of ketone groups (*i.e.* increas-ing n). Very little information about such molecules, much less their metal derivatives, is found in the literature. Some relatively high molecu-lar weight β-polyketones have been reported. For example, Harris and co-workers (*1, 2;* also *see* references cited in *2*) prepared β-polyketones in a homologous series up to and including a 1,3,5,7,9,11,13,15-octaketone, and there are reports of β-polyketone polymers with molecular weights of about 2000 (*3*).

When one considers the chelates prepared from the β-polyketones, it is important to realize that these ligands are readily enolizable and that they form polyanionic species in the presence of a base (*e.g.* Reaction 1). The conjugated sp^2 type carbons in the polyanions impart a rigid planarity to the system so that chelation of metal ions results in planar, polynuclear molecules in which the metals share bridging oxygens. Thus, by use of a homologous series of ligands, a homologous series of chelates may be

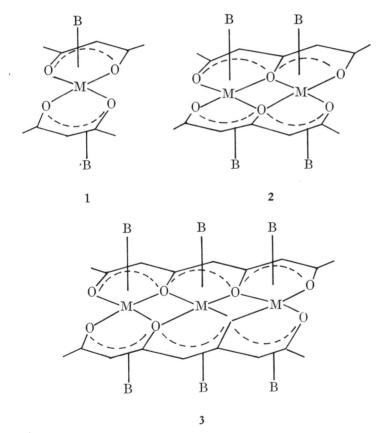

prepared in which the major structural features are constant. The first three members of such a series of divalent metal complexes are **1**, **2**, and **3**. In these structures B represents an adduct ligand, *e.g.* H_2O or pyridine,

which may or may not be present depending on the metal ion. There are a few papers on 1,3,5-triketonate chelates (*4, 5, 6, 7, 8, 9, 10*) but only one report on trinuclear 1,3,5,7-tetraketonate chelates (*11*). To our knowledge, no higher homologs have been reported.

Recently single-crystal x-ray structure determinations for binuclear 1,3,5-triketonates of Cu(II) (*12, 13*), Ni(II) (*7*), and Co(II) (*6*) have confirmed the general structures 1, 2, and 3 and the fact that the ligands are essentially planar. Indeed, in the Ni(II) and Co(II) chelates where B is pyridine, the backbone carbons, the ketonic oxygens, and the metal ions are coplanar within experimental error. The C–C and C–O distances reveal that the π electrons in the ligands are delocalized and further support the belief that the carbons are best represented as sp^2. The other important structural features are: (a) the bridging oxygen-to-metal distance is very similar to the terminal oxygen-to-metal distance, (b) the O_b–M–O_b angle is about 75°, and (c) the M–O_b–M angle is about 102°.

The primary interest in the β-polyketonate chelates so far has been to investigate magnetic exchange between the metal ions within the same molecule. Temperature-dependent magnetic susceptibility was measured on binuclear triketonates of d^1–d^1 (*8*), d^3–d^3 (*14*), d^7–d^7 (*6*), d^8–d^8 (*7*), and d^9–d^9 (*4, 5, 9, 12, 13*) systems. In all cases, quite strong antiferromagnetic exchange is observed with the strongest being the d^9–d^9 Cu(II) chelates. The similar geometry of the metal ions and bridging oxygens implies that differences in the strengths of exchange are caused by differences in the symmetry of the exchanging electrons (*15*). Since the unpaired (d_{xy}) electron in Cu(II) is directed at the ketonic oxygen and is able to participate directly in σ bonding, it is logical that d^9–d^9 exchange is the strongest yet observed in the first-row transition metal ions. More quantitative information about unpaired electron symmetry and exchange strength awaits preparation of new complexes, more magnetic measurements, and additional structural information.

Another area of considerable interest in polynuclear metal complex chemistry for which the β-polyketones seem well suited is that of mixed metal and mixed valence complexes. When one is seeking fundamental data on how the molecular structure and properties of such compounds relate to bulk properties, it is not sufficient to be content with uncertainties concerning purity and the positions of the different ions. In this paper we report on a systematic approach to the preparation of some heteronuclear, molecular chelates, *i.e.* molecules containing two or more

4

different metal ions. The ligands used in this investigation are Schiff-base derivatives of 1,3,5-triketones and ethylenediamine. These molecules contain two different coordination sites, one with two nitrogen and two oxygen donors, the other with four oxygen donors (4). The binuclear chelates of these ligands have rather unusual properties whether the two metal ions are similar or different.

Experimental

The 1,3,5-triketones were prepared by the modified Claisen-type condensation described by Miles, Harris, and Hauser (16). Abbreviations are based on the trivial names: 1-phenyl-1,3,5-hexantrione is benzoylacetylacetone (H_2BAA) and 2,2-dimethyl-3,5,7-octanetrione is pivaloylacetylacetone (H_2PAA).

$H_4(BAA)_2en$. The condensation of ethylenediamine and H_2BAA was effected by the addition of 2.0 ml (0.03 mole) ethylenediamine to 11.0 g (0.054 mole) H_2BAA in a minimum of methanol. The solution was heated to boiling and cooled at $-5°C$. The crude crystals were collected and recrystallized from acetone. The yellow crystals melt at 171°C. Yields are typically 80–90%. Anal.: calcd for $C_{26}H_{28}N_2O_4$: C 72.19, H 6.52, N 6.45; found: C 71.84, H 6.45, N 6.66.

$H_4(PAA)en$. The preparation of this ligand was similar to that of $H_4(BAA)_2en$ with yields of about 90%. The off-white, mica-like crystals melt at 145°–146°C. Anal.: calcd for $C_{22}H_{36}N_2O_4$: C 67.34, H 9.20, N 7.14; found C 66.98, H 9.40, N 7.12.

$NiH_2(BAA)_2en$. This mononuclear complex was prepared by mixing equimolar quantities of $H_4(BAA)_2en$ dissolved in hot acetone and $Ni(C_2H_3O_2)_2 \cdot 4H_2O$ dissolved in a small amount of hot water. When the reaction mixture was cooled, the crude product precipitated and it was then recrystallized from acetone. The red crystals melt at 174°C. Anal.: calcd for $C_{26}H_{26}N_2O_4Ni$: C 63.84, H 5.38, N 5.73, Ni 12.0; found: C 63.56, H 5.32, N 5.73, Ni 11.9.

$Cu_2(BAA)_2en$. Two solutions were prepared: one contained 2.0 g (0.01 mole) $H_2(BAA)_2en$ in 40 ml acetone–160 ml $CHCl_3$, the other 4.0 g (0.02 mole) $Cu(C_2H_3O_2)_2 \cdot H_2O$ in 125 ml methanol. The warm ligand solution was then added to the warm, stirred metal ion solution during about 30 min. The reaction mixture was heated 1 hr until the volume decreased to about 100 ml. The precipitated product was filtered, air dried, and washed three times with 30 ml portions of acetone and once with 30 ml $CHCl_3$. The green product does not melt below 335°C. Anal.: calcd for $C_{26}H_{24}N_2O_4Cu_2$: C 56.21, H 4.35, N 5.04, Cu 22.88; found: C 56.22, H 4.43, N 5.21, Cu 22.6.

$Cu_2(PAA)_2en$. The crude product was prepared by mixing stoichiometric quantities of $H_4(PAA)_2en$ and $Cu(C_2H_3O_2)_2 \cdot H_2O$ (1:2 molar ratio) in hot methanol. After several minutes of heating and then cooling, the crude product precipitated; it was then recrystallized from hot $CHCl_3$. The small green crystals melt at 264°C. Anal.: calcd for $C_{22}H_{32}N_2O_4Cu_2$: C 51.20, H 6.21, N 5.43; found: C 50.46, H 6.13, N 5.38.

NiCu(BAA)$_2$en. The mononuclear complex NiH$_2$(BAA)$_2$en was used as a ligand by dissolving it in acetone and slowly adding an equimolar quantity of Cu(C$_2$H$_3$O$_2$)$_2$ · H$_2$O dissolved in a minimum volume of H$_2$O. After concentration and cooling, the heteronuclear chelate precipitated. The crude product was recrystallized from pyridine, washed with cold acetone, and dried under vacuum. Yield is about 85% based on NiH$_2$(BAA)$_2$en. The dark red-brown product does not melt below 335°C. Anal.: calcd for C$_{26}$H$_{24}$N$_2$O$_4$NiCu: C 56.70, H 4.39, N 5.09, Cu 11.54; found: C 56.85, H 4.47, N 5.12, Cu 11.50.

NiZn(BAA)$_2$en · 2H$_2$O. This compound was prepared by the same method that was used for NiCu(BAA)$_2$en. The crude product was recrystallized from benzene to give a golden brown powder with about 60% yield. The compound does not melt below 300°C. Anal.: calcd for C$_{26}$H$_{28}$N$_2$O$_6$NiZn: C 53.06, H 4.70, N 4.76, Ni 9.97; found: C 53.41, H 4.30, N 4.96, Ni 10.2. Recrystallization from pyridine gave large single crystals of a pyridine adduct, NiZn(py)(BAA)$_2$en · 2py, in which one pyridine is coordinated to the Zn atom. The uncoordinated pyridines are easily removed by washing with acetone. Anal.: calcd for C$_{26}$H$_{24}$N$_2$O$_4$NiZn(C$_5$H$_5$N): C 58.94, H 4.63, N 6.65; found: C 58.51, H 4.72, N 6.61.

NiUO$_2$(BAA)$_2$en. This compound was prepared by the same method that was used for NiCu(BAA)$_2$en. The crude product was recrystallized from acetone to give a dark red powder which does not melt below 330°C. Anal.: calcd for C$_{26}$H$_{24}$N$_2$O$_6$NiU: C 41.24, H 3.19, N 3.70, U 31.43, Ni 7.75; found: C 40.68, H 3.82, N 3.50, U 31.4, Ni 7.4.

NiVO(BAA)$_2$en. The same procedure was used to prepare this compound. Recrystallization from acetone yielded dark red crystals which do not melt below 300°C. Anal.: calcd for C$_{26}$H$_{24}$N$_2$O$_5$NiV: C 56.35, H 4.37, N 5.06; found: C 56.15, H 4.43, N 5.06.

Spectra. Spectra were recorded using the following spectrometers: absorption spectra, a Cary-14; mass spectra, an Atlas C-4 mass spectrometer; ESR, a Varian E-4; and NMR, a Varian A60A. The fluorescence spectra were recorded using an Amico Bowman spectrofluorometer.

Magnetic Susceptibility Measurements. The magnetic susceptibility of the solids was measured by the Faraday method with Hg[Co(SCN)$_4$] used as the calibrant. Corrections for diamagnetism were made using Pascal's constants. Solution measurements were made in CHCl$_3$ by NMR techniques (17).

X-ray Structure Determination. A three-dimensional single-crystal structure was determined on crystals obtained by recrystallization of NiZn(BAA)$_2$en from pyridine. The crystals were monoclinic I 2/c with lattice parameters of $a = 28.403(6)$ A, $b = 8.465(3)$ A, $c = 30.220(9)$ A, $\beta = 105.86(2)°$, and Z = 8. Intensity data were collected by the θ-2θ scan technique with graphite-monochromated Mo–Kα radiation on a Syntex P2$_1$ diffractometer. Of the 5055 data with sin $\theta/\lambda < 0.54$, 2559 had $I > 3\sigma(I)$ and these were used in the solution and refinement of the structure. The structure was solved by Patterson–Fourier methods and refined by full-matrix (isotropic) and block-diagonal (anisotropic) least squares refinement. The conventional discrepancy factors at the present state of refinement are $R = 0.056$ and $R_w = 0.071$.

Results and Discussion

The diamine Schiff-base derivatives of 1,3,5-triketones are an interesting and versatile class of ligands. From the derivatives prepared so far, it appears that the diamine condensations that occur are fairly analogous to those of the 1,3-diketones. For example, amines condense very readily at the carbonyl carbon of an acetyl group but not at the carbonyl carbon of a benzoyl or pivaloyl group whether it be a diketone or a triketone. Therefore, in an unsymmetrically substituted 1,3,5-triketone in which the 1-substituent is $-CH_3$ and the 5-substituent is C_6H_5 or *tert*-C_4H_9, a diamine condenses exclusively at the methyl end. Under a variety of experimental conditions, condensation was not observed at the central carbonyl carbon. As a result of this specificity, binucleating ligands can be prepared in which there are two coordinatively different sites for metal ions. Since one site furnishes two imine-type nitrogens and two oxygens and the other furnishes four oxygen donors, the metals occupying these sites experience different ligand field effects. Several studies have demonstrated that magnetic exchange between metal ions in the triketonates is strongly antiferromagnetic (*4–15*). Therefore, in the Schiff-base triketonate, it is possible to investigate the properties of binuclear complexes in which strongly coupled metals are in significantly different ligand fields.

Heteronuclear Complexes. The presence of two coordinatively distinct sites in the ligand implies that certain metal ions have a preference

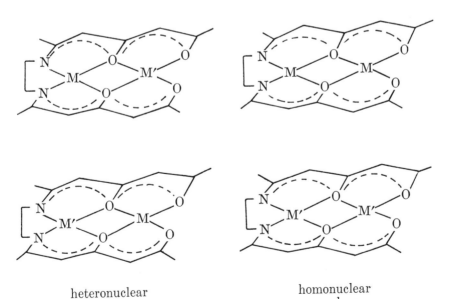

heteronuclear
positional isomers

homonuclear
complexes

for one of the sites and different metal ions a preference for the other site. By using these preferences, we can prepare pure compounds in which two different metal ions are contained within the same molecule. In addition, the positions of the metal ions within the molecule are predictable because of the coordination specificity. If conditions are not completely favorable, however, impurities in the form of positional isomers and homonuclear chelates may be formed. The characterization and identification of a pure product when all possibilities exist is not trivial, and even structure determination is not unambiguous when the two metals have similar atomic numbers. Thus, the approach taken to the systematic preparation of pure heteronuclear complexes must be very deliberate, and characterization must include several different physical measurements.

Our approach to the preparation of pure heteronuclear chelates is to establish the selectivity of the two coordination sites in a ligand such as $H_4(BAA)_2$en by determining the structure of some mononuclear chelates. Once these are unambiguously characterized, they are used as ligands to bind a second metal ion with a proven preference for the available site. The two metals are chosen so as to minimize competition for the same site within the molecule.

MONONUCLEAR CHELATES OF $H_4(BAA)_2$EN. Three well characterized complexes with the general formula $M[H_2(BAA)_2$en], where M is Ni(II), Cu(II), and VO(II), were prepared. Single-crystal x-ray structure determinations of the Cu(II) (18) and VO(II) (19) complexes revealed that the metal ions occupy the site with four oxygen donors under the preparative conditions used. Although the presence of a small amount of the other positional isomer cannot be discounted, sample homogeneity and crystallizability are indications that a significant quantity is not present in the crystalline product which is isolated in high yields.

A structure determination of the other mononuclear chelate, Ni[H$_2$-(BAA)$_2$en], was deemed not necessary because of its unambiguous characterization by spectral and magnetic measurements. The visible spectrum of Ni[H$_2$(BAA)$_2$en] is essentially identical to that of Ni(Acac)$_2$en which is known to contain square planar Ni(II). The IR spectrum contains an intense unchelated carbonyl bond at 1700 cm^{-1} which indicates that the nickel atom is coordinated to two nitrogens and two oxygens. In addition, Ni[H$_2$(BAA)$_2$en] is diamagnetic with X_g obs. $= -266 \times 10^{-6}$ cgs while χ_g calculated from Pascal's constants is -246×10^{-6} cgs, which is further proof of square planar geometry about the Ni.

The spectral properties of Ni[H$_2$(BAA)$_2$en] are not a sensitive measure of purity with respect to the two positional isomers, but the diamagnetism of the compound may be used to judge purity. If the Ni were coordinated to four oxygens, its environment would be very similar to that found in bis(benzoylacetonato)nickel(II), which is paramagnetic by virtue of the fact that it attains higher coordination numbers through solvent coordination or oligomerization (20). Even a small amount of a paramagnetic Ni(II) species would be easily detected by susceptibility measurements. On the basis of the experiments performed so far, it appears that pure isomers of mononuclear complexes are readily prepared under a given set of experimental conditions and that the coordination sites exhibit a significant degree of specificity toward different metal ions.

PREPARATION AND CHARACTERIZATION OF HETERONUCLEAR CHELATES. Once it was established that the two sites in Schiff-base derivatives of 1,3,5-triketones may specifically coordinate different metal ions, it became feasible to prepare pure mixed metal complexes in which the position of the two metal ions is known. As a matter of preparative convenience in the early stages of this work, we chose to use $Ni[H_2(BAA)_2en]$ as a ligand to bind a different metal ion which has a preference for the available oxygen donor site (Reaction 2). The major rationale for using the

$$M = Cu, VO, Zn, UO_2$$

Ni complex as a ligand is as follows: (a) this mononuclear complex is appreciably soluble in most organic solvents including hydrocarbons; (b) we observed that this complex has extremely little tendency to isomerize to the other positional isomer in which Ni is bonded to four oxygens; and (c) the diamagnetism of $Ni[H_2(BAA)_2en]$ is a convenient property for judging whether or not the metal ions change position during the reaction since the Ni is expected to be paramagnetic in the oxygen donor position.

Mass Spectra. The heteronuclear chelates $Zn(py)(BAA)_2en$, $NiCu(BAA)_2en$, and $NiVO(BAA)_2en$ are sufficiently volatile that mass spectra can be obtained. In each case, strong parent ion peaks are observed minus any adducted ligands. The high mass regions of the $NiZn(BAA)_2en$ and $NiVO(BAA)_2en$ spectra are surprisingly simple—only the parent ion, the parent ion minus two mass units, and the doubly charged ions arising from these two species are observed in appreciable concentration above $m/e = 105$. With both these chelates, the loss of two mass units by the parent ion is a very favorable process (Reaction 3). The most logical explanation of the strong P^+-2 peaks is that two hydrogen atoms are lost from the ethylenediamine group which results in a completely unsaturated molecular ion. This explanation for the P^+-2 peak is being investigated further by isotope studies.

M	m/e
Zn	550, 552, 554, 556
VO	553, 555

M	m/e
Zn	548, 550, 552, 554
VO	551, 553

$$-e^-$$

$$p^{2+}$$

$$[p-2H]^{2+}$$

(3)

M	m/e
Zn	275, 276, 277, 278
VO	276.5, 277.5

M	m/e
Zn	274, 275, 276, 277
VO	275.5, 276.5

It is obvious that the peaks around mass 275 are caused by doubly charged ions since the half-mass and full-mass peaks are all quite intense. The importance of the doubly charged ions is certainly the result of the presence of two metal centers, each of which could logically stabilize a positive charge. To pursue this point, the mass spectrum of the mononuclear chelate $Ni[H_2(BAA)_2en]$ was recorded. There is no evidence of formation of doubly charged ions by this compound, which constitutes strong indirect evidence that the maximum positive charge is determined by the number of metal atoms present.

The spectrum of $NiCu(BAA)_2en$ is similar to those of Zn and VO, but it is more complex in that P^+-Cu and related fragments are quite intense. The primary reason for recording the mass spectra was to support the conclusion that these compounds are pure heteronuclear chelates rather than mixtures of hetero- and homonuclear molecules. Since no peaks attributable to species such as $Ni_2(BAA)_2en$ and $Zn_2(BAA)_2en$ were observed, the mass spectra support the conclusion that the samples consist of pure heteronuclear molecules.

Magnetic Properties. Although the analytical and mass spectral data indicate that the heteronuclear chelates are pure, they give no clue as to whether some of the metal ions have reversed their positions during preparation. Since one of our purposes is to be able to prepare pure mixed metal compounds in which positions of the metal ions are known, such a reversal would constitute an impurity in the bulk sample. The magnetic properties of these Ni mixed metal compounds do, however, enable one to determine the metal ion positions with some certainty

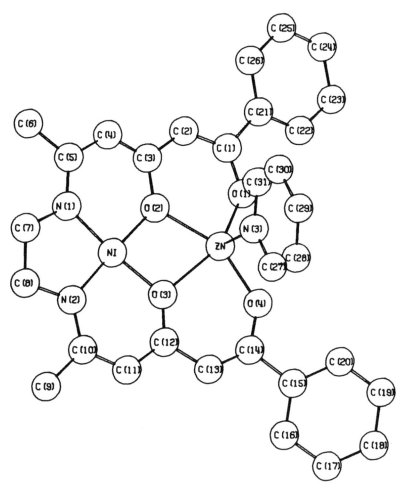

Figure 1. Molecular structure of NiZn(py)(BAA)₂en

since Ni(II) is square planar and diamagnetic when coordinated to the two nitrogens and the two central oxygens. In the alternative position. the Ni(II) is expected to be paramagnetic, as discussed above.

The magnetic susceptibilities of all the heteronuclear complexes prepared from Ni[H$_2$(BAA)$_2$en] are consistent with a diamagnetic, square planar Ni(II). The magnetic moments of NiCu(BAA)$_2$en and NiVO-(BAA)$_2$en correspond to one unpaired electron per molecule, NiZn(py)-(BAA)$_2$en is diamagnetic, and NiUO$_2$(BAA)$_2$en has a small paramagnetism that may result from some orbital contribution.

Structure of NiZn(py)(BAA)$_2$en. Single crystals were prepared by crystallizing NiZn(H$_2$O)(BAA)$_2$en from pyridine, and the structure was proven to be as depicted in Figure 1. It is interesting that although x-ray techniques cannot distinguish between Ni and Zn atoms, the four-coordinate atom must be Ni since the compound would be paramagnetic if Ni were five-coordinate. In addition, the coordination sphere of the five-coordinate atom is very similar to that found in Zn(Acac)$_2$py (21).

The Ni and its four donor atoms (two nitrogens and two oxygens) form a plane in which no atom is displaced more than 0.03 A from planarity. The Zn is in a typically five-coordinate environment in which the metal atom is displaced 0.32 A from the least-squares plane of the four donor oxygens. The average distances in the coordination spheres of the two metal atoms are depicted in Figure 2, and the important angles in the metal coordination spheres are shown in Figure 3.

The geometry of the M⟨O/O⟩M ring in the Schiff-base ligand (BAA)$_2$en^{4-} does not differ greatly from that found in the homonuclear complexes that contain simple triketonate ligands (*see* Table I). The

Table I. Comparisons of M⟨O/O⟩M Ring Structure in Several Binuclear Complexes

Complex[a]	Metal–Metal Distance, A	O–M–O Angle, °	M–O–M Angle, °	Ref.
Ni$_2$(DBA)$_2$(py)$_4$	3.17	78.5	101.5	7
Co$_2$(DBA)$_2$(py)$_4$	3.27	77.3	102.7	6
Cu$_2$(BAA)$_2$(py)$_2$	3.06	75.9	103.6	12
Cu$_2$(DAA)$_2$(py)$_2$	3.05	77	103	13
NiZn(py)(BAA)$_2$en	3.12	79.7, 68.8[b]	104.9	

[a] DBA is the dianion of 1,5-diphenyl-1,3,5-pentanetrione whose trivial name is dibenzoylacetone. DAA is the dianion of 2,4,6-heptanetrione whose trivial name is diacetylacetone.

[b] The angle 68.8° is for O–Zn–O, and 79.7° is for O–Ni–O.

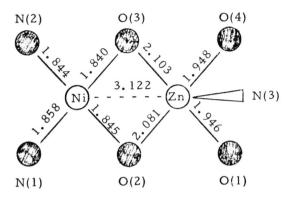

Figure 2. Distances (in Å) in the coordination spheres of Ni and Zn

structural similarities lead one to conclude that any of several metal ions may be introduced without significantly distorting the part of the molecule in which metal–metal interactions take place. Therefore, from the standpoint of magnetic superexchange studies, a great many homo- and heteronuclear complexes can be investigated under conditions of very similar coordination geometry. This fact should prove very useful in the quantitative experimental testing of current theory describing superexchange mechanisms.

Finally, the structure of $NiZn(py)(BAA)_2en$ lends convincing structural support to the contention that the coordinatively different sites in the ligand determine predictably the positions of the two metal atoms. This information, together with macroscopic measurements (particularly magnetic susceptibility) to indicate bulk purity, demonstrates that it is

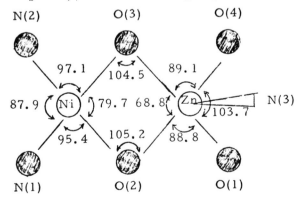

Figure 3. Bond angles in the coordination spheres of Ni and Zn

possible to take an extremely systematic approach to the synthesis of heteronuclear molecular complexes.

Homonuclear Chelates of $H_4(PAA)_2en$. In the above discussion, it was shown that the reaction of 1:1 molar ratios of Cu(II) and H_4-$(BAA)_2en$ at elevated temperatures in MeOH yields $Cu[H_2(BAA)_2en]$ with Cu(II) coordinated to four oxygens. Under different preparative conditions, however, other products are obtained (*see* Reactions 4 and 5).

$$H_4(PAA)_2en + 2Cu(OAc)_2 \xrightarrow[\text{hot}]{\text{CHCl}_3\text{–MeOH}} Cu_2(PAA)_2en + 4HOAc \quad (4)$$

$$H_4(PAA)_2en + 1Cu(OAc)_2 \xrightarrow[0°C]{\text{CHCl}_3\text{–MeOH}} \quad (5)$$

purple solution

Analogous compounds were prepared in which the *tert*-butyl group is replaced by phenyl. These new compounds have unusual properties which should prove useful in the study of polynuclear complexes.

MAGNETIC PROPERTIES. The binuclear Cu(II) compounds of Schiff-base triketonate ligands all exhibit extremely strong antiferromagnetism. The solid and solution magnetic properties of $Cu_2(PAA)_2en$ and the related triketonate $Cu_2(PAA)_2$ are compared in Table II. The high

Table II.   Magnetic Properties of $Cu_2(PAA)_2$ and $Cu_2(PAA)_2en$

Parameter	T, °K	$Cu_2(PAA)_2$	$Cu_2(PAA)_2en$
Magnetic moment (BM)			
solid	300	0.22	0.42
	77	0.00	0.21
solution[a] (in $CHCl_3$)	ambient	0.18	0.00
Magnetic resonance in $CHCl_3$ or $CDCl_3$			
EPR		very weak signal	no signal
NMR		broad absorptions	sharp absorptions

[a] Determined by NMR.

resolution NMR spectrum and the absence of an EPR signal for Cu_2-$(PAA)_2en$ prove that the complex is diamagnetic at room temperature.

Thus, the antiferromagnetism is strong enough to pair completely the electrons on the two Cu(II) ions. This means that the ground state singlet–excited state triplet separation is at least 1400 cm^{-1} by the Bleaney–Bowers method (22). Strong antiferromagnetism in planar binuclear Cu(II) complexes is not uncommon, but to our knowledge no such complexes have been reported in which high resolution NMR spectra are obtained and there is no EPR signal. Since trace amounts of paramagnetic Cu(II) broaden NMR peaks tremendously and are easily observed by EPR, not only is the magnetic exchange in $Cu_2(PAA)_2en$ very strong, but the compound is unusually free of Cu(II) impurities. Even at very high instrumental gains, no EPR signal is observed.

The absence of an EPR signal at room temperature in $CHCl_3$ makes possible a series of unusual mixed metal EPR experiments in which a metal other than Cu is doped into $Cu_2(PAA)_2en$. Introduction of a trace of the new metal ion into the diamagnetic $Cu_2(PAA)_2en$ disrupts the magnetic exchange in the substituted molecules, thereby giving rise to a Cu(II) EPR signal that is attributable to the mixed metal molecules. The metal ion doped into $Cu_2(PAA)_2en$ may be diamagnetic, paramagnetic, EPR active, or EPR inactive. In each case, a signal appears whenever the doped metal replaces a Cu(II) in $Cu_2(PAA)_2en$.

Introducing Zn(II) into $Cu_2(PAA)_2en$, for example, gives the signal in Figure 4 (the spectrum was recorded in $CHCl_3$ at room temperature). The sample was prepared by mixing a $CHCl_3$ solution of $H_2(PAA)_4en$ and a methanol solution that contained Cu(II) acetate and Zn(II) acetate in 95:5 molar ratio; the product was then collected and recrystallized from $CHCl_3$. The spectrum gives an added bonus in that superhyperfine splitting of 15–20 g is observed. Since this is the magnitude of copper–nitrogen splitting and it is almost identical to the purple isomer of $Cu[H_2(PAA)_2en]$, it can be stated with some confidence that the oxygen-coordinated Cu(II) is the one replaced by Zn(II). The green isomer of $Cu[H_2(BAA)_2en]$, which was demonstrated structurally to contain oxygen-coordinated Cu(II), has a simple EPR spectrum without superhyperfine splitting. Analogous experiments with a variety of other metal ions should furnish much information about heteronuclear complexes with potentially strong superexchange interactions.

SPECTRAL PROPERTIES. *Absorption Spectra.* The spectral properties of binuclear Cu(II) complexes which exhibit strong magnetic exchange are not well known. Experimentally there is some advantage to studying the polyketonates and their Schiff-base derivatives since comparisons can be made between the polynuclear molecules and the well known mononuclear diketonates. For example, $Cu_2(PAA)_2en$ can be envisioned as resulting from a fusion of diketonate and Schiff-base diketonate chelates (Reaction 6). Since the ligand fields and metal ion geometries

$$(6)$$

before and after fusion are quite similar, spectral changes in going from mono- to binuclear molecules can be attributed to the formation of the oxygen-bridged system and the resultant change in energy levels. The visible spectra of mono- and binuclear complexes are compared in Table

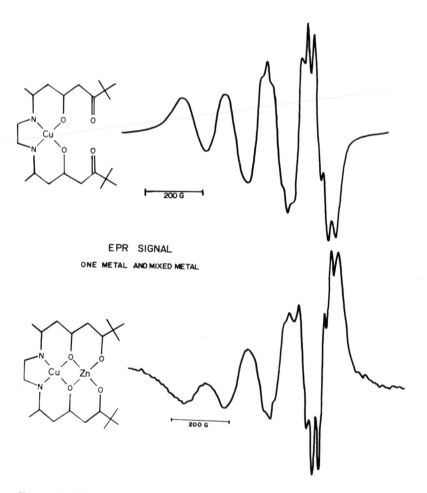

EPR SIGNAL

ONE METAL AND MIXED METAL

Figure 4. First derivative EPR spectrum of Zn(II)-doped $Cu_2(PAA)_2en$ compared with that of the purple isomer $Cu[H_2(PAA)_2en]$

III. It is clear that new absorptions and/or shifts in spectral transitions take place in going from the mono- to the binuclear complexes. Although

Table III.   The Visible Spectra of Some Related Mono- and Binuclear Cu(II) Complexes (in CHCl₃)

Complex	Wavelength, nm (ϵ_{max})		
Cu(Acac)₂			650 (38)[a]
Cu(Acac₂)en		540 (198)[a]	
Cu₂(PAA)₂en	512 (75)	650 (160)	745 (50)
Cu₂(PAA₂)	~475 (75)	645 (282)	735 (65)

[a] From Ref. 23.

it is premature to consider making assignments, these data do indicate that additional spectral studies are warranted.

Emission Spectra. The emission spectra from transition metal complexes can be an extremely useful aid in studying molecular electronic states. The binuclear complex Cu₂(PAA)₂en has a sharp fluorescence emission at 427 nm. This is unusual since Cu(II) compounds rarely, if ever, are observed to emit. It is significant that neither Cu(Acac)₂ nor Cu(Acac)₂en emits under the experimental conditions used to study Cu₂(PAA)₂en. Indeed, there are no reports in the literature of emission from either of these chelates under any experimental conditions. The fluorescence spectra of the protonated ligand, H₄(PAA)₂en, and of Cu₂(PAA)₂en are compared in Table IV. The excitation wavelength in

Table IV.   Fluorescence Spectra of H₄(PAA)₂en and Cu₂(PAA)₂en (in 3-Methylpentane at 77°K)

Substance	Excitation, nm	Emission, nm	Comparative Half-band Widths at Half Height, nm
H₄(PAA)₂en	370	~385	~4
		~400	~7
		~420	~12
Cu₂(PAA)₂en	365	427	1.5
	385	427	1.5

both the ligand and the complex corresponds to a strong absorption, presumably an intraligand $\pi \rightarrow \pi^*$ band. When Cu₂(PAA)₂en is excited at either 365 or 385 nm, sharp emission occurs at 427 nm. No phosphorescence was observed down to about 800 nm.

ELECTROCHEMISTRY. The cyclic voltammogram for the reduction of Cu₂(PAA)₂en in dimethylformamide (DMF) is presented in Figure 5. The reduction occurs at $E_{1/2} = -0.61$ V *vs.* SCE, and it is reversible.

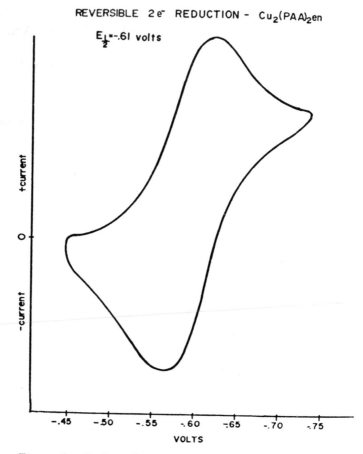

*Figure 5. Cyclic voltammogram for the reduction of Cu₂-
(PAA)₂en in DMF*

On the basis of electrolysis experiments, it was determined that the process involves the addition of two electrons per molecule. This is quite uncharacteristic of simple Cu(II) compounds which generally reduce irreversibly in a one-step, two-electron reduction giving Cu(0) (24). Several of the bis(1,3,5-triketonato)dicopper(II) chelates were studied polarographically in our laboratories (25). These reductions are all irreversible, but they are similar to that of $Cu_2(PAA)_2en$ in that they consist of two-electron reductions.

During the electrolysis in which a solution of $Cu_2(PAA)_2en$ in DMF was qualitatively reduced, the color changed from green to yellow but no precipitate formed. Thus it seems likely that a stable Cu(I) complex was formed. The reversibility of the reduction indicates that perhaps species such as $[Cu_2(PAA)_2en]^{2-}$ or $Cu_2[H_2(PAA)_2en]$ are formed in solution. The nature of these solutions will be investigated further since the electrochemical data suggest that chelates such as $Cu_2(PAA)_2en$ may be useful electron transfer reagents.

Literature Cited

1. Wittek, P. J., Harris, T. M., *J. Am. Chem. Soc.* (1973) **95**, 6865.
2. Murphy, G. P., Harris, T. M., *J. Am. Chem. Soc.* (1972) **94**, 8253.
3. Oda, R., Munemiya, S., Okano, M., *Makromol. Chem.* (1961) **43**, 149.
4. Baker, D., Dudley, C. W., Oldham, C., *J. Chem. Soc. A* (1970) 2605.
5. Murtha, D. P., Lintvedt, R. L., *Inorg. Chem.* (1970) 1532.
6. Kuszaj, J. M., Tomlonovic, B., Murtha, D. P., Lintvedt, R. L., Glick, M. D., *Inorg. Chem.* (1973) **12**, 1297.
7. Lintvedt, R. L., Borer, L. L., Murtha, D. P., Kuszaj, J. M., Glick, M. D., *Inorg. Chem.* (1974) **13**, 18.
8. Lintvedt, R. L., Mack, J., "Abstracts of Papers," 6th Central Regional Meeting, ACS, April 1974, INORG 613.
9. Sagara, F., Kobayashi, H., Ueno, K., *Bull. Chem. Soc. Jpn.* (1968) **41**, 266.
10. Taguchi, Y., Sagara, F., Kobayashi, H., Ueno, K., *Bull. Chem. Soc. Jpn.* (1970) **43**, 2470.
11. Andrelczyk, B., Lintvedt, R. L., *J. Am. Chem. Soc.* (1972) **94**, 8633.
12. Tomlonovic, B., Glick, M. D., Lintvedt, R. L., unpublished data.
13. Blake, A. B., Fraser, L. R., *J. Chem. Soc. Dalton Trans.* (1974) 2554.
14. Borer, L. L., Ph.D. Thesis, Wayne State University, 1972.
15. Glick, M. D., Lintvedt, R. L., *Prog. Inorg. Chem.* (1975) in press.
16. Miles, M. L., Harris, T. M., Hauser, C. R., *J. Org. Chem.* (1965) **30**, 1007.
17. Evans, D. F., *J. Chem. Soc.* (1959) 2003.
18. Gavel, D., Kuszaj, J., Lintvedt, R. L., Glick, M. D., "Abstracts of Papers," 166th National Meeting, ACS, Aug. 1973, INORG 156.
19. Gavel, D., Glick, M. D., Lintvedt, R. L., unpublished data.
20. Fackler, Jr., J. P., *J. Am. Chem. Soc.* (1962) **84**, 24.
21. Belford, R. L., Chasteen, N. D., Hitchman, M. A., Hon, P. K., Pfluger, C. E., Paul, I. C., *Inorg. Chem.* (1969) **8**, 1312.
22. Bleaney, B., Bowers, K. D., *Proc. R. Soc. London* (1952) **A214**, 451.
23. Holm, R. H., *J. Am. Chem. Soc.* (1960) **82**, 5632.
24. Lintvedt, R. L., Russell, H. D., Holtzclaw, H. F., *Inorg. Chem.* (1966) **5**, 1603.
25. Handy, R. F., Ph.D. Thesis, Wayne State University, 1972.

RECEIVED January 30, 1975.

INDEX

S

The text of this book is set in 10 point Caledonia with two points of leading. The chapter numerals are set in 30 point Garamond; the chapter titles are set in 18 point Garamond Bold.

The book is printed in offset on Text White Opaque, 50-pound. The cover is Joanna Book Binding blue linen.

Jacket design by Norman Favin.
Editing and production by Mary Rakow.

The book was composed by the Mills-Frizell-Evans Co. and by Service Composition Co., Baltimore, Md., printed and bound by The Maple Press Co., York, Pa.